Sourcebook for
Farm Energy Alternatives

Sourcebook For Farm Energy Alternatives

James D. Ritchie

McGraw-Hill Book Company

New York   St. Louis   San Francisco   Auckland
Bogotá   Hamburg   Johannesburg   London   Madrid   Mexico
Montreal   New Delhi   Panama   Paris   São Paulo
Singapore   Sydney   Tokyo   Toronto

Library of Congress Cataloging in Publication Data

Ritchie, James D.
    Sourcebook for farm energy alternatives.

    Includes index.
    1.Renewable energy sources.   2.Energy
conservation.   3.Agriculture and energy.
I.Title.
TJ163.3.R57   333.79   82-7738
ISBN   0-07-052951-5   AACR2

1234567890   KGPKGP   89876543

ISBN 0-07-052951-5

The editors for this book were Patricia Allen-Browne and Susan
Thomas, the designer was Mark E. Safran, and the production supervi-
sor was Paul A. Malchow. It was set in Roma by ComCom.

Printed and bound by The Kingsport Press.

Contents

An Economic Introduction to Energy *by Harold F. Breimyer*   xi

Preface   xvii

In Gratitude   xxi

1. **Conservation: The Cheapest Energy Source**   **1-1**

    Managing to Save Energy   1-3
    Planning Field Operations to Save Fuel   1-3
    Planting and Harvesting   1-4
    Saving on Irrigation   1-5
    Managing Machinery and Fuel   1-5
    Reducing Fuel Storage Losses   1-5
    Saving on Grain Drying   1-5
    Farmstead Management   1-6
    Saving Electricity   1-6
    Livestock Management   1-6
    Saving Energy in the Milk House   1-7
    Saving on Ventilation   1-7

2. **Use What You've Got First**   **2-1**

    Getting More Work from Purchased Energy   2-1
    Recovering Hog Heat for Winter Warmth   2-3
    Using Surplus Heat in Dairy Operations   2-4
    Constructing an All-Weather Stock-Watering Tank   2-6

3. **Livestock Buildings**   **3-1**

    Energy for Farm Buildings   3-3

Calculating Heating System Size   3-4
Estimating Heating Fuel Needs   3-6

4. **Insulation and Ventilation**   **4-1**

How Much Insulation?   4-2
How Much Ventilation?   4-7
Ventilation Controllers   4-13
Troubleshooting Ventilation Systems   4-14

5. **Solar Assistance**   **5-1**

Measuring Solar Radiation   5-1
Shading Effect of Buildings   5-2
Active or Passive?   5-4
Collector Efficiency   5-5
Absorber Surfaces   5-8
Cover Glazing   5-10
How Much Collector Area?   5-12
Storing Solar Heat   5-13
Figuring Economic Returns   5-17
Choosing a Solar Contractor   5-19
Solar-Assisted Heat Pumps   5-21

6. **Solar Hot Water**   **6-1**

Types of Solar Water Heaters   6-1
Economics of Solar Hot Water   6-4
Combination Hot-Water Systems   6-5
Versatility of Hot–Water Systems   6-6
In Search of Qualified Installers   6-9
Build a Solar Pond?   6-11

7. **Other Heating and Cooling Options**   **7-1**

Wood Heat   7-1
The Wood Supply   7-2
What Kind of Equipment?   7-5
A Word about Safety   7-9
Energy from Mother Earth   7-9
Controlling the Moisture   7-10
Summertime Cooling   7-10

8. **Energy-Efficient Cropping**   **8-1**

Fertilizer   8-2

Crop Drying   8-3
Field Operations   8-4
Irrigation   8-5
Pesticides   8-5

9. **Low–Temperature Grain Drying**   **9-1**

Over-Drying Is Costly   9-1
Use Drying Energy Efficiently   9-2
Combination Drying   9-4
Grain Depth in Bins   9-5

10. **Solar Grain Drying**   **10-1**

The Solar Option   10-2
Will Solar Pay?   10-3
Solar for Drying Only?   10-6
Integrated Collectors   10-7
To Stir or Not To Stir?   10-7

11. **Drying Grain with Crop Wastes**   **11-1**

Corn Stalks   11-1
Corncobs   11-3
How Much Residue Should Be Used?   11-3

12. **Irrigation Cost Cutting**   **12-1**

Irrigation Methods   12-2
Improving Pumping-Plant Efficiency   12-4
How Much Water—and When?   12-4
Water "Banking"   12-5
Solar-Powered Irrigation   12-6
Wind-Powered Irrigation   12-10

13. **Farm-Grown Fuels**   **13-1**

Liquid Fuels   13-2
How Much Fuel Do You Need?   13-3
The National Biomass "Bank"   13-5
Biogas Production   13-7
Vegetable Oils   13-7
It's Not All Gravy   13-8
Beware of Hucksters   13-9

14. Ethanol   **14-1**

What Is Fermentation Ethanol?   14-2
Pros and Cons of On-Farm Ethanol Production   14-4
Should You Produce Ethanol?   14-6
Feedstocks   14-8
Basic Ethanol Production   14-10
Designing the Plant   14-14
Improving Plant Efficiency   14-19
How Much Can You Automate?   14-21
On the Drawing Board   14-22
Storing, Handling, and Using Ethanol   14-26
A Word about Gasohol   14-28

15. Fuels for Diesels   **15-1**

Extracting Vegetable Oils   15-5
What Will It Cost?   15-6
How about Alcohol in Diesels?   15-7

16. Methane: Cow Power to Burn   **16-1**

What is Methane?   16-2
Producing Gas by Anaerobic Digestion   16-3
Digester Design   16-5
Operating and Maintaining a Digester   16-8
Using Biogas   16-10
Producing Gas by Pyrolysis   16-11

17. Hydrogen: Fuel of the Future?   **17-1**

Objections to Hydrogen   17-1
Energy Potential of Hydrogen   17-2
Storing the Fuel   17-6

18. Producing Electricity on the Farm   **18-1**

Economics of Homemade Electricity   18-1
Make a Load Analysis   18-2

19. Power Blowin' in the Wind   **19-1**

Evaluating the Site   19-3

Which Plant Design?   19-5
The Generator   19-6
Other Uses of Wind Power   19-9

20. Hydroelectric Power   **20-1**

Evaluating the Site   20-2
Measuring Stream Flow   20-3
Measuring Head   20-3
What Kind of Turbine?   20-8

21. Sun-Made Electricity and Standby Power   **21-1**

Photovoltaic Cells   21-1
Standby Electrical Generators   21-1

22. Keeping Out of Trouble   **22-1**

Is It Legal?   22-1
Making Decisions   22-3

23. There's Money Available, But . . .   **23-1**

Income Tax Credits   23-1
Loans   23-3
Grants-in-Aid   23-4
State Programs   23-4

24. Integrating Energy Systems   **24-1**

The Missouri Energy Complex   24-2

25. Using the By-Products   **25-1**

Whole Stillage   25-2
Anaerobic Digestion By-Products   25-2
Vegetable Oil Meal   25-9

Appendix A   Metric and Other SI Units for Agriculture   **A-1**

Appendix B   Tables, Tips, and Rules of Thumb   **B-1**

Appendix C   For More Energy Information   **C-1**

**Appendix D   Directory of Manufacturers and Technical Service
                        Firms (U.S. and Canadian)**   **D-1**

Appendix E   Abbreviations Used in This Book   **E-1**

Appendix F   Glossary   **F-1**

Index   **I-1**

An Economic Introduction to Energy

Harold F. Breimyer
Professor of Agricultural Economics
University of Missouri

The task of the educator in the field of energy is becoming easier. To use a phrase from a generation ago, the "march of time" is making people attentive. Sometimes, current events make them alarmed. We who try to educate on the subject no longer have much trouble convincing audiences of the seriousness of the energy problem.

Our task now is to assess its relative magnitude. How short are we of energy? How serious is the threat of shortage? To what extent will economic signals produce a reasonable balance between energy supplies and needs, and to what extent must other techniques such as subsidy or mandatory controls be employed? What kinds of adjustments will we find necessary, in our living and in our businesses, including farming?

To begin with a ball-park perspective, the national energy scene is somewhere between 68°F thermostats and doomsday. The range between the extremes of simple, easy accommodation and threat of chaos is very wide. When we try to narrow that range and be more specific, we become guarded in our language.

To consider the worst possibility first, there is no reason for fright, for thinking that doomsday is at hand. Fortunately, the United States has very considerable fossil fuel resources. Even though the reserves of petroleum and natural gas are gradually being depleted, those of coal are ample, and we will probably continue to use modest quantities of uranium for nuclear power.

What may be equally meaningful, we begin from a level of energy use so great that in principle we could cut back quite a bit and still live well. We have been using about twice as much energy per capita as other prosperous nations, and—again in principle—we could reduce without incurring privation.

But having offered those somewhat reassuring comments, I add two strict caveats. One is that our institutions, our economic system, even our psychology are now geared to plentiful and cheap energy. Everyone knows we resist transition.

Moreover, in some respects the nature of our country and our economy works against an easy cutback in energy use. We are a big country, geographically. We didn't locate our population in the center of the country where everyone is close together. We have one big population center on the east coast and another on the west coast, with 3000 miles and two mountain ranges between them. Our coal is in one place—in fact, in many places —and our iron ore lies elsewhere. Our grain belt is far from either population center. And so on.

Also, our climate is fairly extreme, requiring heat in the winter (even occasionally in Miami), and although we do not *require* air conditioning in summer, many of our citizens have become addicted to it.

The second caveat is that although we could make adjustments if allowed to accomplish them gradually, a sudden cutoff in delivery of oil would throw us into shock. In such an event, we probably would have to use the militia to prevent disorder. This would be especially true if we do not prepare well for the contingency of a sharp cutoff.

Our nation's energy resources have been inventoried with some care. We know quite a bit about our immediately available oil, gas, and coal. A sum-up will be offered later. We know only a little about the less available materials, and the exotic sources are still a matter of great uncertainty.

One of the most misunderstood aspects of energy is its basic economics. The economics of delivery of a stock energy source is sharply different from that of a currently produced material. The stock energy sources are essentially oil, gas, coal, and uranium-235. These are, in effect, stored resources. They are stored in the ground. For most of them, the cost of delivery is relatively small. The delivery cost looks particularly small alongside the monopoly price that OPEC (Organization of Petroleum-Exporting Countries) is charging for oil.

What I am leading up to is this: Holders of the stock resources play a speculative game. They deliver or not according to how they forecast future price trends. If they think prices will be higher in the future, they hold back. If they think prices will be weak in the future, they are quick to pump or mine, and to deliver. In simpler words, if the prospective price is higher, less will be delivered; if the prospective price is lower, more will be supplied.

I call this "The Perverse Economics of Petroleum." This speculative pattern in delivery of a stock resource differs from the economics of production and sale of a flow commodity, such as most of the products of agriculture. Farmers know well that if the price of their flow products, such as hogs or soybeans, looks good they will produce *more*. If the price outlook is dimmer, they *cut back.*

But there is a further difference. If farmers produce more hogs or soybeans, they do not impair their ability to continue to produce in future years. With a stock resource such as oil, the opposite is the case: The more we use now, the less will be available for delivery in the future.

A lot of oil suppliers have learned this lesson in the past few years. Saudi Arabia is now reluctant to deliver large quantities. Some other OPEC countries are torn between keeping up an income flow and holding back oil in keeping with their perverse economics of petroleum. In the United States, hundreds of oil wells have been capped, as oil is held for a higher price expected in the future.

Of course, some people say that when prices of oil go up we will tap many new sources of energy, such as oil shales. The studies I have seen give little encouragement to that point of view. Oil sands and shales are not that promising a source. Nor will solar or wind or water or any other alternative source yield truly substantial quantities of energy within the next 20 years.

The other side of the equation is the nature of the demand for fossil fuels. It's simple. Almost all users of fossil fuel energy regard their needs as essential. This is true of the industrial firm for which oil, gas, or coal is a fuel or feedstock. The cost is probably only a minor fraction of the firm's total operating cost. The same idea holds nearly as true for consumers who believe their automobiles must be kept in motion as much in the future as in the past. The various attitudes add up to a demand that is extremely inelastic.

The harsh fact of the matter is that the perverse economics of supply and inelasticity of demand add up to extreme difficulty in bringing balance and stability to the energy scene. And these comments reinforce my earlier remarks about the danger of chaos if our inflow of oil from foreign sources were to be shut off suddenly.

I turn now to energy in agriculture.

Agriculture has a dual personality with respect to energy—it uses it, and it produces it. The farm technology that has been adopted in this century is essentially energy-intensive. Yet the products of agriculture contain energy. That's essentially what agriculture is; it is the process of converting the energy of the sun into products that human beings can use. For the most part they have been food products, but can also be fibers, tobacco, and some vile things that may be illegal.

So we ask the "biomass" question: Is it feasible to use the products of agriculture for conversion into industrial energy?

That question will be developed further in pages ahead. On the positive side, though, is the fact that most crop production is *energy-efficient*. That is to say, even though present cropping methods are energy-intensive, most yield more energy than they use. To my knowledge, all field crops have a positive energy ratio: The energy content of the harvest is greater than the energy content of all the inputs employed. For many vegetables this is not true, but for wheat, corn, soybeans, and such crops, the energy ratio ranges from 2:1 to 5:1 or higher.

In the future, energy efficiency in crop farming may actually increase, as higher costs of energy force more conservative energy use. Recent increases in energy prices and the prospect of further ones have made farmers sit up, take notice, and adjust.

I have usually thought of energy as constituting roughly a third of the variable costs of producing field crops. But my colleague Kenneth Schneeberger quotes data from records kept by 200 Missouri farmers who are above-average managers. For nonirrigated corn and soybeans, data for 1979 show the following costs per acre:

	Corn	Soybeans
Fuel and lubrication	$ 16.00	$13.60
Fertilizers and lime	40.50	7.50
Chemicals	15.00	13.50
Drying	4.40	
Total energy-related costs	75.90	34.60
Other variable costs	41.60	33.50
Total variable costs	$117.50	$68.10

The conclusion follows that energy prices will henceforth influence farmers' choices of crops and cultural practices to some extent. But to continue balancing the positive and the negative, we note that agriculture is fortunate in that its total energy requirement is relatively modest. The statistic is now familiar that production of food on farms absorbs only 3 percent of all the energy used in the economy. Adding cotton and tobacco puts on another percent or so, but these are relatively small numbers. Moreover, of the energy used in farming operations, the biggest share goes to commercial fertilizer, particularly nitrogen fertilizer. Power for field equipment stands second.

Hidden in the average data are certain operations that *are* energy-intensive. Deep-well irrigation is extremely vulnerable to high-cost energy. Most midwestern irrigation does not involve such deep pumping, but any irrigation other than gravity opens up energy-cost considerations.

I'll now touch briefly on the best available information on the extent of our mineral energy resources. The subject has been a bone of contention for at least 10 years. The two studies published recently on which I draw are the report of the private research institution Resources for the Future and a similar report of the Harvard Business School.

To sum up, conventional petroleum and natural gas are essentially resources of the twentieth century. If we include what Resources for the Future calls "inferred" resources, and even add a certain amount of undiscovered resources, the available supply will last only two to three decades.

There is some difference of opinion about the supply of "clean" uranium, but at the reduced rate of building nuclear plants, it seems likely that the mineral will last pretty well through the next century.

Coal is, of course, a more plentiful resource, and even with sharply increased mining, supplies ought to last two centuries—perhaps three. Likewise, if we should eventually get nuclear energy from the breeder reactor, it will prove a very large energy source. But that is a possibility for the next century, not for this one.

The next category of minerals is the so-called "unconventional" mineral fuels. These are the oils and gases from tar sands, shales, and such. These lend themselves to science fiction, and perhaps will eventually be a large source. The biggest problem is that a great deal of energy is required to get oil out of shales. In addition, the slag from mining western shales could cover the state of Utah. None of my information sources foresees any appreciable amount of energy from "unconventional" mineral fuels before the year 2000. After that, it's a guess. The maximum potential is, of course, impressive.

Synfuels is a misnomer. Such fuels are not synthetic but are simply converted forms of basic mineral fuels. Most talked about is gasification or liquefaction of coal. Almost all national projects are confined to the east and west coasts, near population centers. Some coal will be gasified or liquefied, but there is substantial energy loss in the process.

Number one on the hit parade right now is *biomass,* and I have supported this idea for a long time. Biomass is a varied category including the burning of wood, the converting of manures to methane, and the making of ethanol for Gasohol from organic materials. Corn and sorghums are among the more expensive organic materials, although readily available and now being used. Lower-valued materials are more economically feasible and offer more promise in the long run.

The subject of biomass is a touchy one. I believe we need to explore every possibility, but there is a lot of pie-in-the-sky thinking just now, and many of the ideas advocated are simply not economical.

The same general judgment applies to solar energy. Solar has some good potential for small-scale use in homes, hog barns, and grain dryers, but it has not yet proved technically and economically feasible for large-scale industrial use. The best hopes for solar energy lie in the next century.

Which brings us to the subject of energy conservation. The axiom is that the cheapest energy is energy saved. It follows, I suppose, that we could easily save 40 percent if we wanted to badly enough. My view is that the first 10 percent of saving comes easy, the second 10 percent comes harder, and the third and fourth 10 percent will be resisted like fury. Getting beyond 20 to 25 percent savings means making major changes in our life-style and our industrial processes, including some modifications in farming techniques. Being human, we aren't eager to make those changes.

In summary, I would say that at the University of Missouri, where I practice my profession, 90 percent of the scientists who do their homework on energy concur in the general proposition that our energy supply is shrinking but that we should take a balanced approach. We must pursue all avenues toward improving our sources of energy. We must also work toward conservation and using less.

Even as we set out in all these directions, we also must be honest with ourselves. There is no panacea, no magic solution on the horizon. Certainly there is no easy way to avoid an almost perilous dependence on imported oil. If a flare-up in the Middle East should substantially reduce our imports of oil, which provide roughly one-third of our oil supply, pandemonium would reign. We would impose mandatory allocation quotas, enforced by the police powers of the state. It would be neither easy nor pleasant.

I wish I could end on a more hopeful note. I repeat my opening theme: We are not promising magic solutions or trouble-free times, but we are also not warning of doomsday. We are saying that the situation is serious, that it is highly vulnerable to political winds (particularly those blowing from the Middle East), and that it ought to lead all of us to do whatever we can, both privately and through the policy route.

We cannot avoid adjusting to the use of less energy in industry, in trade and commerce, and in the way we live and conduct our daily affairs.

Preface

As a farmer, farm builder, or agribusinessman, you need not be told that we Americans have a ravenous appetite for energy—all kinds of energy. And it's no surprise what energy prices are doing to your costs, as illustrated in Fig. 1 on the next page.

In fact, we Americans use so much energy that we have to talk about it in terms of quadrillions of Btu's, or *quads.* One Btu, you'll recall, is the amount of heat energy required to raise the temperature of one pound of water by one degree Fahrenheit. One quad of energy is equal to the energy available from 500,000 railcars of coal, assuming that each car hauls 83 tons. Or, to put it another way, burning a wooden kitchen match produces about 1 Btu of heat, whereas 1 quad of energy is the equivalent of burning the crude oil cargoes of 75 supertankers.

Farming is an energy-intensive activity. Energy used in agriculture is consumed directly as fuel and electricity or indirectly as fertilizers, pesticides, and feed additives—many of which are derived from petroleum. (See Fig. 2.) However, of the nearly 80 quads of energy consumed in the United States each year, production agriculture uses only 2.5 quads—just over 3 percent of the total.

With consumption of just more than 3 percent of the total U.S. energy diet, agriculture and forestry produce food and fiber for some 225 million Americans and a goodly number of people in other parts of the world.

However, the efficiencies of U.S. agriculture have more often been measured in terms of *people* than units of energy. U.S. farms have grown bigger and fewer, as individual farmers have begun to manage more crop acres, larger numbers of livestock, bigger flocks of poultry. One farmer now produces enough food for 70, 80, or 100 people (depending on whose figures you read and how this statistic is measured), but he does it primarily because of technologies based on an assumption of cheap energy.

Look at what has happened in farming in the past 30 years. Commercial fertilizer use has increased more than fivefold, to the point that fertilizing accounts for more of the energy used in agriculture than any other single farm operation. Since 1950, the *number* of tractors used on U.S. farms has increased less than 30 percent, but tractor *horsepower* today is nearly 2½ times that in use 30 years ago.

In the past few years, the use of diesel engines has grown to the point that we're burning nearly as much diesel oil as gasoline in farming operations. And, while diesel is more efficient than gasoline on a gallon-for-gallon basis, refineries get less diesel fuel than gasoline from a barrel of crude oil.

Agriculture is not only consuming more energy than ever before, but also consuming it at a much higher unit cost. The critical importance of energy to agriculture—and to the nation's grocery bill—was brought home to all U.S. citizens by the Arab oil embargo of 1973–1974.

Unhappily, the outlook for the next five years and beyond is not particularly bright where energy supplies and prices are concerned. Take another look at Fig. 1. The price of oil is expected to rise at an average of 17 percent per year, doubling again in less than five years. Natural gas will go up by even more, according to many economic analysts— something on the order of 21 percent per year. And electricity will rise in cost by about 12 percent per year.

Whether these predictions hold true or not, cheap energy will not return. Not for farmers or for nonfarmers. We will have to learn to get along with less, regardless of the alternative sources of energy that are developed. We will have to make more efficient use of the energy we do consume, and substitute *low-grade* sources of energy (solar, geothermal, wood, etc.) for *high-grade* sources (petroleum, natural gas, and coal) wherever possible. And we'll have to resign ourselves to paying higher and higher costs for the energy we purchase.

But the picture is not totally bleak, where agriculture is concerned. Big changes are coming, sure. But farmers will have the opportunity to set the pace in energy conservation and alternative energy development. Higher prices already have encouraged farmers to adopt many conservation practices, such as increased insulation of buildings, reduced field tillage, closer regulation of ventilation in livestock and poultry housing, and better maintenance of equipment.

In the years ahead, combinations of production inputs will change. Farmers will lean toward new technology that tends to make energy use more efficient. Changes will be dramatic, but farming is not likely to return (as some doomsayers prophesy) to animal power to replace engine-driven equipment. Not unless someone magically produces the 18 to 20 million horses and mules that would be needed, and the 50 to 60 million acres of forage land to feed them.

Given enough time to develop the technology to use them, U.S. farms have the raw

PRICE OF SELECTED INPUTS IN AGRICULTURAL PRODUCTION

INDEX OF INPUT PRICES

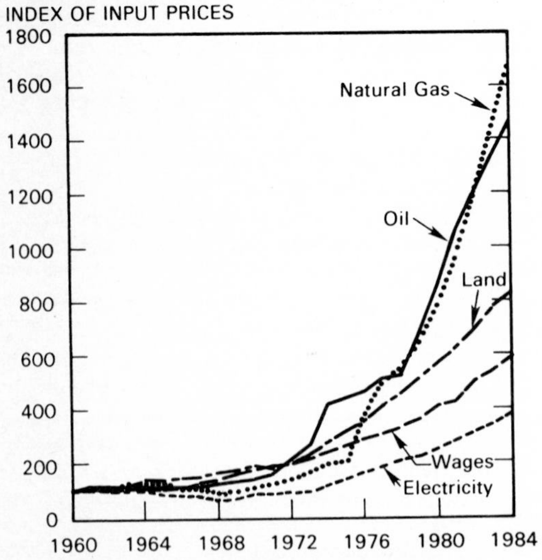

Fig. 1 Dollar cost of energy is outpacing other farm inputs. *(Source: Data Resources, Inc.)*

ENERGY USED IN AGRICULTURAL PRODUCTION

Total
2,022 Trillion BTU's

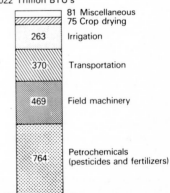

Fig. 2 Actual farm production accounts for only 3 percent of U.S. total energy use. Fertilizers and pesticides represent more than a third of the energy used in agricultural production. *(Source: U.S. Department of Agriculture.)*

resources to provide huge quantities of alternative energy, for use in agriculture and elsewhere in the economy. These are renewable resources from the sun, from wind, or from growing plants (so-called "biomass" energy, such as fuel-grade alcohol, vegetable oils, methane gas, or directly burned crop residues and wood).

And much of the technology needed to make agriculture a major energy producer is here now—albeit at the "Model T" stage in some areas. Admittedly, much of the technology is out of feasible economic reach for most farmers at the present, but that will change rapidly. Farmers are accustomed to adapting new technology to their operations, and you'll read about some of the ingenious alternative energy installations now in use, in the chapters ahead.

Of key importance to agriculture will be the comprehensive energy policy that is finally hammered out in Washington. How nearly will farm policy and energy policy follow the same track, toward the same goals? Take the U.S. grain situation. For the past several years, when farmers have grown more grain than the market would take at a given price, the federal government has had some kind of support program to put a floor under grain prices. During much of that time, the price stayed pretty close to the floor (except as export demand took surplus production), and grain exports have proved to be as vulnerable to political whim as are petroleum imports.

Now, there's an upsurge of interest, encouraged by both the U.S. Department of Agriculture and the U.S. Department of Energy, in producing fuel from grain. And, to the extent that fuel production results in "disappearance" of grain surpluses, the price increase benefits all grain producers—whether they make alcohol or not. History indicates that a 5 percent reduction in supply generally brings about a 10 percent increase in grain prices.

But, if we're really serious at the national policy level about both supporting farm prices and reducing our dependence on imported fossil fuels, why aren't we burning corn—cobs and all—as boiler fuel to generate electricity? Pound for pound, corn has the potential to produce nearly as many Btu's as soft coal does, *and* it sends fewer emissions into the air. The petroleum fuels saved by burning corn in electrical generating plants could be put to other uses for which only petroleum energy is suitable.

In fact, with the proper encouragement from policymakers and planners, farmers could well become the energy heroes of the 21st century. Left to their own devices, many individual farmers will no doubt become relatively self-sufficient in energy in the next 25 to 30 years. With the right kind of direction from the powers that be, farmers could produce a great deal of surplus energy for use by the rest of society.

But that prospect is well down the road, and it's a road strewn with "ifs" and "buts." Those dollars we're spending for foreign oil are buying time, but that time is ever more expensive—and, like the world's oil, in ever shorter supply.

Of immediate concern to your farm business are your energy bills for this year and next. Of critical concern to all of us is the very real possibility that, through some political pique or whim, our overseas oil suppliers will turn off the tap. How do you make your farm operation less vulnerable to shrinking supplies and ballooning prices of energy? Where are your limited dollars best spent in energy conservation and alternative energy systems?

To help answer those questions is the goal of this book. It will not promise, as some publications on energy alternatives appear to do, to tell you how to solder together a few feet of copper tubing and an old clothes boiler so that you can home-brew your way to energy independence. (This is not to cast any aspersions on "junkyard" technology; if Cyrus McCormick had not tinkered with his homemade harvesting machine, we might not have the self-propelled combine today. For that matter, many farmers are confirmed and inventive tinkerers who build or modify equipment that is more suitable to their purpose than anything they can buy from a machinery manufacturer.) But throughout the chapters ahead, you'll find emphasis on economic returns, and payback on energy dollars invested. You'll find many examples of equipment and operating techniques that have worked—and are working—for other farmers, again with cost and return information wherever possible. And, you'll find the best technical information the author could obtain from farmers, farm builders, engineers, and chemists across the country.

Much of our purpose in this book is to present energy options and discuss them in clear economic terms. Let's use, as an example, a dairyman with a herd of high-producing milk cows and reasonably good forage- and crop-producing land. Can he afford to set aside 10 acres of that land to produce feedstock to distill 1500 gallons of fuel alcohol, when that same 10 acres could grow forage to produce another $8000 worth of milk? A better energy-saving option for him might be to install a heat exchanger to recover the heat removed from the milk by a cooling-tank compressor. With the recovered heat, the milk parlor and washdown water could be heated.

The dairyman may very well choose to do both, however, and we help show him how. But there's a standard economic axiom that says everything has opportunity costs. The dairyman cannot use those 10 acres—and the 24 hours in each day—to maximum benefit for both forage and fuel.

The business of farming, like any other business, is the management of limited resources. Each farmer has only so much land, labor, capital, energy, and management skill to invest in his business operation. That's why much of our emphasis in this book is on helping farmers and farm builders assess their energy options in terms of costs and returns. You'll find that economics has at least equal billing with hardware in the pages ahead.

DISCLAIMER

While the text of this book takes on masculine gender in most references for convenience, there is no intent on the author's part to slight or ignore women, who are involved mightily in food and fiber production. In fact, farm women were "liberated" to the point of working shoulder to shoulder with men from the time America was first settled. Their role has not diminished.

IN GRATITUDE

In a way, professional writers deal in energy: the energy of ideas. We're the transmission lines that distribute ideas from where they are generated to where they are consumed. A great many people generated ideas and information for this book.

To the dozens of people in other professions—farmers, builders, engineers, chemists, economists—whose ideas and experiences are transmitted on the pages ahead, this writer is profoundly grateful.

Special thanks are due Richard E. Phillips, agricultural engineer, who views energy and building innovations not solely on the basis of whether they will work, but on whether they will work *practically*. Also, to Harold F. Breimyer, who so astutely depicts the current energy situation and its economic implications, my thanks. Dr. Breimyer's analysis in the introduction to this book sets the tone and focus of those chapters that follow.

And last, but by no means least, a fervent but not-often-uttered "thank you" to my wife, Kathy, whose patience and attention to myriad detail have contributed much to whatever craft I may have developed at transmitting words and ideas.

James D. Ritchie
Versailles, Missouri

Sourcebook for
Farm Energy Alternatives

Conservation: The Cheapest Energy Source

In the foregoing pages, Dr. Harold F. Breimyer offers this axiom: "The cheapest energy is energy saved," and adds that we in the United States could fairly easily save 40 percent of our total energy consumption, if we wanted to badly enough.

No doubt, big potential for energy savings exists on most U.S. farms. And there's more and more motive for saving. The gallons of fuel and kilowatthours of electricity you don't buy add up to more dollars, as the cost of energy increases.

Because energy has been so cheap for so long, most of us have developed wasteful habits—at least inefficient habits. So what if I leave a light on in the machine shop? Five 60-W bulbs would have to burn for 3½ hours to use 1 kWh of electricity, which costs less than a dime in most rural areas.

Who cares if you forgot something on this morning's trip to town, and have to burn another three gallons of gasoline for a second round trip this afternoon? The price of gasoline is up, but those three gallons of gasoline only cost a few dollars.

Big deal, huh?

Well, it's not so much when you compare that 1 kWh of electricity with the amount of power you might use to operate an irrigation rig or run the exhaust fans all year in a confinement livestock building. An extra three gallons of gas to go after a forgotten washer or setscrew is just a drop in the bucket, compared with the fuel used to grow 500 acres of corn.

But it *is* a big deal when you add up all the careless and wasteful habits we've acquired. For most of us, effective energy conservation involves developing a changed attitude, a new energy ethic. In the past few years, the fact has been brought home that the world's oil barrel does have a bottom, and we're closer to the bottom than we formerly suspected.

Today America must import foreign oil to meet its own demand for *high-grade energy* (energy used for sophisticated purposes, such as fueling internal-combustion engines). And yet half of today's high-grade energy sources (gas and oil) are diverted to produce *low-grade energy,* the kind used to heat spaces and water. Wood can be burned to heat spaces but cannot be used to fuel our modern self-propelled vehicles. We need to use more of such alternative resources to meet our need for low-grade energy and to reserve higher-grade forms of energy for those tasks that only high-grade energy can do.

In other words, you can help heat a hog house with solar energy, but no one has come up with a practical way to run a tractor directly on solar power. And, as we continue to make the shift from high-grade to low-grade sources of heat, *conservation* must be a key in any solution to our overall energy problems.

Still, energy conservation is not without practical limits in mechanized agriculture. You can cut out a tillage trip or two, or combine fieldwork operations to save tractor fuel, but if you do not perform enough tillage to prepare an adequate seedbed, crop yields will suffer. With minimum tillage systems, there are compensating higher requirements for pesticides. The same thing is true of fertilizer usage. Or just about anything else. At some point, saving on energy inputs starts to cut into production and into gross income.

Much the same is true of livestock operations. If your physical plant and production schedule are geared to farrowing pigs, raising broilers, or milking cows on a year-round basis, you'll no doubt choose to spend the energy dollars needed to ventilate buildings in summer and heat them in winter. Ongoing fixed costs in many confinement animal buildings argue against letting them sit idle, even for a few weeks of the year.

However, that doesn't mean that energy bills are monthly dues you are permanently locked into paying. Look at each form of energy you use as a budgetary expense that can be controlled, at least within limits.

All farms use a combination of energy sources. You pay for energy on the basis of the units consumed: kilowatthours of electricity, gallons of gasoline and diesel fuel, cubic feet or therms (1000 ft³) of natural gas, pounds or tons of fertilizer, etc. Energy conservation will not do much to lower the *per unit* price of energy.

But, whether you shift to one or more so-called "alternative" sources of energy or stick with conventional sources, the units of energy you manage to do without will continue to save more and more total dollars.

The way to start is to prepare a month-by-month *energy profile* of your farm. Record how much (how many units) of each energy source you purchased per month for the past year. How many gallons of diesel or gasoline were bought in January? In February? How many kilowatthours of electricity? How much LP gas? Fertilizer? Pesticides? It shouldn't take long to round up the receipts of energy bills paid. Once you have them in hand, analyze your energy costs to see how much you have been paying for each source, and when.

Each farm, because of its geographical location and mix of enterprises, has different patterns of energy consumption at different times of the year. Chances are, your use of some energy sources does not coincide with your purchase of that source. For example, fertilizer and farm chemicals often are ordered and paid for in winter, although these energy items are not used until later in the season.

Your next step in analyzing your energy profile is to estimate as closely as possible when and how the energy was used. At some times, this may be a fairly rough estimate. If you buy and store hundreds of gallons of tractor fuel, you may not meter how much is consumed for each farming operation.

It helps your energy-use planning, however, to make the best estimate possible of the use of each source. As you look through your energy bills for the past year, which months are high? Which are low? In which season do you use the most electricity? Gasoline? Diesel fuel? What operation is the major consumer of fuel? Deepdraft plowing? Combining grain? Irrigation?

The next step is to ask: Where can I save units of energy and dollars, without hampering my farm operation or output? Can I cut out a tillage trip during spring fieldwork, or go to more efficient tillage implements? (Remember, the soil is *your* energy "ace in the hole"; any energy-conserving operation that also results in soil conservation has a double benefit.)

Would a shelterbelt of windbreak trees pay off? (It would, in many cases, and for years to come.) Could you increase the use of high-moisture feeds and use less heat processing in a crop-livestock operation? How about the use of more legumes in crop rotations, to reduce the amount of commercial fertilizer needed? Or, perennial crops that would let you avoid annual tillage costs?

Can you make seasonal changes in livestock operations to save energy and money? For example, could you shift calving season by a month or two so that cows can make more use of crop residue after weaning? Perhaps a hog operation could go to spring-and-fall pig farrowings, to reduce the need for energy for summer cooling and winter heating.

The point of all this is: The energy situation has gotten personal. It affects you and me, and the way we're accustomed to living and doing business. But it doesn't affect you and me in the same ways. Your farm operation probably is not duplicated exactly anywhere else in the country. Not every tailor-made energy-saving solution will fit your farm.

There are promising alternative energy sources on the horizon: solar, alcohol fuels, wind, methane, etc. But for the short run, these hold scant promise of replacing much conventional energy on a big scale, although some farmers (again on an individual, personal basis) are approaching energy self-sufficiency right now and more will in the months ahead.

Given enough time and encouragement, U.S. agriculture can become virtually self-sufficient in energy. But it won't happen next year, or in the next 10 years. The most immediate action we all can take is to cut down, tune up, and tighten up to save. The amount saved by each conservation measure may be a little or a lot, but each unit of energy saved buys time. Time for you to shift your cropping enterprise to a more energy-efficient combination, time to build or remodel farm buildings to make use of alternative sources of energy, time for you to evaluate your potential as a producer of energy, as well as a producer of food and fiber.

MANAGING TO SAVE ENERGY

Concern over fuel cost and supplies is causing farmers to alter their management techniques. The changes vary from making fewer trips to town through building new energy-efficient buildings to cutting out a couple of tillage trips across the field, but few farm families are completely untouched by more costly energy.

In a poll of 1000 Missouri farm families conducted in 1980, here were some of the ways farmers reported they are conserving energy in their farm operations and homesteads:

• Nearly 65 percent said they were reducing the number of times they drive across the field for tillage; a fourth of the farmers were going to all-out minimum-tillage methods in crop operations.

• Half of the families polled had installed wood-burning stoves for home heating.

• More than a third had added insulation to the farm home or other buildings.

Next to *indirect* energy purchases in the form of fertilizer and pesticides, the biggest energy expenses on crop farms are for tillage, grain drying, and irrigation.

The opportunities for saving energy vary from farm to farm, but among the following energy-conserving tips you may find several more you will want to consider for your own operation.

PLANNING FIELD OPERATIONS TO SAVE FUEL

Before fieldwork begins, start planning field operations. Even a rough plan will indicate bottlenecks and reveal operations that consume fuel but do not contribute to increased production or profits.

• Schedule machine movement from farmstead to field and between fields for minimum road time.

• Match equipment to tractor size. Use a smaller tractor for light loads, or throttle back on light loads if a larger tractor is used.

- Do not let tractor engines idle for long periods of time. A medium-sized tractor uses ¼ to ½ gal of fuel per hour when idling and does no work.
- Make wide, level turnrows to reduce time lost in turning.
- Keep tillage tools sharp, clean, and properly adjusted.
- Plow when soil moisture and conditions are most favorable.
- Till only as deeply as necessary to maintain high yields. Deepdraft plowing is a major energy user in crop operations.
- Use weights to keep tractor tire slippage at 15 percent or less.
- Reduce tillage where possible. Cut out any tillage operation that does not increase or maintain crop yields. Combine as many operations as possible in one pass across the field.
- Select tools that consume less fuel, if final results are similar. For example, use a chisel rather than a moldboard plow.

PLANTING AND HARVESTING

With most farm crops, the optimum planting season is fairly short, as is the harvesting season. With time at a premium on both ends of the crop-growing season, options to scale down on use of either planting or harvesting equipment may be limited. However, there are ways to make these operations more energy efficient:

- Plan rapid planter-box filling and combine-hopper unloading to reduce engine idling time.
- Complete harvest before lodging of crops causes slow, fuel-gobbling harvests.
- Inoculate legume seed (soybeans, peanuts, clovers, peas) with the proper rhizobia to stimulate nitrogen fixing from the atmosphere.
- Use fast-dumping grain wagons to reduce tractor idling time at the grain bin.
- Keep cutter knives and shear bar sharp and adjusted to proper clearance on forage harvesters.
- Don't chop silage too finely. When you double the length of cut, you reduce power requirements by a third. Chop just fine enough so that the material packs tightly in the silo.
- Don't use a recutter screen in the forage harvester unless it is absolutely necessary. A screen can increase fuel use by 20 percent.

Saving on Grain Drying   Grain drying is a common practice on most crop farms. Fuel requirements per bushel to dry corn from a moisture content of greater than 25 percent to an acceptable 15.5 percent can be greater than fuel requirements of field operations to produce the corn. For this reason, when thinking about ways to save energy, take a long look at crop drying operations early on.

- Delay harvesting until corn is at 22 percent moisture to reduce the amount of water to be removed by artificial drying. However, keep in mind that when moisture content of unharvested corn falls below 18 percent, field losses due to shelling and shattering mount rapidly. Leaving 1 bu out of every 10 in the field to save a little drying fuel is false economy.

- Investigate adapted early- and medium-season crop varieties. If these will yield nearly as much in a normal year as full-season varieties, consider planting at least part of your acreage with earlier varieties that will be ready to harvest sooner.
- Use natural air or low-temperature heat for drying.
- Use conventional methods to dry the crop down to about 20 percent moisture, then finish drying with unheated air. By this time, weather is generally cooler, so corn with 20 percent moisture can be held longer without damage.
- Store the crop as "wet" grain, either in airtight structures or with acid preservatives to prevent molds and spoilage.
- Harvest as ear corn, if a corn picker is available, and store in ventilated cribs, particularly if the grain is to be fed on the farm. However, ear corn usually must be shelled or ground before being sold or fed, and losses to rodents, birds, and insects can be high in open bins.

SAVING ON IRRIGATION

Irrigation, particularly in western and southwestern regions, makes a big demand on energy. On some farms, two to three times as much fuel is used to pump water as to till, plant, and harvest the crop. Anything that improves crop production per unit of irrigation water pumped will improve the energy efficiency of an irrigated crop operation.

- Favor low-pressure irrigation systems, such as furrow irrigation or drip-trickle systems. They use less energy than high-pressure systems, such as traveling guns and center-pivot sprinklers, and they also lose less water to evaporation.
- Use a precise method of scheduling irrigation to complement rainfall and to just meet crop moisture needs.
- Install reuse (tailwater) systems on surface irrigation.

MANAGING MACHINERY AND FUEL

Machinery that is maintained and operated correctly not only saves time and fuel but also does a better, more reliable job.

- Read (and understand) the operator's manuals for your equipment.
- Double-check machinery adjustments and settings.
- Keep all engines well tuned; check and replace air filters regularly.
- Keep cutting edges sharp and clean; keep moving parts well lubricated.
- Don't overfuel an engine. Use only the type of fuel recommended for the engine.

Reducing Fuel Storage Losses   Evaporation losses as high as 9.6 gal/month have been noted in dark-painted 300-gal gasoline tanks stored above ground. By painting the tank white and locating it in the shade, you can reduce evaporative losses to 2.4 gal/month. A pressure-relief valve on the white, shaded tank can further reduce losses to about 1.3 gal/month. Underground tanks can reduce losses still further.

- Don't store gasoline on the farm for more than six months.
- Clean or replace fuel filters regularly.

FARMSTEAD MANAGEMENT

Operating a farm business involves making regular trips with an automobile or pickup, both on and off the farm. This is a necessary part of running a farm but can consume a great deal of fuel in a year's time. There are ways to trim down:

- Keep a supply of normal operating and maintenance items on hand, to eliminate the need for rush trips to town.
- Check equipment carefully and make needed repairs during the off-season, to avoid emergency trips for repair parts.
- Use the lowest-fuel-consuming vehicle whenever possible. A four-wheel-drive pickup with a 250-hp engine consumes a lot more gasoline per trip than does a four-cylinder sedan.
- Keep a shopping list of parts and supplies so that as many jobs as possible can be accomplished with each trip.

Saving Electricity Electricity is still among the better buys in energy. One kWh of electricity will shear 40 sheep, pump 1000 gal of water, wash 70 lb of laundry, or operate a radio for 15 hours. But that hard-working power still costs money, and there are ways to save it:

- Turn off lights, motors, and heaters when not needed. Repair leaking faucets and water-lines to reduce pump operation.
- Insulate to prevent heat loss when heating with electricity.
- Use natural ventilation where possible. Control heaters, ventilating fans, and cooling equipment with thermostats to limit operation.
- Grind feed no finer than required. Keep hammers and screens in hammer mills—and burrs in burr mills—in good condition to reduce power requirements.

Livestock Management Generally, livestock farming is a relatively low consumer of energy. Nationally, the production of meat animals, poultry, and dairy products consumes only about 225 trillion Btu each year—not counting the energy used to produce feed grains and forage crops for these animals, of course.

A pickup truck is probably the hardest-working vehicle on many cattle farms and ranches. Some farmers find that herd-tending duties can be handled just as well and more cheaply by using an all-terrain motorbike, or even that traditional means of cowboy transportation: the horse. Here are some other energy-saving suggestions for cow country:

- Use self-limiting supplements in self-feeders when feeding cattle on range or pastures, to cut the number of miles driven to take care of feeding chores. Self-fed liquid and dry rations can be put out in volume with one trip.
- Store hay as near the feeding area as possible, and manage hay hauling to reduce long engine idling periods and partial loads.
- Consider feeding dry cows every other day or every third day, to cut down on tractor or pickup trips. Depending on your hay feeding setup, you may be able to feed cows every second day with little wasted hay and little effect on condition or performance.
- New methods of pasture fencing can save time, materials, and energy during construction. The so-called "New Zealand" fence has 10 strands of nonbarbed wire suspended on posts set 50 ft apart, and can be driven over with a pickup at any point, saving a long detour to the gate.

- In feedlots, a major consumer of energy is steam-flaking equipment. Consider going to dry-rolled or high-moisture feed methods as an energy-saving alternative.

Saving Energy in the Milk House   Energy costs are no longer a minor item in a dairy farmer's expense column. In Arizona, the average dairy cow runs up an annual utility bill of $26. In Iowa, where winters are longer and colder, the energy bill is even higher—about $35 per cow. As a result of climbing costs, dairy farmers are turning on fewer lights, readjusting their thermostats, and making sure their milk-house equipment is operating smoothly in an effort to cut down on fuel bills.

- Look for more energy-efficient ways to use waste heat, process feed, and dispose of manure.
- Milk cooling accounts for 10 to 17 percent of the energy used in a dairy operation, and it uses an expensive kind of energy: electricity. Direct expansion coolers are more energy-efficient than ice-bank coolers.
- New insulating materials with washable surfaces can reduce the heating requirements for concrete and concrete-block milk parlors.

Saving on Ventilation   While ventilation is necessary in a swine confinement building, studies show that most swine housing is overventilated and is therefore wasting energy. During winter, as much as 90 percent of the thermal energy required to heat a farrowing house goes into heating the ventilation air.

- Keep windows to a minimum in "warm" confinement buildings. Glass windows can lose 10 times more heat than a well-insulated wall. Hogs do not need a lot of sunlight, and an insulated wall costs less than a window to install in the first place.
- Plan zone heat for baby pigs in the farrowing house. This can be either floor heat or some type of radiant heat. Both electric floor mats and hot-water piping work well.

Even with the rapid evolution of alternative energy systems, conservation will be a key word for all farmers from now on. Livestock producers will need to improve building design and feed-processing efficiency. Crop growers will perform fewer tillage operations and go to more fuel-efficient machinery.

The primary motive for conservation, of course, is energy cost savings. No one doubts that energy prices will continue to go up and will be an ever-larger part of farm operating costs.

Your energy costs, plus the fact that you can cut back on energy use only so much without jeopardizing production, mean you will need to generate even greater efficiencies in farm production. As far as energy is concerned, the surest short-term way to become more energy-efficient is to manage to use less of it wherever possible.

2

Use What You Have First

Designing an energy-independent farm operation would be a simpler adventure if you could start with bare acres and a clean sheet of paper—and a wad of money.

A solar-heated farrowing house and pig nursery could go on the south slope of a hill; a methane digester would logically be located downhill from the hog buildings so that effluent could be fed into the gas plant by gravity; an ethanol "still" could be sandwiched neatly between the grain bins (equipped with solar dryers, of course) and the methane digester—handy to both feedstocks and a source of heating fuel.

Fortune seldom hands us such neat packages, however. For most, coping with mushrooming energy costs is a holding action. Most farmers must make do with what they have, at least until they can ease their way into alternative energy.

Many farmers are making rather modest investments now that will give them a head start in the years ahead, when most experts expect energy to be scarcer and more expensive. There are many ways to add energy-saving features to present crop and livestock facilities, without dipping too deeply into the farm's capital resources or cutting back on production. When new equipment or structures are planned, energy-saving features can be designed in from the beginning. An example is the open-front hog building shown in Figs. 2-1 and 2-2.

GETTING MORE WORK FROM
PURCHASED ENERGY

A good money-saving approach is to cut back on energy use, to get more work from each unit of energy purchased, while moving to renewable energy systems. In Chap. 1, we outlined several energy-saving management steps that can be put into practice at little or no extra expense. Now, we'll look at some relatively low-investment ways to make better use of energy sources already available on many farms.

There are still going to be energy bills to pay. The first step, regardless of the energy source, is to invest in those systems that will give the greatest return for each dollar spent. That means you will have to evaluate how well each conservation and renewable-energy measure would suit your own particular farm operation.

If you still have handy those energy bills you paid in the past 12 months, take another look at them now. For easy figuring, let's assume you spent $12,000 for fuel and electricity last year. If you multiply that amount by 10 (for an average interest rate of 10 percent) and capitalize your energy dollars out, you can get an idea of what kind of investment you could

FIG. 2-1   Open-front "cool" livestock buildings can usually be designed for natural ventilation, as is this modified Nebraska-type swine finishing building.

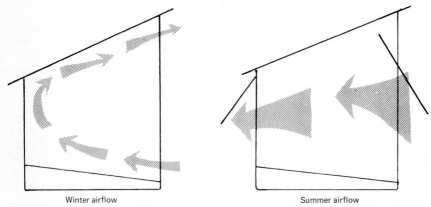

Winter airflow   Summer airflow

FIG. 2-2   Panel and shutter design controls winter and summer airflow in naturally ventilated livestock buildings. In winter, air enters through openings low in the front wall, then moves back along the sloped floor, picking up heat and moisture as it goes. The heated air rises, and is directed back out the front by the ceiling angle. In summer, larger panels are opened in both front and back for straight-through ventilation.

justify on the basis of present energy costs, to produce all your fuel and electrical power needs.

Let's say you could invest $120,000 in alternative energy systems with a life expectancy of 10 years or more, and that this investment would provide all your fuel and power needs for the next 10 years. The $120,000, spread over 10 years, would come to $12,000 per year.

It isn't quite that cut-and-dried, of course. There are tax incentives that let you get back part of the cost of alternative energy systems from Uncle Sam, and you can take investment tax credit on many of them (we'll go into that subject more thoroughly in a later chapter). And, if conventional energy costs double in the next 10 years, a system that makes you energy-independent would pay for itself much sooner.

On the other hand, if you must borrow all or part of the money, interest costs extend the time before the investment is paid back.

The above is a rather idealized example, as you can see. For most farmers, investing in equipment to produce *all* the fuel and electrical power needed is somewhat less than practical. Priorities must be put on limited dollars, whether the dollars are owned or borrowed.

The wisest course of action, generally, is to invest *first* in the materials, systems, and equipment that will give the greatest rate of return. As you review your energy expenses for the past year, notice which farm operations were the big consumers of diesel fuel, gasoline, LP gas (propane), and electricity. The energy source that requires the biggest cash outlay deserves your first scrutiny.

Suppose you are a dairy farmer with a 100-cow herd and are spending $2500 per year to cool milk, heat the milking parlor, and heat washdown water. Could you invest in heat exchangers to recover the waste heat from the milk-cooling tank and use it to heat water? Say you could invest $2500 to modify your present cooling-tank compressor. If the invest-ment saves $250 per year, that's a 10 percent return—not counting possible tax benefits or interest payments.

Obviously, that would be a better investment than spending $10,000 for a bank of solar collectors that only saves $250 per year. It's easier, too. That heat from the refrigeration unit removes heat from an existing source: milk. If it isn't being used now, it's exhausted outdoors and wasted.

The point of all this is: There are sources of energy on virtually every farm that are not now being used, or not being used to full advantage. In many cases, this energy can be put to work with a smaller capital investment than it takes to buy or build alternative energy equipment. The balance of this chapter goes into detail on several of these "use-what-you've-got-first" applications, but by no means exhausts the opportunities to get more mileage out of each energy unit purchased.

RECOVERING HOG HEAT FOR WINTER WARMTH

A sow and her litter of pigs produce about 600 Btu of sensible heat per hour, as a normal by-product of metabolism. However, they also produce a great deal of moisture, by respiration and from the evaporation of body wastes and spilled drinking water. The moisture must be removed by ventilation to provide a healthful environment for the animals, and fresh air must be brought into the building.

As the moisture is removed by a conventional ventilation system, it carries out much of the heat. Agricultural engineers estimate that up to 70 percent of the heating required in "warm" livestock buildings goes to heat ventilating air.

It doesn't have to be that way. A heat exchanger built into a specially designed ventila-tion system lets a Missouri pork producer heat his 20-sow farrowing house largely by reclaiming heat that otherwise would be exhausted from the building.

The heat exchanger is an assembly of 16 corrugated roofing sheets, each 10 ft long. A diagram of the unit is shown in Fig. 2-3. The completed unit measures 10 ft by 30 in by 7 in. Between each pair of metal sheets, in alternating fashion, warm air from inside the building is exhausted and cold air is pulled from the farrowing house attic. By passing cold air in one direction and warm air in the other direction, with only a thin layer of metal between, the heat exchanger recovers up to 70 percent of the building's heat.

"The beauty of it is, the colder it gets outside, the more heat we recover," says Bob George, agricultural engineer who helped the hogman, A. J. Cutbirth, design the unit. "The more temperature difference that exists between outside air and the temperature inside the building, the more efficient the heat exchanger becomes. It's better than solar heating, because it works 24 hours a day—not just when the sun shines."

A key element in making the heat exchanger work is a good air-filtering system. An

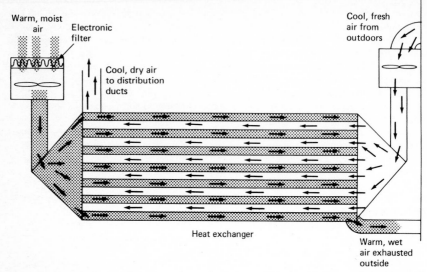

Warm, moist air

Electronic filter

Cool, fresh air from outdoors

Cool, dry air to distribution ducts

Heat exchanger

Warm, wet air exhausted outside

FIG. 2-3   How the heat exchanger works to salvage some of the heat that normally is exhausted to the atmosphere.

electronic filter on Cutbirth's device removes 99 percent of all dirt particles and much of the moisture from inside air. Without the filter, dust pulled through the exhaust blower would build up in the exchanger and gradually decrease its ability to recover the building's heat.

For warm-weather ventilation, larger exhaust fans are installed in the farrowing house wall. Zone heat for newborn pigs is provided by heat lamps, the only supplemental heat used in the building.

"At present energy costs, the heating bill for a 20-sow farrowing house can run $200 or more per month during the coldest part of winter in Cutbirth's region," says George. "A well-designed heat exchanger could easily save $500 per year on energy needed to heat the building."

At that rate, Cutbirth will repay the $700 cost of the materials in the heat exchanger in 1½ years. Commercial heat exchangers of similar design are now on the market, ready to be fastened to the wall and connected to exhaust and intake blowers. These units cost somewhat more than Cutbirth's homemade version, but should save enough energy to recover the original investment in three to four years.

The heat-exchanger idea can be used in any confinement livestock or poultry building where artificial ventilation is needed to remove moisture and odors.

USING SURPLUS HEAT IN DAIRY OPERATIONS

Milk cooling and water heating consume half to three-fourths of the total electricity used on a dairy farm. A 50-cow dairy operation may use 8000 kWh or more per month for water heating—or the equivalent in natural gas or LP gas.

Milk must be cooled from about 95°F as it comes from the cow down to about 38°F to keep in storage. Cooling 1 lb of milk by 1°F removes about 1 Btu of heat. For every 1000 lb of milk produced at each milking, about 60,000 Btu is removed when milk is cooled. In addition, the electrical energy in running the refrigeration cycle will produce 8000 to 15,000 Btu. So, each 1000 lb of milk cooled produces 65,000 to 75,000 Btu, the equivalent of burning 1 gal of LP gas at 70 percent efficiency.

Heat must be extracted from the milk; heat must be added to water and the milk-house space. Why not take the heat from the milk and add it to the water?

This can be accomplished with heat exchangers. The equipment converts an air-cooled refrigeration condenser to a water-cooled unit. Heat exchangers can be added to most present cooling-tank equipment or purchased with a new bulk tank. As a bonus, a water-cooled unit operates at 5 to 10 percent greater efficiency than does an air-cooled refrigeration system.

How much such a move will save on energy bills depends on what you are now paying for energy to heat water, and how much hot water you use. A heat-recovery system such as that outlined above, when used to replace an electric water heater, can save $39 per month if 100 gal of hot water is used each day and electricity costs 6 cents per kilowatt-hour. See Fig. 2-4.

More efficient cooling-tank–heat-exchanger systems will produce 1 gal of 140°F water for each 1 gal of milk cooled from 95 to 38°F (assuming a water supply that enters the system at 60°F or warmer from the well).

You can save even more on energy costs by installing a collection tank that lets you reuse some of the washing, rinsing, and sanitizing solution to wash down the floors of milking areas or holding pens. In some cases, where the washing solution is high in milk fat or the rinse solution has high acid content, it may be desirable to follow up with a final

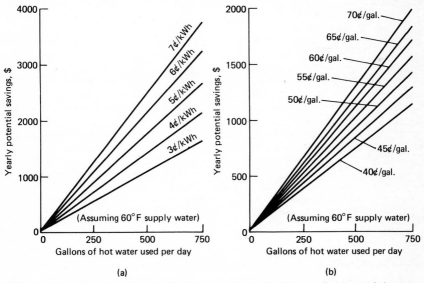

FIG. 2-4   Potential savings from the Fre-Heater, with *(a)* electric water heating and *(b)* LP-gas water heating, assuming water usage does not exceed Fre-Heater output. The amount of savings depends on the quantity of hot water used, the amount of milk cooled, and the fuel costs to heat water. *(Graph: Paul Mueller Company.)*

rinse of clear water. Still, it's another way to save water and the energy required to heat it.

Some dairy farmers who have installed these heat-recovery systems find that they suddenly have a surplus of hot water. Where it's convenient to do so, some have piped the hot water through insulated lines to their residence, to supplement the hot water required by the family.

CONSTRUCTING AN ALL-WEATHER STOCK-WATERING TANK

Livestock need a supply of clean, fresh water for efficient growth and performance—water that is neither too cold in winter nor too warm in summer.

While stock-tank heaters are not big energy users on most farms, the fuel they consume adds to the annual energy bill. An earth-insulated concrete watering tank, set below a pond or reservoir dam and equipped with a trickle overflow, can cut down on daily management chores and the need for external energy to prevent winter freezing.

This kind of freeze-free watering tank also permits livestock to be fenced away from the pond water source and thus eliminates the risk of animals falling through ice and drowning.

The following construction details, provided by the University of Missouri, describe how to build one such all-weather tank. The idea permits of several variations.

Concrete used should be of high quality. If you buy ready-mix, specify a 3000 psi (pounds per square inch) "break" mix containing 6 percent air entrainment, to provide resistance to surface scaling and to the effects of acids in soil and manure. If you mix your own, use washed gravel that contains no aggregate larger than 1 in in diameter. Use six sacks per cubic yard of concrete or cement that has an air-entrainment additive. Use just enough water to make the mix workable.

Construction Steps:

1. Excavate topsoil and all organic matter from the tank site.
2. Install water supply and overflow piping. The overflow should drain to an area on the surface where livestock will not trample the water into a mudhole. Place a screen over the overflow outlet, to keep out small animals and insects.
3. Backfill and tamp the ditches where drainpiping is laid. Place gravel to a depth of 5 or 6 in on the site where the tank will be built.
4. Cut and shape $3/8$-in reinforcing rods for the tank, as shown in Fig. 2-5. Wire all junctions of the reinforcing rods.
5. Assemble the forms as in Fig. 2-6. Apply a coating of grease or oil to the inside form corners to make disassembly easier. Use wing nuts on corner bolts on the inside of the forms so that the forms can be removed without destroying the lumber.
6. Do plumbing for the supply and overflow pipe inside the tank. Note that the overflow should be fitted so that it can be unscrewed to drain the tank completely. Plug pipe ends to keep out any spilled concrete.
7. Build tank cover forms on a flat surface and place reinforcement rods in them.

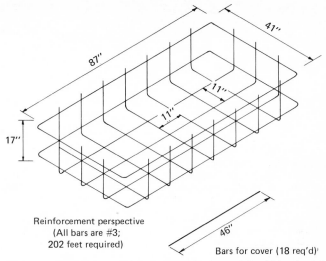

Reinforcement perspective
(All bars are #3;
202 feet required)

Bars for cover (18 req'd)

FIG. 2-5 Number 3 (⅜-in-diameter) reinforcing rod should be shaped to these dimensions for the freezeproof water tank. Tie joints and overlaps securely with wire.

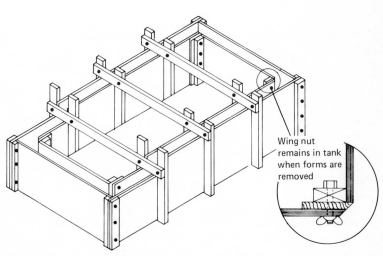

Wing nut remains in tank when forms are removed

FIG. 2-6 Isometric view of assembled forms, ready for pouring concrete. *(University of Missouri.)*

8. Pour concrete in the forms, starting with the wall sections. Work the concrete just enough to make sure there are no voids in it. A depression around the overflow drainpipe should be shaped by hand, to allow for draining the tank for occasional cleaning.

9. Let the concrete set for three days—five is better—before removing the forms.

10. Install the shutoff valve and float valve, and trim the overflow pipe to the proper length.

11. After cover slabs have cured for at least seven days, unform them and install them on top of the tank.

12. Install a retaining wall bulkhead of treated posts and lumber, to retain the earth backfill that will insulate the tank.

13. Form and pour a slab in front of the tank. A step, as shown in the cross-sectional view in Fig. 2-7, helps to discourage cattle from fighting at the tank and prevents the deposit of manure in the tank.

14. Backfill earth around and over the back of the tank. The fill should be at least 24 in deep, for freeze protection.

15. Open the water supply valve to the tank and adjust the float valve so that the normal water level is about $\frac{1}{4}$ in below the top of the overflow pipe.

16. Drill two $\frac{5}{16}$-in holes in the overflow pipe, one just below the normal water level of the tank, the other just below that. Plug the holes with two $\frac{1}{4}$-in brass machine screws. In extremely cold weather, one or both of these screws can be removed to allow a small amount of water to flow through the overflow pipe. This brings in more water from deep in the pond and prevents freezing.

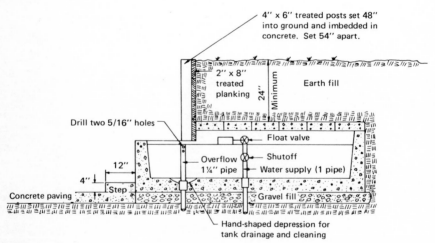

FIG. 2-7 Cross-sectional view of completed tank and earth cover. The hand-shaped depression allows for complete drainage of the tank. *(University of Missouri.)*

If built properly and dismantled carefully, the forms can be disassembled and reused many times.

None of the above energy-saving features will divorce your farm from the fuel truck or the electricity power line. But every move you make now to reduce your dependence on outside sources of energy will pay bigger and bigger dividends in months and years ahead.

3

Livestock Buildings

According to estimates by the farm building industry, the typical farm building contractor will erect more than 40 structures this year, at an average cost of more than $40,000 per building. More than half of these buildings (8 out of 10 of those built to house livestock) will be insulated to some degree.

If you belong to the 50 percent of farmers who will build, remodel, or add to a building in the next 12 months, the economics of incorporating energy-saving features has never been better. Of course, the time, effort, and materials that go into new construction or retrofit projects also represent energy consumed—and dollars spent. But planned quality construction is, in the long run, the most economical construction.

Energy-wise construction starts with an accurate assessment of the climatic region, expected temperature extremes, winter and summer sun angles (see Figs. 3-1 and 3-2), building location, and building use. Consider the following when planning the location of a livestock building.

- Drainage should be away from the home and other buildings. Manure-laden runoff should not enter waterways leading to streams or lakes.

- Livestock buildings should be placed downwind (with respect to prevailing breezes) from the home.

- Future plans for expansion should be considered. Leave room for more buildings to be located in relation to feed storage and roadways.

- Feeding and marketing access should be planned as part of the building project, as should lots, pens, sorting areas, and loading/unloading facilities.

- Other factors, such as the accessibility of water and utility lines, should also play a role in the selection of building location and design.

Once the building's location is selected, make sure the site is prepared for fast, efficient construction. A well-organized system for handling building materials saves time if you're doing the building yourself, and saves labor costs if you're hiring someone to do it.

Plan the concrete work for footings, foundations, floors, and paved areas for both strength and ease of maintenance. The design of much concrete work will be determined in livestock buildings by the system of waste disposal: gutter flushing, manure pits, "self-cleaning" sloped floors, etc.

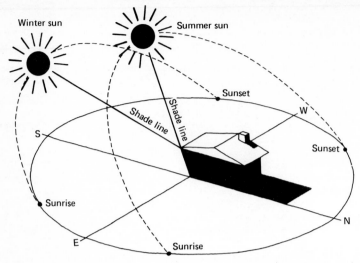

FIG. 3-1 Building design should take into account winter and summer sun angles for the region. Above shows position of the sun at noon for latitude 40° north.

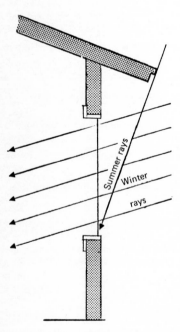

FIG. 3-2 Roof overhangs can be designed to shade windows in summer, yet allow sun's rays to penetrate the building in winter. This is the simplest form of solar heating.

ENERGY FOR FARM BUILDINGS

The following information on the subject of energy in farm buildings, particularly in livestock housing, is based largely on material and counsel from Richard E. Phillips, agricultural engineer, University of Missouri.

The main purpose of animal housing is to provide for an altered environment that will promote more efficient growth and production of livestock and poultry. The one element that is most often artificially modified is temperature, either by removing or by adding heat. Moisture and odor control are somewhat related considerations.

All creatures (including human beings) have a temperature range at which they are most efficient at production. Within a given species, this range may change with age and size of the animal. Young animals typically require higher temperatures for well-being and to perform at maximum levels.

Some buildings also may have higher temperature requirements than are dictated solely by animal performance. For example, cattle perform well at temperatures down to 0°F. However, water lines freeze and the people who must manage the cows are decidedly uncomfortable at temperatures around 0°F.

Animal confinement buildings generally fall into three categories of temperature control:

"Warm" buildings are maintained at relatively high constant temperatures to meet the needs of animals at certain stages of life or growth. Supplemental heat is usually required in cooler weather. Buildings for farrowing pigs, brooding poultry, and growing young calves fall into this category.

"Cool" buildings are designed to house animals at high enough density so that, with proper insulation, radiated body heat will ordinarily keep winter temperatures in the optimum performance range. An example would be semiconfinement structures for finishing hogs.

"Cold" buildings are designed to protect animals from the direct blasts of extreme weather, with little control over the temperature inside the structure. These buildings may contain some insulation to control condensation in winter and heat gain in summer, but no supplemental heat is used. Sheep housing, dairy free stalls, confinement beef feedlots, and housing for swine breeding herds generally fall into this category.

As noted earlier, warm-blooded animals produce a certain amount of body heat. In some housing systems, it is possible to retain this body heat with structural insulation and use it to maintain building temperatures in the desired range, with little or no outside energy sources used. However, ventilation is a critical feature of successful livestock structures. It serves four principal functions:

1. It provides a source of oxygen to meet respiration needs of the animals.

2. It removes excess body heat produced by animals and helps maintain desired temperature ranges.

3. It dilutes airborne disease organisms with fresh air, reducing the potential for disease spread.

4. Perhaps most important, it partially removes excess moisture from the building.

Estimated animal heat production and recommended winter *minimum* ventilation rates when supplemental heat is required are listed in Table 3-1. In warmer weather, normal rates and installed ventilation capacity would be much higher.

TABLE 3-1   Estimated Animal Heat Production and Recommended Winter Minimum Ventilation Rates

Animal	Sensible heat produced, Btu/h	Minimum ventilation, cfm
Dairy animals: .		
Adult	2,900 per 1000 lb	25 per 1000 lb
6 to 12-month calf	675 per animal	8–10 per 100 lb
Swine:		
Sow and litter	600	20
50-lb pig	220 per animal	2–3 per animal
100-lb pig	310 per animal	5 per animal
200-lb pig	460 per animal	10 per animal
250-lb hog	520 per animal	10 per animal
Poultry:		
Laying hen	3.2 per lb	0.125 per lb
5-week-old broiler	7 per lb	0.125 per lb
10- to 20-lb turkey	3.5–4 per lb	0.125 per lb
Sheep:		
Adult, with fleece	2 per lb	4
Adult, sheared	3.8 per lb	4

CALCULATING HEATING SYSTEM SIZE

There are several ways to estimate the size of equipment needed to provide supplemental heat for a building and to estimate the amount of energy used. Residential heating engineers commonly use *heating degree days* to estimate the seasonal heating requirement. Heating degree days are determined by subtracting the average outside temperature from a base of 65°F—which assumes that no supplemental heat will be needed when temperatures are at or above 65°F.

Let's say the average temperature for October 30 is 45°F. The degree days for that particular day would be 20, or 65 minus 45. Degree days are cumulative. To find the degree days for a month or year, simply add together all the degree days for that period. A chart of average heating degree days per month for selected U.S. locations is included in App. B (Table B-3).

Heating degree days can give you a good idea of how many days you can expect to need supplemental heat in a dwelling or heated livestock building and about how much additional heat must be supplied for each month during the heating season. The next exercise preliminary to installing a heating system in your livestock or poultry building is to calculate heat loss, and Richard Phillips takes us through the steps:

"For calculation purposes, heat loss [of a building] is divided into two components, conductive and infiltrative, or ventilation loss," he says. "Conductive loss is that heat which is lost directly through walls and ceilings and is usually calculated separately for each component, as well as for windows and doors. The values then are summed for total conductive loss."

Here's the formula for calculating conductive heat loss:

$$HL = \frac{A}{R} \times D$$

Where HL = heat loss, Btu/h
      A = area of component (door, wall, etc.), ft²

R = insulative R value of component

D = temperature difference between inside and outside, when outside temperature is at expected minimum

"Infiltrative or ventilation heat loss is the amount of heat required to warm up the air that either leaks into the building naturally or is drawn in by the ventilating system operating at the minimum level," notes Phillips.

For most insulated, well-constructed buildings, normal leakage is calculated at one complete air change per hour. This is compared with the minimum ventilation rate, and the larger of the two is used to determine infiltrative heat loss, by this formula:

$$HL = 0.018 \times D \times V$$

where 0.018 is constant, D is the temperature difference between inside and outside at the lowest temperature, and V is the air exchange rate in cubic feet *per hour*. If the normal infiltration leakage value is used, V would be the volume of the building in cubic feet. If the minimum ventilation rate is greater than the leakage rate, V would express the minimum ventilation rate in cubic feet per hour. HL again represents heat loss in Btu's per hour.

The total heat loss of a building is the sum of conductive and infiltrative losses. Often, this is used directly to determine the size of supplemental heater required for a building. A more accurate method is to take the total heat loss figure and subtract the expected animal heat production from it.

For example, calculate the size of supplemental heating equipment needed for a 16-sow farrowing house that is 24 by 50 by 8 ft. The building is insulated with R-13 in the walls and R-23 in the ceiling.* There are no windows and two 3- by 7-ft doors with an R-2.17 value each. The inside temperature of the building is to be kept at 60°F. What size heating system will be needed on a day when the outside temperature is −10°F?

First, compute the conductive heat loss:

Component	Area		R value	Difference	Heat loss
Walls	8 × (24 + 24 + 50 + 50)	÷	13	× 70	= 6,375
Ceilings	24 × 50	÷	23	× 70	= 3,652
Doors	3 × 7 + 3 × 7	÷	2.17	× 70	= 1,354
				Total conductive loss	= 11,381

Then, compute the infiltrative heat loss:

Minimum ventilation rate for 16 sows is 16 times 20 cfm (from Table 3-1) times 60 to give the rate in cubic feet per hour = 19,200. Since ventilation is larger than the normal infiltration rate of one air change per hour, use it to estimate infiltrative heat loss:

$$HL = 0.018 \times 70°F \times 19,200 = 24,192 \text{ Btu/h heat loss}$$

Now, add together conductive and infiltrative heat losses to determine the total heat loss in Btu's per hour:

$$11,381 + 24,192 = 35,573$$

As noted in Table 3-1, one sow and her litter produce about 600 Btu of sensible heat per hour. If the farrowing house is full, the 16 sows and litters will produce 9600 Btu/h

---

*How well a material resists heat flow is commonly measured by what are called *R values*. The more resistance to heat flow, the higher the R value.

to help heat the building. Subtracting this number from the 35,573 Btu/h heat loss gives 25,973 Btu/h as the size of supplemental heating system required.

Of course, this is based on a fully occupied building. In many cases, it's advisable to build in some reserve heating *capacity* to ensure against the possibility of the building being less than full when the coldest weather sets in.

ESTIMATING HEATING FUEL NEEDS

When it comes to estimating *energy requirements* for building heat, a more reliable figure to use is the long-term average daily temperature in your area, rather than the expected extreme low temperature. You'll need a heating system with a capacity to meet the worst temperature, whereas the actual weather is usually somewhat less demanding. In figuring actual heat energy needs, consider the heat animals produce within the building.

For example, if the average temperature in January is 30°F, how many Btu's of supplemental heat energy will be required for the entire month for the farrowing house described above, assuming that the house is occupied by 16 sows and litters of pigs? By working through the same exercise as above (except with a 30° temperature difference, rather than 70°), we come up with a total heat loss of 15,180 Btu/h.

Subtracting the 9600 Btu/h body heat generated by the animals in the building leaves 5560 Btu/h heat loss. That's 133,440 Btu/d, or 4,136,640 Btu of supplemental heat needed for January with an average daily temperature of 30°F.

How much fuel oil would you need to buy to meet January's heating requirements in the example above? How much LP gas? Electricity?

Most of the energy used to heat livestock buildings is supplied by petroleum fuels or electricity, although alternative sources such as solar, wood, and geothermal (earth) systems are catching on rapidly. Energy is sold in various units: gallons of fuel oil or LP gas, cubic feet of natural gas, kilowatthours of electricity. But a livestock producer actually buys Btu's of energy, and the selling units are not much of an indication of the actual energy content of the fuel purchased, in every case.

Refer to Table 3-2 for comparisons of different fuels and heating systems. How much LP gas would it take to provide the 4 million Btu of heat needed by the farrowing house described earlier, for the month of January? If the heat is provided by equipment that is 80 percent efficient, about 56 gal of LP gas would be needed.

Suppose LP gas sells for 60 cents per gallon. What would the equivalent costs for other heating fuels be? Figure 3-3 compares costs of various fuels. Lay a ruler or other straight-edge horizontally across the figure to intersect the LP-gas column at 60 cents per gallon. An equivalent amount of heat energy from other sources would have the following costs:

Natural gas	65 cents per cubic foot
No. 2 oil	80 cents per gallon
Electricity	2.75 cents per kilowatthour
Coal	$130 per ton
Wood	$148 per cord

This compares fuels on the basis of actual Btu's of heat energy delivered, at the heating system efficiencies shown in Table 3-2. However, some fuels—such as LP gas and electricity—incorporate more convenience of use and ease of regulation than others, such as coal or wood. This will also be a factor in your decision, as will the long-term outlook for a particular fuel's availability.

There are other considerations when choosing a heating system and fuel source for a

TABLE 3-2   Comparison of Fuels and Heating Systems

Source	Selling unit	Potential heat per unit, Btu	Average system efficiency, %	Actual heat per unit, Btu
Natural gas	100 ft³	100,000	80	80,000
LP gas	Gallon	92,000	80	73,600
No. 2 fuel oil	Gallon	140,000	70	98,000
Electricity	kWh	3,413	100*	3,413
Coal	Ton	25,000,000	65	16,250,000
Wood	Cord†	27,500,000	65	17,900,000

*Resistance heaters (baseboard, radiant, ceiling cables, etc.) are 100% efficient. For central electric furnaces, deduct 10% for duct losses.
†Based on 128 ft³ of air-dried hardwood.

warm livestock or poultry building. As outdoor temperatures increase, the ventilation requirements for moisture removal increase. As *indoor* temperatures increase, the ventilation requirements for moisture removal decrease rapidly, because air at 70°F will hold nearly twice as much *absolute* humidity as air at 60°F.

Radiant-type heaters (such as electric heat lamps) heat animals and objects in the building but do not add much heat to the air. And, as warm air can carry more moisture than cooler air, the type of heating system installed will in part influence ventilation levels. The 20-cfm minimum ventilation rate per sow and litter, as noted in Table 3-1, is

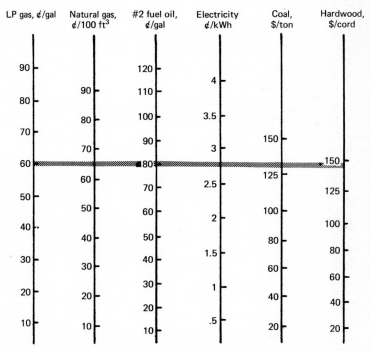

FIG. 3-3   Equivalent costs for various farm fuels on the basis of available Btu's of heat energy. Shaded line indicates the example problem discussed in this chapter.

adequate for indoor temperatures of about 70°F. If the indoor temperature goes down to 50°F, less than half the moisture produced by the animals will be removed by a 20-cfm ventilation rate.

Obviously, as indoor temperature is increased, conductive heat loss increases—because of the greater difference between indoor and outdoor temperatures. However, ventilation heat loss decreases rapidly at higher inside temperatures and can result in an overall reduction in the rate of total heat loss.

This means that at higher indoor temperatures buildings can be dehumidified with less ventilating energy, since heat increases the moisture-carrying capacity of the air. With some buildings and some types of livestock, it may be more economical to remove part of the moisture with condenser-type dehumidifiers, as long as enough ventilation is provided to meet the animals' need for oxygen and to provide enough fresh air to diffuse disease-causing organisms in the building.

In the chapters ahead, we discuss alternative sources of energy—solar, underground, wood—that can cut the conventional energy requirements of confinement buildings. The big question, as a producer considers these alternative sources, is: How much investment is justified by savings on conventional energy?

Bear in mind that many alternative energy systems will still require conventional heating plants as a backup—often heating plants of as large capacity as if the alternative system were not installed. Also, some alternative energy systems may require a change from one conventional energy source to another. Solar heating, for example, is not very compatible with electric radiant heat. Solar energy typically heats either air or water, which distributes the heat through a building, thus requiring either ducts or piping. Much the same is true of wood.

You'll also want to consider the expected rate of increase in conventional energy costs. Trying to predict energy costs with accuracy is a hazardous undertaking these days. But your net savings on an alternative energy system often will depend, in large part, on how these costs change over the life of the investment. In the past decade, costs of conventional sources of energy have climbed at an annual average rate of slightly more than 10 percent, with the sharpest increases in petroleum fuels coming in the latter part of the decade.

Finally, you will want to select an alternative energy system that has an expected useful life nearly equal to the expected life of the building.

4

Insulation and Ventilation

A key element in the successful environmental control of livestock and poultry structures is insulation. However, extra insulation does not return the same benefit in animal housing as it does in human dwellings, because of the ventilation requirement, as noted earlier. The "point of diminishing returns" is reached sooner when you add more insulation to a farrowing house, for example, because much of the heating requirement is for cold, fresh air pulled in by ventilating fans.

In an average "warm" confinement livestock or poultry building, the largest fraction of structural heat loss is conductive loss through the building's perimeter. Adding 2 in of a foam insulation—such as Styrofoam®—can cut the total structural loss by about 20 percent. Adding still more insulation to bring walls to R-25 and ceilings to R-34 can result in another 20 percent savings.

However, the additional energy savings do not justify piling on insulation beyond a certain optimum. A 20 percent savings in conductive heat loss through the walls, ceilings, doors, and windows may result in only a 5 to 10 percent saving in the heating bill, again because of the ventilation requirements of closely spaced livestock or poultry in the building.

Therefore, a building designed for energy efficiency incorporates a balance of the factors that bear on energy consumption: insulation, ventilation, size and shape of the building envelope, floor type and material, arrangement of objects within the building, and so on.

"Most swine farrowing buildings are overventilated and are wasting heating energy in winter," says Dwaine Bundy, agricultural engineer, at Iowa State University. "As much as 90 percent of the thermal energy required to heat a farrowing building during winter goes into heating the ventilation air."

Bundy has compared effects of different waste-handling systems, floor systems, and air distribution patterns on the ventilation rates required in farrowing houses. Much of what he found can be applied to other "warm" livestock and poultry confinement buildings.

For example, his three-year study showed that a ventilation rate as low as 10 cfm per sow and litter can maintain good air quality in a well-designed total building package. He also found that commercial ventilation companies often do not design their systems with enough range in capacity. Many, for instance, had a minimum ventilation rate as high as 40 cfm. Others, conversely, were not large enough to provide adequate air flow during hot weather.

"One reason companies make ventilation systems so high in minimum capacity is that air distribution inlets were not well planned," he says. "As long as energy costs were low, this was less important. Now, they are rethinking inlet distribution."

In fact, energy costs are forcing more coordination of all features that affect energy consumption in buildings. Fully slatted floors, for example, are not needed in farrowing buildings; slats are necessary only near the sow's head—under the feeder and waterer—and under the dunging area.

"Where there's no moisture to evaporate, you don't need slats," says Bundy. He also notes big differences in winter ventilation requirements for buildings with different floor systems.

Both ammonia levels and relative humidity remained in acceptable ranges with partially slatted floors over manure pits and over gutters that were flushed twice weekly, even at ventilation rates as low as 10 cfm per sow and litter. In buildings with solid concrete floors, the ammonia level was five times higher and the relative humidity was dangerously high when the ventilation system was operated at such a low rate.

In short, a confinement livestock production system must be designed and operated not only to make the most economical use of each energy unit consumed, but also to provide an environment that is most suitable for the health, gain, and feed efficiency of animals housed. Optimum energy efficiency is not necessarily achieved by layering on more insulation (although insulation is a vital element) or by shutting down ventilation blowers in winter.

HOW MUCH INSULATION?

Most building materials possess a degree of insulating ability, but for our purposes, the term *insulation* refers to those materials which have relatively high resistance to heat flow.

Heat moves from one location to another by conduction, convection, or radiation—or a combination of these methods. Heat always moves from a warm area or surface to a cooler area or surface. Conducted heat passes from one location to another by warming the material that separates warm from cool areas. Convected heat is transmitted by a moving fluid, such as circulating air. Radiated heat involves the passage of heat through a space without warming the space, as when the sun's rays heat the earth or objects without appreciably heating the air. In a farm building, heat is lost and gained by all three methods.

The transfer of heat through an insulated wall is rather complicated and may involve all three of the methods of heat transfer. But simple methods and materials can economically and effectively resist the flow of heat. Air itself, depending on its position, how thick the layer is, and how well it is trapped between surfaces, can be a good insulating material.

A reflective material, such as metallic foil, placed so that there are air spaces around it, can reflect nearly all the radiant heat that strikes it, although the foil has low R value when it comes to resisting conductive heat loss. Also, only part of the heat to be retained in a building is radiant heat, so other insulating materials are needed to resist the flow of heat by conduction and convection.

The amount of insulation to use for the most *economical* resistance to heat flow depends on several factors, and at best is a compromise between the cost of additional insulation and the savings in heat energy. For one thing, insulation does more than merely help conserve heat in cold weather. Insulating materials also help reduce the rate of heat *gained* inside the building in hot weather. Temperatures of walls and roofs exposed to direct sunlight can be 60°F higher than air temperature. Insulation helps prevent this heat from migrating into the building's interior. Adequate insulation, in tandem with ventilation to remove excess moisture from the building, can also control condensation ("sweating") on inside surfaces. In a poorly insulated building, inside wall and ceiling surfaces become cold in winter. If these surfaces become cold enough, the air next to them becomes saturated (as air cools down, it is able to hold less moisture) and moisture condenses.

There is, as mentioned, a strong relationship between increased insulation and energy savings, but the economics of adding more insulation depend upon the difference between the investment in added materials and the value of energy saved. That holds true even in climates with bitterly cold winters. Swedish engineers Krister Sallvik, Christer Nilsson, and Sven Nimmermark computed the value of added insulation to swine and poultry buildings, taking into account the expected increase in energy costs over the life of the buildings. For a farrowing-nursery building in central Sweden (designed for a minimum outside temperature of $-21°C$) the engineers determined that the optimum, or maximum-profit, thickness of insulation was 90 mm in walls and 150 mm in ceilings, if the energy price increased by 6 percent annually.

As a livestock producer or farm builder selects the amount and kind of insulation for a particular building, a number of questions need to be taken into account. How is the insulation to be used? Is it purely for heat-flow resistance, or must the material have some structural qualities? How adaptable is the insulation material to the particular use in the building? What is its cost in relation to that of other materials? Can it be applied easily? How about moisture resistance and fire safety?

Table 4-1 lists R values for commonly used construction materials. In general, the heavier and denser a material, the poorer its insulating properties. The most common insulating materials are bulky, porous, and lightweight and contain a multitude of air spaces. Many common insulating materials—fiberglass, cellulose, mineral wool, vermiculite—are packed in bags as loose fill, especially adapted for use in the ceilings of new and existing buildings. These loose-fill types can also be used in side walls, but special provisions must be made to prevent the material from settling and leaving part of the wall uninsulated. A separate moisture barrier must be provided with loose-fill insulation to prevent condensation and moisture migration into the insulation.

Other insulating materials are made in batts and blankets of widths to fit in the spaces between standard framing member intervals of 16 and 24 in. Insulation in this form is most adaptable to side-wall application, but needs a vapor barrier, unless the batts or blankets are manufactured with a vapor shield attached.

Rigid insulation board provides some structural strength as well as heat resistance. Some such products have a vapor barrier on the inside surfaces; others must be protected with a separate vapor barrier. Rigid insulating board commonly is used for sheathing and roof insulation in "cool" and "cold" animal buildings.

Rigid foam insulation is made of several materials—polystyrene, polyurethane, foam glass—and typically comes in the form of blocks and boards. Compared with fiberglass and mineral wools, this material is fairly expensive to use for general insulation, but is well suited for perimeter or floor insulation to control condensation around the edges of concrete floors and prevent heat loss through concrete foundations and floors.

A growing practice is to place insulation on the outside of concrete (between the concrete and the earth fill or earth subgrade) to allow the mass of masonry to act as a "heat sink," particularly in solar-heated buildings, as shown in Fig. 4-1. However, some rigid foams deteriorate when in contact with moisture. Also, some are flammable and should not be used without a fireproof covering.

Reflective insulation—usually metal foil—is efficient at resisting the downward flow of heat and is effective in keeping out summer heat. However, dust and corrosion quickly reduce the value of reflective insulation; some care should be taken in locating this type of material.

Plastic foams, so-called "foamed-in-place" insulating materials, are made by foaming organic plastics with air or inert gases. Most common plastic foams are polystyrene, polyurethane, and urea-formaldehyde. These materials are often used to insulate existing wall cavities and to insulate interior liners for farm buildings, such as a concrete-block wall that separates a farrowing area from a nursery or growing area in a building.

New developments in concrete construction have given rise to several precast and

TABLE 4-1   Insulating Values for Various Materials

Material	Thickness, in	R value
Insulation:		
Glass fiber batts	1	3.7
Rock wool batts	1	3.7
Rigid fiberglass board	1	2.5
Extruded polystyrene, plain	1	4.0
Extruded polystyrene, expanded	1	5.0
Milled paper, wood pulp	1	3.7
Sawdust, wood shavings	1	2.2
Rock wool, loose fill	1	3.0
Fiberglass, loose fill	1	3.0
Vermiculite	1	2.3
Expanded polyurethane	1	6.25
Urea-formaldehyde foam	1	5.0
Building Materials:		
Common brick	4	1.0
Poured concrete	1	0.08
3-core concrete block	8	2.0
Beveled wood siding	½	0.8
Plywood siding	⅜	0.6
Metal siding, insulating backing	⅜	1.8
Hardwood lumber	1	0.9
Softwood lumber	1	1.25
Gypsum board	½	0.45
Hardboard	½	0.7
Particle board	1	1.1
Building Fixtures:		
Solid-core wood door	1¾	2.17
Solid-core wood door with metal storm door	—	3.23
Metal exterior doors:		
Urethane core	1¾	2.5
Polystyrene core	1¾	2.1
Windows:		
Single pane	—	0.88
Double pane	—	1.79
Insulating glass (³⁄₁₆-in space)	—	1.45
Insulating glass (½-in space)	—	1.72

"plastered" concrete techniques that incorporate insulation in the masonry. Concrete is a poor insulator but possesses the ability to store heat energy and release it slowly. By incorporating insulating materials in the concrete, or by constructing a sandwich-type wall with inner and outer layers of concrete around a core of rigid insulation, both structural strength and heat resistance are achieved.

The only fair way to compare insulation materials for their relative heat resistance is by the cost per R value. Often, insulation contractors and manufacturers quote R values on an *installed* basis, thus taking credit for the insulating qualities of the siding, interior wall covering, etc., in their claims for the product. You can spot this sales gimmick. If the same material is quoted at different R values for its use in walls, ceilings, and floors, you can be pretty sure the quote is on the basis of the insulation plus other building materials installed.

Recent federal legislation requires that insulation be labeled with the R value per unit

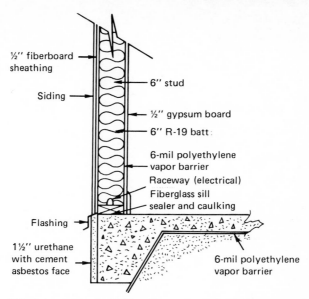

½" fiberboard sheathing

Siding

6" stud

½" gypsum board

6" R-19 batt

6-mil polyethylene vapor barrier
Raceway (electrical)
Fiberglass sill sealer and caulking

Flashing

1½" urethane with cement asbestos face

6-mil polyethylene vapor barrier

FIG. 4-1 One method of installing perimeter insulation on a concrete slab floor. *(Owens-Corning.)*

of thickness. If the insulation supplier can't or won't provide this information, you may want to shop elsewhere.

To calculate the cost of an insulating material on the basis of its resistance to heat flow, compute the cost per net square foot of the material. Suppose you want to install more insulation in a building attic, to achieve a total R value of R-19. You price one kind of insulation that has an R value of 3.2 per inch of thickness and costs 7 cents per square foot. Another type has an R value of 3.7 per inch of thickness, but costs 8 cents per square foot. Which is the better buy at an insulation thickness to give R-19?

You'd need 5.9 in of thickness of an insulation of the first type to total R-19. At 7 cents per square foot for a 1-in-thick layer, that's 41.3 cents per square foot. With the second type of insulation, you'd need only 5.1 in of thickness to get R-19. At 8 cents per square foot, your cost would be 40.8 cents per square foot.

However, the amount of insulation needed in a farm building will depend on the expected outside minimum temperature, the number and size of animals housed, and the inside temperature to be maintained. The thickness of insulation (thus the resistance to heat transfer) is a problem to be solved by the designer of each building, after considering the initial cost of the material and its installation, the effects of condensation, operating costs for heating (and cooling) based on present and projected energy costs, and the specific nature of the building and its inhabitants being insulated.

Minimum insulation recommended for the five temperature zones of the contiguous 48 states is shown in Fig. 4-2.

However, in most cases, additional insulation in "warm" or supplementally heated livestock and poultry buildings would pay an economic return on the extra cost of the materials. Don't *under*insulate. Insulation is the most economical part of the construction in a new building and is fairly easy to install in most existing structures. If it's inadequate or poorly installed, you'll be paying the bills for it as long as you operate the facility.

FIG. 4-2 Recommended R values as insulation minimums in each of the five temperature zones:

Zone	Roof	Walls
1	8	4
2	8	6
3	12	8
4	16	12
5	20	16

The R value is the easiest to calculate of several ways to measure insulation's heat resistance and is the measurement most often used by the building industry. However, there are other means and symbols of expressing the insulating value of a material.

K value gives the amount of heat in Btu's per hour that will pass through a piece of insulating material one inch thick and one foot square when the temperature differential between the two sides is one degree Fahrenheit. *C value* is similar to K value, except that it expresses the insulation value for a total thickness of material—again in Btu's per hour at the same standards. *U value* is the overall coefficient of heat transmission, or the amount of heat in Btu's that will pass through a complete wall, ceiling, or floor construction in one hour per degree of temperature difference between the air on the warm side and the air on the cool side of the structure.

Do a careful job of installing insulation. Follow the manufacturer's directions. He knows his product and how to install it for best performance. Always install a continuous vapor barrier on the "warm" side of insulating materials. Moisture is ever present in animal housing; if it isn't controlled, it can ruin the insulating value of materials. See Fig. 4-3.

For slab floors, insulation often is more effective at reducing heat loss when placed under the concrete slab than when placed vertically on the foundation wall. For best resistance to heat flow, use both. Methods are shown in Fig. 4-4. Where floors are designed to be used as part of the storage for heat energy—as in some solar heating systems—it may be economical to install insulation under the entire floor area. In other cases, except where

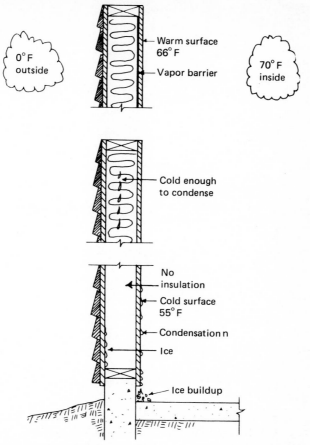

FIG. 4-3 Condensation occurs when warm, moist air encounters a cold surface. Insulation helps control "sweating" by keeping wall and ceiling surfaces warmer. *(Midwest Plans Service.)*

the soil stays wet for long periods, a 4-ft-wide strip of insulation under the perimeter of the slab floor may be adequate.

One final note on insulation: Check with your insurance agent and lender before installing "foamed-in-place" plastic insulation materials. Some of these products are highly flammable and can give off toxic fumes when burning. Insurers and lenders may be reluctant to insure or finance buildings with these materials installed. In any case, plastic foams should be protected with a fire-retardant covering over both inside and outside surfaces.

HOW MUCH VENTILATION?

A mechanical ventilation system is essential for virtually all "warm" animal confinement housing and is needed for some "cool" buildings as well. Earlier, we pointed out the major

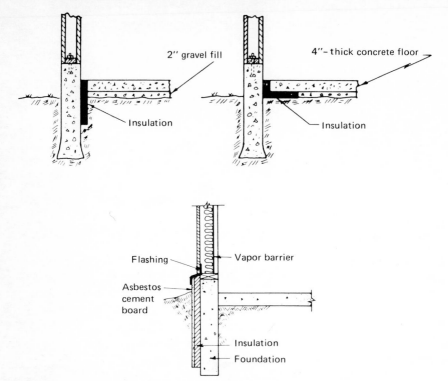

FIG. 4-4 Optional methods of insulating concrete slabs. *(Agricultural Engineering Department, University of Missouri.)*

reasons for ventilation: to control moisture, to provide fresh air needed by animals, to move enough air to dilute disease organisms in the air, and to control or moderate extremes of temperature.

In most buildings, if temperature control and respiration needs are met, moisture buildup and airborne-disease control will also be at satisfactory levels. With a well-designed, well-operated system, cool, dry air is drawn into the building. As the air warms inside the building, it picks up moisture. The warmed, wetter air is exhausted from the building. The objective is to design a ventilation system that provides *just enough* cool outside air to accomplish these functions, but not so much cool air that an unnecessary load is placed on the supplemental heating plant.

Ventilation air requirements vary from low levels in winter, when the system is removing heat supplied by supplemental heating equipment, to many times those levels in summer, when the main purpose of ventilation is to remove *excess* heat. The ideal ventilation system would be infinitely variable, from minimum to maximum capacity.

Unhappily, the systems that would be ideal for most confinement animal housing are priced too high to be used for that purpose. The best compromise is more often a system that will provide three or more levels of air movement through the building. Table 4-2 lists ventilation requirements for several species of livestock and poultry commonly housed in "warm" buildings. Use Table 4-2 as a guide for sizing continu-

TABLE 4-2   Ventilation Requirements for Various Farm Animals

Animal	Desired indoor temperature, °F	Ventilation rate per listed animal unit, cfm		
		Winter minimum	Winter normal	Summer
Sow and litter	60–80	20	20	210
40-lb pig	70	2	15	36
100-lb pig	60	5	20	48
150-lb pig	60	7	25	72
250-lb pig	60	10	35	120
Dairy cow	55	35	100	200
Dairy calf	55	10	50	75
Beef (per 1000 lb)	55	15	100	200
Poultry (per lb)	55–60	0.125	0.5	1

SOURCE: Richard E. Phillips, agricultural engineer, University of Missouri.

ous-run *winter minimum* ventilation capacities, unless your operation has special requirements.

Rates shown in Table 4-2 are totals for each level and are not additive. In other words, a total of 120 cfm of air per 250-lb pig is required for summer ventilation. This might be provided by a winter minimum fan moving 10 cfm, a winter average fan providing another 25 cfm and a summer fan providing 85 cfm—all operating to make the needed summer level.

Basing ventilation systems on animal weight or per-head populations partially eliminates animal density in the building as a factor in the performance of the system.

A single-speed fan with an interval timer, as shown in Fig. 4-5, allows a wide range of ventilation capacity, but requires manual resetting of the timer for each adjustment. A 1000-cfm fan set to run 3 min and shut off for 7 min provides an average air flow of 300 cfm. However, such an intermittent system can cause rapid temperature changes in a building and should be equipped with anti-backdraft curtains on air inlets.

Variable-speed fans can control airflow from a minimum of about 20 percent to full

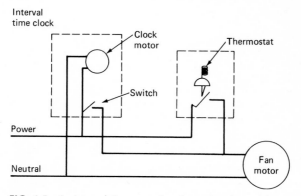

FIG. 4-5   An interval timer runs fans intermittently to deliver a minimum volume of air. Usually, a thermostat is wired to override the time clock so that fans run continuously at higher temperatures.

capacity. Solid-state controllers regulate voltage going to capacitor-started, capacitor-run motors. When fan controls are set at 60°F, the fan operates at almost full capacity when the temperature reaches 65°F and cuts back to minimum speed when temperature drops to about 55°F.

Whatever the type of fans used, the lowest or minimum level of capacity should be designed to provide just enough airflow to meet respiration requirements when the fan runs continuously. In most buildings, at least one fan should operate all the time when the inside temperature is above 35°F. See Fig. 4-6.

In summer, the primary mission of a ventilating system is to remove excess heat. In many buildings, this ventilating air requirement is provided by fans controlled by thermostats, set to operate any time the inside temperature is above 75°F. In addition (or instead), large panels may be opened to allow for natural air circulation. Most modern buildings incorporate features for both natural and forced ventilation for summer cooling.

In between winter minimum and summer maximum ventilating capacities, additional fans are thermostatically set to start in increments of about 5°F, to prevent sudden temperature drops. These fans provide additional air movement to control temperature and moisture as the weather warms up, as on warmer winter days and in spring and autumn.

Most livestock ventilating systems are of the *negative-pressure* exhaust type, where the fan expels warm, moist air, and cool, fresh air is "sucked" into the building through slots, louvers, or dampers. See Fig. 4-7 for typical exhaust fan locations.

This arrangement subjects fans to dust, moisture, and corrosive acids in the exhausted air. For that reason, fan blades, housings, and shutters should be built of heavy-gauge, corrosion-resistant material. The fan motor should be completely enclosed, with sealed bearings. Choose fans rated for livestock building service by the Air Moving and Conditioning Association (AMCA) or an equivalent standards agency. To ventilate most "warm" animal buildings, fan capacity should be listed at $\frac{1}{10}$- to $\frac{1}{8}$-in static pressure.

Obviously, as fans in a modern building are designed only to exhaust air, they cannot be relied upon to distribute the incoming air. The uniformity of air distribution within a building is a function of the location, design, and adjustment of air inlets. Thus placement of air inlet points is a most critical feature of exhaust-type ventilation systems. Air inlets should be designed so that air entering the building will be distributed evenly throughout the structure and mix with air already inside the building.

With ventilating systems that "push" air into the building, as with some alternative heating systems, the incoming air must be distributed by a different method. Figures 4-8 and 4-9 show one option.

However, as most ventilation systems for livestock buildings are of the negative-pressure type, we will focus primarily on how to size and locate distribution inlets for air being "pulled" into a building. A general rule of thumb for sizing air inlets is to multiply the total winter fan capacity by $\frac{1}{4}$ to give the area in square inches of the air intake.

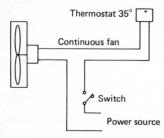

Thermostat 35°

Continuous fan

Switch

Power source

FIG. 4-6   A winter minimum ventilating fan may be set to shut off automatically if inside temperatures drop below 35°F. *(Midwest Plans Service.)*

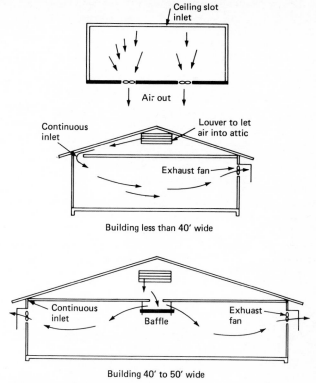

FIG. 4-7   Typical exhaust fan locations. Use a center diffusing inlet when the building is more than 30 ft wide; use two or more air inlets for buildings over 48 ft wide. *(Midwest Plans Service.)*

An adjustable slot is the most commonly designed air inlet in livestock buildings, and usually is located along the longitudinal dimension of the structure, at the eave or in the ceiling, as shown in Fig. 4-10.

How wide should an air inlet slot be made in a building 60 ft long, when the building houses 600 pigs that weigh 250 lb each?* Referring to Table 4-2, the minimum winter ventilation rate per pig is 10 cfm, for a total winter minimum of 6000 cfm.

$$\text{Air intake size} = 6000 \times \tfrac{1}{4} = 1500 \text{ in}^2$$
$$\text{Length of slot} = (60 \text{ ft} \times 12) \text{ in} \times 2 \text{ slots} = 1440 \text{ in}$$
$$\text{Width of slot} = 1500 \text{ in}^2 \div 1440 = 1 \text{ in}$$

For this particular situation, a 1-in slot the length of each long wall is adequate air inlet area. If a center slot intake and suspended baffle system is used, there should be a slot down each edge of the baffle, when the baffle is adjusted to 1 in below the ceiling surface.

*This sample problem and solution are taken from the *Swine Handbook of Housing and Equipment,* published by Midwest Plans Service.

FIG. 4-8   Using polyethylene duct tubing is a relatively inexpensive way to distribute heated air in a pressure ventilating system. *(Photo: U.S. Department of Agriculture.)*

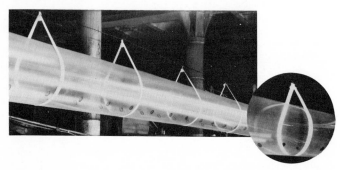

FIG. 4-9   Better-made polyethylene duct tubing has openings placed to provide more air velocity and turbulence. *(Photo: Sto-Cote Products, Inc.)*

Use these air inlet slot locations for different size buildings:

- For buildings up to 30 ft wide, use either a center ceiling slot or a slot along one outside wall.
- For buildings 30 to 48 ft wide, use slots at the ceiling along both outside walls.
- For buildings over 48 ft wide, use slots at the ceiling along both outside walls and one or more center ceiling slots.

In buildings where manure is stored in pits below the floor, some ventilation of the space between the liquid manure and the floor surface is needed to prevent the buildup of odors and possibly toxic gases. This can be accomplished by installing a low-volume fan to run continuously to pull air from the area above the pit.

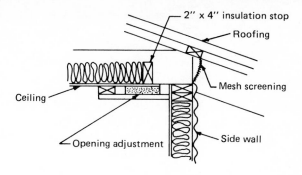

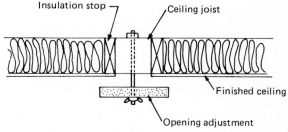

FIG. 4-10   Sectional views of two common slot inlet designs and opening control devices. *(University of Missouri.)*

VENTILATION CONTROLLERS

The most frequently used controls for ventilating fans are *reverse-acting* thermostats. These are temperature-activated on/off switches that energize the circuit when the temperature rises, as opposed to heating-type thermostats that energize the circuit when temperature drops. Controllers must endure the same dust and corrosive elements as the fans they control, and should be chosen for their reliability under these conditions. In addition, fan motors should be equipped with overload and overheat protection devices that will shut the motor off before it is damaged.

Fan controllers of other types are often used in livestock ventilation systems.

• Low-temperature cutoffs can be installed on winter minimum ventilating fans to prevent freezing in the building in the event of unusually cold weather or heating system failure.

• Time clocks, or percentage timers, provide a variable winter minimum level of ventilation, which can be increased as animal size or density is increased.

• Humidistats control fans by sensing the moisture level in the air inside a building.

Humidistats are a potentially great energy-saving device, but reliability of most to date has been disappointing.

- Throttling controllers provide high and low ventilation rates by adjusting fan intake, usually by thermostatically controlled motorized dampers or louvers.

- Reversing fans provide exhaust in winter and pressure in summer, either by electrical reversal of the fan motor or by pivoting the fan 180° in its housing. Except in special circumstances, this is not a particularly efficient ventilation control method.

- Two-stage thermostats provide high and low ventilation rates, when used with two-speed fan motors. The thermostat automatically switches between high and low speeds, according to the temperature inside the building.

- Solid-state controllers are used with variable-speed fan motors to provide infinite variation in air delivery across the capacity of the fan. This type of controller can be used to good advantage in housing where requirements vary frequently, because of fluctuations in outside temperature or animal population.

TROUBLESHOOTING VENTILATION SYSTEMS

Even well-designed, well-operated ventilation systems can malfunction—electrical power is lost, equipment does break down, etc. If a power failure might result in livestock or poultry losses, it's advisable to install an emergency alarm or power generation system, in addition to the normal control mechanisms. Emergency alarms usually employ battery-operated magnetic relays connected to fan circuits, or a solenoid valve that drops open to operate a compressed-air horn. Automatic backup generator systems are usually economically out of reach of most livestock and poultry operators, although a portable emergency generator—such as an alternator driven by a tractor PTO—may provide practical insurance.

Once again, we turn to the counsel of Richard Phillips, Missouri agricultural engineer, on troubleshooting livestock ventilation equipment. Below, Phillips lists the symptoms and most likely causes of ventilation system failures.

Symptom: Building is colder than desired, even though outside temperature is not severe. (*Note:* A properly designed, well-operated livestock ventilation system produces an environment that is desirable for the animals, not necessarily for people. Make sure the complaint is legitimate by checking the temperature needed for the type of animals housed in the building.)

Likely Causes:

1. Fan controls are set to maintain too low a temperature, or temperature calibration is wrong on the controller. Locate a thermometer next to the sensing unit on the thermostat and check its operation.

2. Building is not properly insulated. Refer to the map of temperature zones and recommended minimum insulation levels in Fig. 4-2.

3. Animal density is too low to make the ventilation system work properly.

4. Air inlets are open too wide, allowing excessive amounts of heat to escape from the building when fans shut off.

5. Insulation is wet. This usually happens when a vapor barrier has not been installed or

is installed on the wrong (cold) side of the insulation. Suspect this problem if interior surfaces have excessive condensation.

Symptom: Building is too warm.

Likely Causes:

1. Fan controls are set to maintain too high a temperature.
2. There is not enough fan capacity for animal units housed.
3. Fans are not delivering rated capacity because of dirt on blades, shrouds, and shutters.
4. Air inlet is too small or clogged with dirt. Be suspicious of inlet restrictions if fans suck you inside the building when you open the door.
5. Obstructions around either the intake or exhaust side of the fan are reducing its capacity.

Symptom: Excessive moisture in building; air feels damp and heavy.

Likely Causes:

1. Ventilation is inadequate, particularly if temperature is too high in the building.
2. There is too much ventilation. Tip-off will be cool temperature. Air is not being allowed to warm up enough to pick up moisture. Check thermostat settings.
3. There are leaks in waterers or plumbing.
4. Gas heaters are unvented. A major by-product of burning gas is water vapor. This problem occurs most often in farrowing houses and calf nurseries that do not have large amounts of absorbent litter material.
5. Livestock diseases that produce diarrhealike symptoms are present and add substantially to the moisture load. With some ailments, a deterioration in environmental conditions may be the first indication of a disease outbreak.

Symptom: Excessive condensation on interior surfaces.

Likely Causes:

1. Extremely cold outside temperature is causing the ventilation system to operate at minimum levels for long periods of time. This condition will correct itself when weather warms up.
2. Windows have only single glazing. Allowing a small amount of air to enter at the top of the windows will often cure the problem.

Symptom: Fans not working.

Likely Causes:

1. A fuse has blown or a circuit breaker has tripped.
2. Overheat protection on fan motor has tripped, possibly because of dirt and dust accumulation on fan motor.
3. There is a defective fan motor or thermostat.

Symptom: Controls not working properly.

Likely Causes:

1. Controls are defective.

2. Control settings are not properly calibrated.

3. Sensing units for controls are not properly located. The sensing units should be placed where they will measure and respond to true room conditions, such as near the center of the building at a height that is just out of reach of animals.

4. There are faulty wiring circuits.

Symptom: Excessive fuel costs, in a building that is well-designed and insulated.

Likely Cause:

1. Controls on either the ventilation or heating equipment (or both) are improperly adjusted. Heating system controls may be set too high, or fan controls may be set to move more air than is necessary. Only the winter minimum ventilating rate should be used when supplemental heating equipment is in operation.

5

Solar Assistance

The sun is not a new energy source for livestock producers. For years, livestock and poultry buildings have been constructed with windows on the south, to pick up heat from the low rays of winter sun. In buildings of long ago, natural ventilation systems often employed the dynamics of sun-heated air to convect currents through a building.

Recently, meat and milk producers have taken a renewed interest in the sun. Climbing costs of conventional energy are spurring farmers to find ways to make more efficient use of the "free" energy it provides.

Still, despite soaring costs of conventional fuels, the major limitation to the development of solar heating is more often economic than technological. It is possible to make capital investments in solar energy equipment that will exceed any potential economic return for the foreseeable future.

With that dash of cold water thrown, we'll go on to discuss several practical applications of solar heating systems for livestock and poultry buildings. You will find a great deal of information in the following pages on costs and returns of various systems described, as well as data to help provide a basis for evaluating the economic return of others.

The place to begin is with the relationship of the sun to various locations on earth at different times of the year. For your specific area, you need to know how much usable radiation falls during the times you need to use solar heating. This figure, along with the estimated efficiency and collector area size of whatever system you happen to be considering, will help determine how many Btu's of heat you can expect to receive from such a system and whether its installation will be worth the cost in terms of Btu's per dollar invested.

Thanks to astronomers and mathematicians, we know precisely where the sun is located at any particular instant. Appendix B contains maps showing the amount of insolation (or solar radiation received) during each month of the year for various areas of the United States.

MEASURING SOLAR RADIATION

How much solar radiation can you expect in your area during December? During January? That depends on the amount of sunshine (day length) and on atmospheric conditions. In some areas, there are so many cloudy days in winter that the amount of radiation reaching a solar collector is relatively small. In other regions, sunny days and chilly nights for many months of the year make solar a more attractive proposition.

Insolation, or the amount of solar energy falling at any one place, is measured most often in *langleys,* as shown on the maps in App. B. One langley equals one gram-calorie per square centimeter per minute.

The average amount of solar radiation reaching the earth's atmosphere per minute (called the *solar constant*) is just under two langleys. This is the equivalent of 442.4 Btu/(h)(ft²), or 1395 W/m².

You can roughly convert gram-calories per square centimeter to Btu's per square foot by multiplying the figure given in langleys by 3.69. Or you can work it out to a more precise calculation. One square foot equals 929 square centimeters; one gram-calorie is the amount of heat needed to raise the temperature of one gram of water by one degree Celsius (whereas one Btu is the heat energy needed to raise the temperature of one *pound* of water by one degree Fahrenheit); and one pound equals 435.59 grams.

The effects of solar energy can be calculated in the design and location of any building —whether it fits the description of a "solar" structure or not—to get the effect of sunshine inside the building in winter and to shade out the sun's rays in summer.

Consider, for example, the effect on an open-front livestock shelter. Generally, this kind of building would be placed to orient the long dimension or ridge of the building in an east-west direction, with the open side to the south. This provides a couple of advantages:

1. The area of the building exposed to late afternoon summer sun is reduced by facing the narrow wall to the west.

2. The long, open south wall admits direct solar radiation in winter, when the sun is low in the sky. At the same time, the roof overhang can be designed to shade out the higher, hotter summer sun.

SHADING EFFECT OF BUILDINGS

It's not that difficult to calculate the solar penetration and shading effect for various buildings. As shown in Fig. 5-1, the horizontal penetration (HP) of sunlight behind the edge of the overhang into an open-front, south-facing building is

$$HP = \frac{\text{height of overhang above ground}}{\text{altitude of sun at time being considered}}$$

where altitude of sun = 90° − latitude + solar declination.

Solar declination refers to the position of the sun at noon relative to the earth's equator, and can be calculated from the equation below to within a half-degree of accuracy, which is close enough for most installations.*

$$\text{Declination} = -23.5 \times \cos{(N + 10)} \times \frac{360}{365}$$

where N = number of days elapsed in year. If the angle of declination of the sun is −5° (the figure you would obtain for 40° north latitude on March 10), then the altitude angle of the sun at noon at that latitude is 90 −40 + (−5), or 45° above the horizon.

In much the same way, the vertical shading (VS) effect can be calculated to determine how much overhang is needed to shade a building's south wall in summer:

*The *ASHRAE Handbook,* published by the American Society of Heating, Refrigeration and Air-Conditioning Engineers, 345 East 47th St., New York, NY 10017, contains a full set of solar-angle equations for the United States.

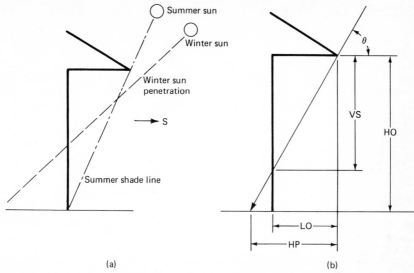

FIG. 5-1 *(a)* The effect of seasonal change in the sun's angle on shade and solar penetration of a building. *(b)* The relationship of sun angle of elevation, solar penetration, and roof overhang height. HP = horizontal penetration (in feet) of sun beyond edge of roof overhang; LO = roof overhang beyond south-facing wall (in feet); HO = vertical distance from floor to eave; VS = roof overhang × angle of the sun (indicates the distance down the wall the overhang shadow will progress at a given time and season). *(Courtesy: Richard Phillips, University of Missouri.)*

$$VS = \text{length of overhang} \times \text{sun's altitude}$$

See Fig. 5-1 for a graphic representation of the effects of seasonal change in the sun's altitude on shade and solar penetration. This is the simplest form of passive solar energy at work, and these considerations require little or no added investment in building costs.

In most regions of North America, solar heating systems can be designed that will do a little or a lot to provide the supplemental heat needed for livestock and poultry buildings. However, the more you expect from a solar system, the more complex and costly the equipment is likely to be. Solar energy may be free, but capturing it can be expensive.

In a "warm" confinement building occupied year-round, *when* is nearly as important as *how much* when it comes to solar heating. For most solar systems, you'll still need a supplemental or backup heating system that uses conventional fuels—to take over heating chores when the sun isn't getting through.

In other words, solar collectors on a farrowing house in Henry County, Illinois, might provide 70 to 80 percent of the supplemental heat needed in April. But it's what the system does on the coldest, cloudy day in January that dictates the size of the supplemental heating equipment installed. That's one reason solar heating equipment is so costly—it usually is *supplementary to* a conventional heating plant, which is little reduced in capacity from what it would be if the solar system were absent.

However, most animal housing that uses supplemental heat can benefit from the use of solar *assistance*. Heated buildings already require fan ventilation, so an air-moving vehicle is available to bring heat from the solar collector into the building. In winter, more heat is usually required to heat incoming ventilation air than is lost through walls, ceiling, and foundation of a well-insulated building. Heating the incoming air only 10 or 20°F will reduce the fuel needed to keep the building warm.

Low-cost solar collectors, installed on the south roof of a building, can provide this preheating function, without adding appreciably to the cost of the building. An even better option is to build the collector vertically, on the south wall (see Fig. 5-2). During the cold months of the year, when the sun angle is low, a vertical south-facing collector picks up more solar energy per square foot than one on a $^4/_{12}$ roof slope. This was proved near Sioux Falls, South Dakota, where a solar unit on the south wall collected about five times more heat per square foot than a similar collector on a south-facing roof with a $^4/_{12}$ pitch. Part of the advantage was due to less frost and snow on the vertical collector.

Later on in this chapter, we'll describe a vertical solar collector that utilizes concrete blocks for heat storage.

ACTIVE OR PASSIVE?

The state of the solar art is generally focused on two broad categories: active and passive systems. Very simply, active systems employ mechanical equipment and transmission fluids—usually air or water—to move the heat from the solar collectors to where it is used to heat building spaces, water, or storage material. Passive solar systems use the orientation of the building, types of materials, design of the system, insulation, and placement of the collector surface in relation to the space or material to be heated.

Passive solar techniques—such as the orientation of an open-front building to the south —are often incorporated in "cool" livestock and poultry buildings. However, as "warm" animal housing typically is equipped with mechanical ventilation equipment, the solar system installed is more often active (usually designed to preheat incoming ventilation air).

Extreme temperatures are not needed for most farm building heat. The so-called "flat-plate" collector is less critical of position than concentrating collectors that generate high temperatures and is much cheaper to build and install. This type of collector is the usual choice for farm space-heating needs. See Fig. 5-3.

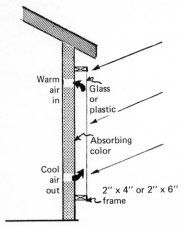

FIG. 5-2 Low-cost solar collectors installed vertically on a building's south wall can pick up part of the heating load.

There are dozens—perhaps hundreds— of designs for flat-plate solar collectors, but they all have essentially the same characteristics and perform the same functions. A flat black plate absorbs energy from the sun; a circulating fluid (usually air or water) picks up heat from the plate and moves it to the point of use or to storage.

An air collector can be as simple as a bare sheet of metal painted dull black to absorb more heat, with air circulated behind the plate. More often, however, collectors are covered with one or more transparent surfaces—glass, plastic, fiberglass, or some other durable glazing—with air circulated between the cover and the black absorber plate, as shown in Fig. 5-4. The cover reduces heat loss to outside air and also helps trap the longer rays from the sun. Insulation commonly is installed

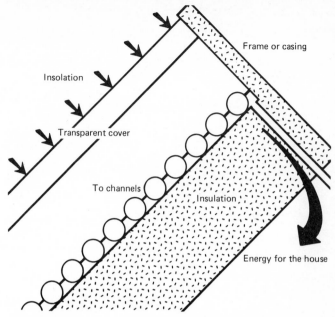

FIG 5-3   Flat-plate solar collectors are essentially glass-covered boxes with heat-absorbing surfaces and insulation. *(U.S. Department of Energy.)*

behind the absorber plate, to prevent conduction losses through the back of the collector.

Water collectors are similar in principle to air collectors, with some differences to accommodate the liquid. In most water collectors, tubes are attached to the absorber plate or are located in the space between the absorber plate and the glazing. In simpler water collectors, the liquid flows across the surface of the absorber plate (built of corrugated metal, in many cases), rather than through tubes.

There are potentially more problems with water collectors than with air collectors. For one thing, if the system springs a leak, water can do more damage to building materials than can escaping warm air. For another, water can freeze in cold weather. However, these problems are not insurmountable. Chapter 6 goes into solar hot-water systems in detail.

COLLECTOR EFFICIENCY

Solar collector efficiency is measured by the percentage of useful heat collected compared with the total amount of solar radiation striking the collector surface. If 20,000 Btu of solar energy hit an 8- by 120-ft collector between noon and 1 P.M. on February 1, and the collector gathers 10,000 Btu of usable heat energy during that hour, the collector is 50 percent efficient. Fifty percent is fairly high efficiency for flat-plate collector designs.

Efficiency can be reduced by heat losses through the front, sides, and back of the collector, by reflection of solar rays off the transparent cover, by radiation from the heated

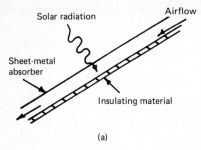

(a)

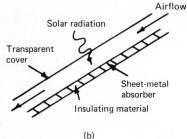

(b)

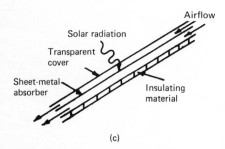

(c)

FIG. 5-4   Three types of solar collector: *(a)* bare-plate, *(b)* covered-plate, *(c)* covered suspended-plate.

flat plate, and by losses in the ducting and transfer medium. Collector efficiency also is affected by the angle of the collector in relation to true south, as well as the angle or "tilt" of the collector from the horizontal.

For maximum efficiency, collectors should be oriented due south. However, flat-plate collectors can accept direct or reflected sunlight from a range of angles. The collector can be aimed from 15° east of south to about 35° west of south, with little loss of performance.

As mentioned earlier, the sun strikes the earth at different angles—depending on its location in relation to the equator and the season of the year. In a solar system designed for winter heating, the collector surface should be nearly perpendicular to the sun's rays in the middle of January—usually the coldest time of the year.

The optimum angle is the sum of the latitude plus 15°. If you're building a farrowing house at Anderson, Indiana, which is about 40° degrees north latitude (40° north of the equator), collectors for winter heating should be angled at about 55° from horizontal. However, the "tilt" can be off by as much as 10° without having any noticeable effect on the capability of a flat-plate collector.

Collector efficiency also depends on the temperature in the collector, compared with the outside temperature. As you recall from the exercise on figuring a building's heat loss, the wider the temperature difference, the higher the rate of heat loss. The same holds true for a solar collector: The greater the temperature difference, the more heat is lost back to the atmosphere.

That's a big reason why flat-plate collectors are often designed to reach fairly low maximum temperatures. The name of the game is collecting Btu's of energy, not necessarily attaining the highest temperatures possible. There are many more Btu's in a bathtubful of 90°F water than in a teakettleful of boiling water.

A collector's efficiency at that instant in the day when all components are operating at their peak may be considerably higher than average day-long efficiency. This should be kept in mind when comparing manufacturers' or designers' claims for their particular equipment. Table 5-1 shows typical efficiencies for various types of flat-plate collectors.

TABLE 5-1   Efficiency of Flat-Plate Collectors

Collector type	Day-long efficiency, %
Bare plate	30
Covered plate, single-glazed	35
Suspended plate*	40
Suspended plate, double-glazed	45

*A suspended-plate collector is designed so that collection air can circulate across both the top and the back surfaces of the absorber plate.

Collector efficiency also depends on the surface of the absorber plate, and reflectors often can be used to increase the amount of radiation striking the collector. Of course, any improvements in collector efficiency through these methods must be weighted against increases in cost—both to install and to maintain the equipment. A less efficient, but cheaper collector may be the more cost-effective for a particular installation.

Solar-heated air provided by a simple, inexpensive collector can be used in a number of farm applications. Marvin Hall, veteran solar advocate at the University of Illinois, and his colleagues have come up with a portable collector that is both inexpensive and flexible. See Fig. 5-5. The 12- by 24-ft suspended-plate collector can be located near grain bins in the fall, to provide 10 or 15°F of heat for the slow drying of grain, then skidded to another location to supplement the heating system for a machine shop or livestock building. Figure 5-6 shows construction details of this collector.*

The collector can be built for about $1000 worth of materials, not counting fans and their controllers, and will provide up to 400,000 Btu of heat per day when fitted with a 5000-cfm blower. That combination gives a peak heat rise of about 15°F through the collector, which is ample for low-temperature grain drying. At this rate, the collector can potentially supply 28 to 30 million Btu over a six-month season. If the collector replaces 385 gal of LP gas at 60 cents per gallon, this makes the "payback" time on the initial cost of materials only four to five years, which is better than most solar expenditures.

Set up to heat a building, the portable collector provides a peak heat rise of 40°F, at an airflow of 1000 cfm. Some places for which the Illinois-type collector might be a good choice are machine shops, dwellings, and animal housing.

*A full set of plans can be obtained from Plans Service, 202 Agricultural Engineering Building, University of Illinois, Urbana, IL 61801. Send $1 and ask for Plan SP 546.

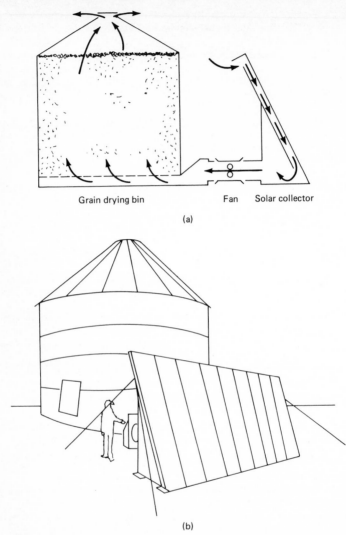

Grain drying bin   Fan   Solar collector

(a)

(b)

FIG. 5-5 *(a)* Schematic and *(b)* perspective of 12- by 24-ft portable collector used to dry grain. The more uses made of heat from a collector, the more economical solar heat becomes. *(University of Illinois.)*

ABSORBER SURFACES

An absorber plate should soak up as much of the sun's radiant energy as possible, lose as little heat as possible to the surrounding atmosphere, and transfer the heat retained to the circulating fluid.

Materials vary in their ability to perform these functions. Generally, dull surfaces absorb

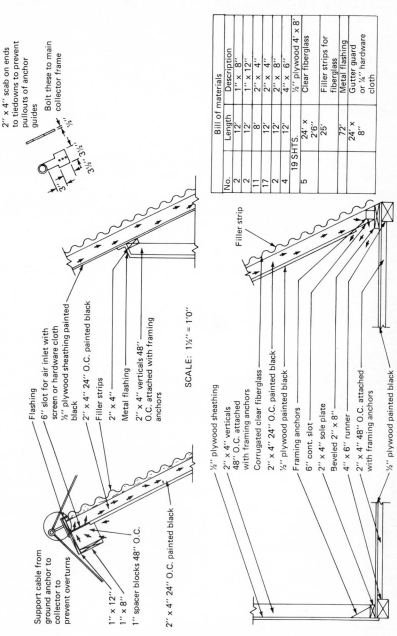

Bill of materials		
No.	Length	Description
2	12'	1" × 8"
2	12'	1" ×12"
11	8'	2" × 4"
17	12'	2" × 4"
2	12'	2" × 8"
4	12'	4" × 6"
19 SHTS.		½" plywood 4' × 8'
5	24' × 2'6"	Clear fiberglass
	25'	Filler strips for fiberglass
	72'	Metal flashing
	24' × 8"	Gutter guard or ¼" hardware cloth

2" × 4" scab on ends to tiedowns to prevent pullouts of anchor guides

Bolt these to main collector frame

SCALE: 1½" = 1'0"

Support cable from ground anchor to collector to prevent overturns

1" × 12"
1" × 8"
1" spacer blocks 48" O.C.
2" × 4" 24" O.C. painted black

Flashing
6" slot for air inlet with screen or hardware cloth
½" plywood sheathing painted black
Filler strips
2" × 4"
Metal flashing
2" × 4" verticals 48" O.C. attached with framing anchors

Filler strip

½" plywood sheathing
2" × 4" verticals 48" O.C. attached with framing anchors
Corrugated clear fiberglass
2" × 4" 24" O.C. painted black
½" plywood painted black
Framing anchors
6" cont. slot
2" × 4" sole plate
Beveled 2" × 8"
4" × 6" runner
2" × 4" 48" O.C. attached with framing anchors
½" plywood painted black

FIG. 5-6 Details of a 12- by 24-foot suspended-plate solar collector. O.C. means measured center-to-center. (*University of Illinois.*)

more heat than smooth ones; dark-colored surfaces absorb more heat than lighter ones. A black surface absorbs more radiant energy than any other color—about 95 percent of the solar radiation that strikes it. A flat black (dull-finish) paint helps reduce reflective losses.

Multiple coatings can reduce the amount of heat radiated from the absorber plate to surroundings. Coatings absorb about the same amount of radiant heat as regular black surfaces, but radiate and reflect less heat. Table 5-2 lists energy absorption and emittance for several materials.

Materials often used for collector plates are copper, aluminum, and steel. Copper is most expensive, but has the highest thermal conductivity. Steel is least expensive, but has the lowest conductivity of the three metals.

With water-heating collectors, where conductivity is more important than with some air collectors, copper tubes bonded to the absorber plate can be spaced further apart than can materials of poorer conductivity. Also, soldered bonds of the pipe to the plate are easier to make with copper than with aluminum. Some absorber plates on the market have bonded-in tubes in a radiatorlike design that gives good thermal contact.

With air collectors, if economics dictate low building costs, such materials as plastic or plywood may be used for absorber plates. These materials do not perform as efficiently as metals, but are often acceptable since the cost per square foot of collector area can be reduced.

In a "solar attic" design, common to many swine farrowing houses in the Midwest, the absorber plate surface may be black-painted plywood, fiberboard, or other standard building materials. Figure 5-7 shows the airflow pattern of a typical "solar attic" building.

COVER GLAZING

Glass is most often used as a cover for flat-plate collectors for space heating of homes and public buildings. Glass is a good transmitter of solar radiation and emits little radiant heat from the absorber surface back to the surroundings. This "greenhouse" effect is why the inside of an automobile gets hot on a sunny day, even when the outside air temperature is low.

Glass has an 87 percent transmittal of solar radiation; that is, 87 percent of the incoming radiation can pass through the glass. The rest is either reflected or absorbed in the glass itself. Special low-lead-, low-iron-content glass retains a greater percentage of the heat

TABLE 5-2   Energy Absorptance and Emittance of Various Materials

Material	Absorptance, %	Infrared emittance, %
Flat black paint	0.97–0.99	0.97–0.99
Concrete and dark-colored stone, paint, or brick	0.65–0.80	0.85–0.95
Light-colored paint, brick, clay, or fireclay	0.50–0.70	0.85–0.95
Glass	0.04–0.40	0.90
Aluminum paint	0.30–0.50	0.40–0.60
Dull brass, copper, or galvanized metal	0.40–0.65	0.20–0.30
Selective surface:		
Copper treated with $NaC10_2$ and NaOH	0.87	0.13
Copper, aluminum, or nickel plate with CuO coating	0.80–0.93	0.09–0.21

SOURCE: University of Missouri.

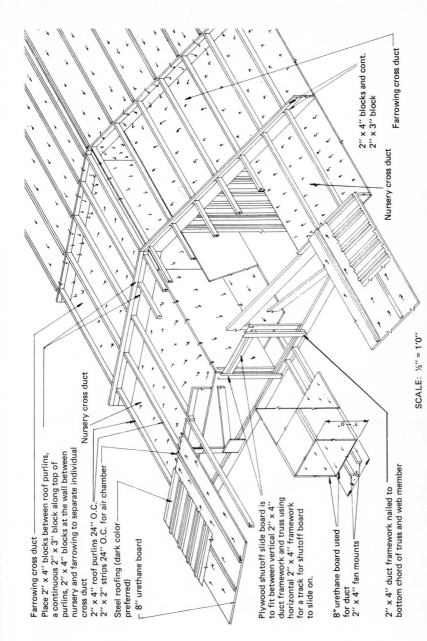

Farrowing cross duct

Place 2" x 4" blocks between roof purlins,
a continuous 2" x 3" block along top of
purlins, 2" x 4" blocks at the wall between
nursery and farrowing to separate individual
cross ducts

2" x 4" roof purlins 24" O.C.
2" x 2" strips 24" O.C. for air chamber

Nursery cross duct

Steel roofing (dark color
preferred)

8" urethane board

Plywood shutoff slide board is
to fit between vertical 2" x 4"
duct framework and truss using
horizontal 2" x 4" framework
for a track for shutoff board
to slide on.

8" urethane board used
for duct
2" x 4" fan mounts

2" x 4" duct framework nailed to
bottom chord of truss and web member

Farrowing cross duct

2" x 4" blocks and cont.
2" x 3" block

Nursery cross duct

SCALE: ½" = 1'0"

FIG. 5-7 32-ft solar-heated farrow-to-finish hog unit. Bare-plate collectors incorporated in dark-colored metal roofing preheat ventilating air for this swine building. O.C. means measured center-to-center. (*University of Illinois.*)

5–11

generated than does ordinary plate glass; however, the cost of special glass is much higher. See Fig. 5-8 for an example of its application.

Plastic films or sheets also are used as covers for collectors, although only specially made plastics will withstand the sun's rays for any length of time without degrading or yellowing. Transparent plastics transmit about 90 percent of the sun's radiation. However, they do not perform as well as glass at trapping the heat; they allow 30 percent or more of the long-wave radiation to be radiated to the surroundings from the absorber plate.

Still, for many low-temperature applications, such as grain drying, or for temporary installations, plastics have some advantages over glass. They are easier and cheaper to install and are less subject to damage from wind and hail. In some designs, two glazings are used on the cover—an outer one of glass and an inner one of plastic.

In animal housing, "greenhouse-type" corrugated fiberglass or fiberglass-reinforced plastics often are used as collector covers, sometimes incorporated as part of the roof. Both are tough, semitransparent materials that perform in much the same way as glass. However, green-tinted fiberglass transmits less solar radiation: about 80 percent, compared with 87 percent for glass.

HOW MUCH COLLECTOR AREA?

A rule of thumb for solar installations used to heat human dwellings calls for 1 ft² of collector surface for each 3 to 4 ft² of house floor area. If you live in a well-built, well-insulated 2000-ft² home in a cold but sunny climate, 700 ft² of collector can supply 60 to 70 percent of the annual heating requirements, as well as much of the domestic hot water needed.

With animal housing, collector surface area is more often a function of the south-facing

FIG. 5-8 Dual solar collectors are studied at Mississippi State Poultry Research Laboratory. Collector in foreground heats ventilation air only; smaller unit heats water that is stored for heating the poultry building at night. *(Photo: U.S. Department of Agriculture.)*

roof or wall area. A 120-ft-long farrowing house, with a $^4/_{12}$ pitch roof that is 14 ft wide and an 8-ft-high sidewall, has the potential for 2640 ft² of collector area—although the cost of building such a collector might not be justified by the savings in supplemental heating fuel.

However, with translucent-fiberglass corrugated panels used to replace conventional metal roofing and siding materials on south-facing surfaces, the cost of building a collector that occupies the south half of a roof or the south-facing wall may be little more than the cost of building without the solar feature.

Agricultural engineer Richard Phillips outlines some design characteristics for flat-plate solar collectors.

Bare-plate Collectors:

1. Design temperature rise through the collector should be 10°F or less.
2. Maximum air velocity in the collector should be less than 750 ft/s.
3. Minimum air space behind the collector should be ¾ in.
4. Absorber plate material should be black plastic sheeting or black-painted sheet metal.

Covered-plate Collectors:

1. Design temperature rise through the collector should be 50 to 75°F.
2. Maximum air velocity should be less than 750 ft/s.
3. Minimum air space behind the collector should be ¾ in.
4. Absorber plates should be of dark-colored metal or dark-painted wood.
5. Cover materials should be of low-iron-content glass, ultraviolet-resistant plastic sheeting, or fiberglass-reinforced plastic with Tedlar® coating. Use double covers over collectors designed to exceed a 75°F temperature rise.
6. Back and sides of the collector should be insulated with R-13 or greater.

Covered-plate Liquid-Transfer Collectors:

1. Protect from freezing if required. Ethylene glycol antifreeze can be used in closed-loop systems.
2. If the liquid has a boiling point below 300°F, provide pressure-relief valves in the system.
3. Design friction head losses to less than 6 ft.

STORING SOLAR HEAT

Even in the mildest North American climate, the winter sun doesn't shine during all the hours of the day. Therefore, if solar heat is to be utilized round the clock, solar energy must be stored.

This situation is aggravated by the fact that the coldest temperatures often occur on cloudy days or at night, when the sun is not contributing but supplemental heat is needed. It's possible to store enough solar energy to meet the heating requirements for several cold, sunless days. However, in practice, a supplemental heating system is usually installed as a backup, and solar energy storage, if used at all, is designed to provide enough heating capacity for just a day or two.

The most common storage medium for solar air collectors is rock or concrete. During the period when the collector is "harvesting" solar heat energy in excess of building needs, the storage material is warmed either directly by the sun or indirectly by heated air passing through it. Then, when the sun is not shining, the rock storage gives up heat either directly to the building or to air that is circulated through it.

Leslie Prier, of Newtonia, Missouri, spent an additional $8600 for a 2000-ft² solar collector wall to preheat ventilation air coming into his 40- by 250-ft turkey brooder house. The 10,000-bird brooder is the first stage in Prier's turkey-raising operation and is the most critical from an energy-consumption standpoint. The building houses poults from a tender one-day-old to about six weeks of age.

For heat, the brooder building needs about 4700 Btu per degree difference per hour, between inside and outside air temperatures. On an average December day, with an inside temperature of 80°F, total energy demand would be about 4.9 million Btu.

Space along the south wall allowed construction of an 8- by 250-ft solar collector, for a total collector area of 2000 ft². On a completely clear day in December, the collector nets nearly 2 million Btu. On the average, the system collects about 500 Btu (ft²)(d), or some 1 million Btu on an average December day.

Prier measured the heat collected on 29 sunny days between November 17 and January 1. The collector added 44.2 million Btu to the ventilation air coming into the building. Prier estimates that 77.1 million Btu of solar energy were radiated toward the collector during that period, so the system's efficiency was about 57 percent. The 44 million Btu harvested is the equivalent heat value of burning 591 gal of LP gas. At the time, LP gas cost nearly 60 cents per gallon in Prier's area. On that basis, the solar collector recovered about $350 of the initial $8000 extra investment, in just the 29 days the system was monitored.

Here's the way the system works. Fresh air is pulled into the collector through a manifold along the lower edge of the system and ducted through a rock heat-storage bin beneath the collector. See Fig. 5-9.

Two fans provide 4000 cfm of outside air, which has been heated by the sun's energy directly or has picked up heat from the rock storage, and distribute the heated fresh air through a perforated 18-in-diameter plastic tube duct. See Fig. 5-10.

Air temperature inside the building is maintained at 70°F. Until poults are three weeks old, they are kept in zones heated to 100°F by 34 pancake-type LP-gas brooders.

The solar collector is covered with Tedlar®-coated fiberglass, over the black-painted plywood absorber plates. The storage bin, beneath the collector, contains 1½-in crushed rock and is designed to store the maximum energy collected in one day. The rock bed performs well at retaining collected heat during the day and releasing it at night. In some instances, ventilation air entering the building from the rock storage is as much as 30°F above outside air temperature.

Solid rock or concrete has a density of about 150 lb/ft³ and a specific heat of 0.2 Btu/(lb)(°F). Crushed rock of 1½-in average diameter has a density of 115 lb/ft³. A cubic foot of crushed rock will store 230 Btu of heat energy for every 10°F of temperature increase. For a practical working range, storage temperatures are seldom more than 50 to 60°F higher than air temperatures, which means that 1 ft³ of crushed rock can store about 1250 Btu. See Fig. 5-11.

Some cautions are in order, however. The air in a livestock or poultry building contains a great deal of dust and moisture. For this reason, most rock storages are designed to introduce fresh air from outside, rather than recirculate air from within the building.

Solar systems employing a liquid transfer medium typically use water, which has a density of 62.5 lb/ft³ and a specific heat of 1 Btu°F. Using the same 50 to 60°F working

FIG. 5-9 Local conditions can affect system design. In areas with heavy snowfall, the collector air inlet might need to be located higher above ground level. *(Photo: Farm Building News.)*

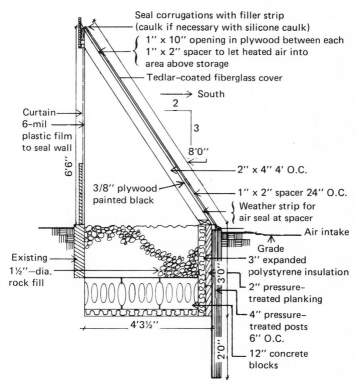

Seal corrugations with filler strip
(caulk if necessary with silicone caulk)

{ 1" x 10" opening in plywood between each
1" x 2" spacer to let heated air into
area above storage

Tedlar–coated fiberglass cover

→ South

2

3

8'0"

Curtain
6-mil
plastic film
to seal wall

6'6"

3/8" plywood
painted black

2" x 4" 4' O.C.

1" x 2" spacer 24" O.C.

{ Weather strip for
air seal at spacer

Air intake

Grade

Existing
1½"–dia.
rock fill

3" expanded
polystyrene insulation

2" pressure-
treated planking

4" pressure-
treated posts
6" O.C.

12" concrete
blocks

4'3½"

3'0"

2'0"

FIG. 5-10 Cross-sectional view of the collector-storage system built by Leslie Prier. O.C. means measured center-to-center. *(University of Missouri.)*

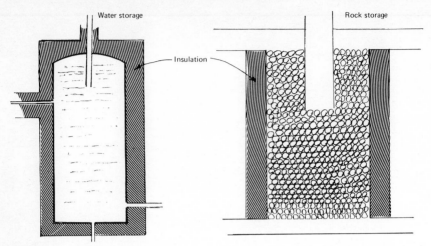

FIG. 5-11 Collected solar heat is commonly stored in either rocks or water. As a rule of thumb, 2 gal of water storage is needed for each 1 ft² of collector area. With rock storage, allow 0.5 ft³ of rock for each 1 ft² of collector.

temperature range as for rock storage, 1 ft³ of water will store about 3400 Btu of heat energy.

However, water is not always popular as a storage medium in livestock buildings, for a couple of reasons. For one, the storage must be watertight and able to withstand design loads that are generally higher than rock storage. This adds to the cost of construction. For another, water freezes. If antifreeze is used, some type of heat exchanger must be used in the water storage to keep the expense of the antifreeze solution low. The alternative is to design an automatic drain-down collector system which would empty water from the collector when the sun is not shining. These techniques are discussed more thoroughly in the next chapter.

Change-of-state storage materials, often called *eutectic salts,* are chemicals that will solidify or "freeze" at temperatures within the normal range of those in a solar heating system. The phase-change process (freezing or thawing) provides an advantage in storing or releasing heat, compared with direct temperature changes through a material that does not change its state.

For example, Glauber's salt ($Na_2 SO_4 \bullet 10H_2O$) releases 108 Btu/lb when it solidifies at 91°F. Obviously, the same amount of heat can be stored in much less space when these materials are used rather than rock or water.

One drawback to using these chemicals has been the cost of packaging—often higher than the basic cost of the phase-change material itself. Another disadvantage is that most of these chemicals can go through only so many freeze/thaw cycles before they lose their ability to change state.

Newly tried materials, such as calcium chloride hexahydrate, which changes phase at 81°F, show promise of being longer-lasting. Also, new packaging methods are being devel-

Material	Crushed rock	Concrete	Water	Glauber's salt
Heat stored, Btu/ft³	1250	1650	3430	12,500

oped. The University of Delaware has come up with a patented, hydrated salt that melts and freezes at about 55°F and is packaged in sausage-shaped plastic containers.

FIGURING ECONOMIC RETURNS

Can you economically use solar energy in your farm buildings? There may come a time when supplies of other fuels become so low as to dictate the use of solar energy in farm structures. Then there will be little choice, perhaps.

At present, however, the economic feasibility of using solar energy depends on how well this source competes with other available fuels. The principal costs of operating a solar system in a "warm" livestock or poultry confinement building are fixed costs, the costs of owning the system. Operating costs may be little or no more than those for conventional heating and ventilation equipment.

The fixed costs are depreciation, interest on investment, repairs, taxes, and insurance. Depreciation is the replacement cost associated with the normal life of the system. It can be calculated in several ways, depending on the tax situation of the owner, but for cost estimates, the straight-line method of depreciation customarily is used. This method calculates annual depreciation by dividing the initial cost by the expected years of useful life.

Most solar systems should have a useful life equal to the life of the building in which they are installed. For example, let's say a solar heating system will have an anticipated useful life of 20 years. What would be the annual depreciation on a system costing $10,000?

$$\text{Depreciation} = \frac{\$10,000}{20 \text{ years}} = \$500 \text{ per year}$$

Interest is the cost of money borrowed—or the lost income that the money would have earned if it had not been invested. The realistic way to figure interest is to use the average commercial interest rate, which at the time of publication ranged in the 16 to 18 percent area.

Repairs will depend on the type of solar heating system installed, how much operating equipment is included in the system, and the quality of the initial construction. There isn't much available information on repair costs, but you may want to include an estimate of 4 to 5 percent of the initial cost per year.

Depending on where your farm is located, real estate or property taxes may be assessed on a solar collection system. Many states have exempted solar equipment from taxes; others have not. Generally, installed solar equipment is assessed as part of the structure and taxed at the same valuation rate.

Insurance premiums may reflect the added cost of insuring a solar system, particularly if the equipment makes the structure more vulnerable to damage—such as glass breakage by hailstorms.

To get an accurate picture of costs, the system should also be charged for any labor required to adjust or operate the equipment, as well as for electricity to operate any extra motors, sensing devices, and controllers associated strictly with the solar system.

The economic clincher for any alternative energy system is the "payback," or the rate of return on the investment. In the case of solar heating equipment, the payback is in dollars saved on purchased heating fuels.

Predictions of how much heating fuel a particular solar energy system will save are, at best, educated guesses. As noted earlier, the actual heat energy delivered by a solar system depends on how much energy is received from the sun and on the efficiency of the collector—and a host of factors can influence the efficiency of the collector.

The time of day the sun shines also affects the amount of useful radiation received. Some solar energy can be collected even on cloudy days, but the best operation is on clear, sunny days. Dust or haze can affect the amount of energy intercepted.

Swine producer Gene Pogue built a 28- by 120-ft nursery building with the entire south-facing roof covered with translucent fiberglass. About 480 ft² of that is a water collection system, with 1800 gal of hot-water storage under the floor and an automatic drain-down system to prevent freezing. Pogue insulated the building to R-24 in the ceiling and R-19 in the sidewalls.

At full capacity of 600 pigs, the building has a heating requirement of 800 Btu of supplemental heat per degree of temperature difference. This is with the space temperature maintained at 75°F and the floor in the pigs' sleeping area maintained at 95°F. This zone heat for the pigs is provided by the solar water collection system, backed up by an LP-gas-fired boiler.

Engineers at Kansas State University have monitored the performance of a unique vertical solar collector (with an area of 380 ft²) coupled with a concrete block wall storage system, and find that each square foot of this sytem produces the heat equivalent of 2 gal of LP gas during a typical heating season. At that rate, the initial investment could be recovered in about 10 years if LP gas is the heating source supplemented or in about seven years if solar replaces electric heat.

The Kansas design is adaptable to any existing building with an expanse of south-facing wall. Main features of the solar wall are a stack of concrete blocks (each 6 by 8 by 16 in, painted black and laid so that openings are from front to back) and a double transparent cover on a frame that allows ventilating air to pass between the covers as it enters the system.

Moving the air between the two layers of glazing allows some heat to be collected that otherwise would be lost. The air then removes heat from the south side of the concrete blocks first, thus cooling the surface to further reduce heat loss from the storage. Inside, a fan connected to a ducting system moves the air to a furnace in the farrowing house.

The Kansas-type solar wall can be built for about $7.50 per square foot of collector space, with $4 of that amount allocated for labor.

Other collection systems vary in cost and in the time it takes to recoup the initial investment. To work through a sample problem, let's assume that a 10- by 80-ft collector is being built in north-central Missouri, at N40° and that the day-long efficiency of the system is 40 percent. How much energy will be captured per square foot of collector during December?

Referring to Fig. B-12 in App. B, we find that the nearest recording weather station received 158 langleys per day, on the average, during the month. To convert this figure to Btu's per square foot, multiply by 3.69.

$$158 \times 3.69 = 583$$

The result is that about 583 Btu of energy can be expected to be collected by each square foot of collector per day.

For the 31 days of December, each square foot of collector area is bombarded with 18,073 Btu. Across the entire 800 ft² of collector surface fall a potential 14,458,400 Btu of solar radiation.

However, the collector is only 40 percent efficient at receiving and trapping those Btu's from the sun, which reduces the total Btu contribution of the 10- by 80-ft collector to 5,783,360 Btu for December.

The commercial fuel equivalents can be computed by referring to Table 3-2.

Use the present prices for the fuel—or fuels—you normally use to heat animal housing

Electricity	1694 kWh
LP gas	78 gal
Natural gas	72 therms (1 therm = 100,000 ft³)
No. 2 oil	59 gal

to compute what this type of solar collector would "pay back" for this one month's operation.

However, the above illustration is a rather simplified approach, which assumes that *all* the solar Btu's possible will be harvested and put to use—and that all of that heat would be *needed*. In a normal season, a Missouri December would probably have a few sunny days when very little supplemental heat would be needed and a string of cold, cloudy days when more supplemental heat would be needed than could be provided by stored solar energy. That's why estimating savings with a solar heating system is fraught with error.

Several reasonable adjustments to the cost of a solar energy system can be made to help reduce the fixed costs and increase the savings. If conventional sources of energy continue to increase in cost, as most experts predict, the savings in fuel generated by a solar heating setup will be more significant as years go by. The same is true assuming that inflation will continue—buying now will save money (dollars), as compared with buying later.

Also, investment tax credits and energy tax credits can be taken for most solar heating systems installed on farm buildings. These are dollar-for-dollar credits that can be deducted directly from taxes due. In addition, some states allow property tax and income tax deductions for solar equipment.

CHOOSING A SOLAR CONTRACTOR

Despite their cost, solar systems have multiplied steadily in the past few years. Something like a quarter million installations (including solar hot-water systems) have gone into homes and other buildings, and the number nearly doubles each year.

The number of competent contractors and manufacturers has not kept pace, however. Probably the biggest cause of problems with solar energy systems is poor installation procedures, followed closely by shoddy material.

Complaints by owners range from obvious goofs (one contractor's crew installed Australian-made collectors and followed the manufacturer's instructions to the letter—never stopping to realize that the sun is in the *north* in Australia) to suspected frauds (a solar manufacturer in a southeastern state used a foam insulation that started to vaporize when collector temperatures reached 150°F).

Still, it isn't too difficult to avoid being burned when you buy solar. Consider the project as carefully as you would any other major purchase. Get as much information as you can on the type of solar system you're interested in: How does it work? Who builds the components? Who can install it properly? How about repair parts? Fortunately, many solar systems installed on animal housing are fairly simple in construction and operation and for the most part use readily available building materials.

Standards are set for manufactured solar components. Choose a collector that has been approved by a federal or state testing program or that has been rated by an independent testing laboratory. Make sure that insulating materials will withstand the temperatures generated. Choose transparent cover materials that will not darken or degrade in sunlight.

The best equipment built will not function properly if it is improperly installed, however. If you're doing the job yourself, complete your homework before you start construction.

If you're hiring the work done, your local association of farm builders is a good place to start looking for a solar contractor. Some utility companies also work with experienced contractors. Agricultural engineers at your state land-grant university should be able to help you locate a builder and give you some good counsel on constructing and operating a solar heating system. The National Frame Builders Association, 1406 Third National Building, Dayton, Ohio, can refer you to qualified contractors in your state, as can the staff of *Farm Building News,* 733 North Van Buren, Milwaukee, Wisconsin.

When contacting any suppliers or installers, find out their performance records. How many systems have they installed? Can you talk to the owners? Do members of the firms have heating, plumbing, electrical, and engineering experience?

The manufacturers of solar collectors should have published information on the Btu's their units produce per square foot of collector per heating season month. About 10,000 Btu/ft² per month is theoretically available at about 40° north latitude, but 3000 to 3500 may be closer to reality.

Check out the warranties offered to cover hardware and installation work. If something goes wrong, who fixes it? For how long? Are repair parts easily obtainable?

Before you accept or pay for a solar heating system, make sure it is taken through a complete operation test, as pictured in Fig. 5-12. All controls should work, and any leaks (in a water system) should be repaired immediately. Make sure the system runs long enough to confirm that the collector does indeed pick up solar energy and that it increases the temperature in the space and in the storage.

FIG. 5-12 Before a solar installation is accepted—and paid for—the owner should witness a run-through of the entire system.

SOLAR-ASSISTED HEAT PUMPS

Heat pumps can be combined with solar collectors, in either air-to-air or water-to-air heat-exchanger systems, to heat farm buildings. Heat pumps are fast gaining popularity, especially in human housing, either as an in-between auxiliary heating system or as the only backup to solar.

A heat pump is basically a reversing refrigeration unit. It can pump heat from the indoors for cooling, just as an air conditioner does, and it can also reverse the cycle to extract heat from outdoor air to heat the indoors.

The problem is that as the outside temperature drops, so does the heat pump's efficiency. At the same time, colder temperatures make space-heating requirements go up. In some newer installations, heat pumps are being used to draw heat from solar-warmed air, rather than from the cold outside air, to increase the efficiency of both the heat pump and the solar system. Heat exchangers from which the heat pump draws its Btu's can be connected to either an air or a water system.

In such systems, the heat energy delivered is the sum of the heat extracted from the air plus the electric energy needed by the compressor of the machine. The efficiency of a heat pump (called the *coefficient of performance,* or COP) is measured by dividing the total Btu's delivered by the Btu's supplied in the electricity which runs the pump.

In conventional operation, a COP of 1.5 to 2 is fairly typical. In other words, the heat pump delivers 1 1/2 to 2 Btu of heat energy for each Btu of electrical energy consumed. By combining a heat pump with a solar collector, the COP of the heat pump can be boosted to 3 or more.

A key advantage of combining a solar collector with a heat pump is that the solar system can be designed to operate more efficiently as well. As mentioned earlier, the efficiency of a solar collector increases as the temperature of the absorber plate (and the circulating fluid) decreases. There is less temperature difference with outside air, thus less heat lost from the system. A properly designed heat-pump solar-energy system will have the collectors working at peak efficiency and the heat pump supplying only the additional heat required.

In a livestock or poultry building, the amount of dust and moisture in the inside air must be considered if a heat pump is to be installed. Heat-exchanger surfaces need to be kept absolutely clean; moisture can cause corrosion and damage to equipment.

6

Solar Hot Water

Heating water is one of the easier, more practical ways to utilize solar energy. In applications where hot water is needed year-round, as in family households or dairy operations, the payback on a solar water heater can be twice as fast as on solar space heating alone.

Many of the principles of collecting and using solar heat discussed in Chap. 5 apply equally to solar water heating. In fact, in many installations, water heating can be an adjunct of solar space heating, utilizing the same collectors to heat both air and water.

Domestic solar water heating is enjoying a brisk business. A big reason, as mentioned, is that domestic hot water is needed year-round, rather than on a seasonal basis. Another is that solar water-heating systems can be built for reasonably low cost, under $2500 in many cases. A third reason is that water-heating units can be retrofitted on existing homes, with less trouble and cost than space-heating collectors.

In the average household, each person uses about 20 gal of hot water each day, for bathing, cooking, laundry, dishwashing, etc. If the water is heated from about 55°F as it comes from a deep well to 120 to 140°F, water heating uses 40,000 to 45,000 Btu per day, or a million to 1½ million Btu each month—for each of the 12 months of the year. With conventionally heated water, domestic hot water typically makes up 15 to 20 percent of the annual household energy bill.

In this book, we're primarily interested in applying alternative energy methods to farming operations, but much of the information that follows can be adapted to farm household use as well.

TYPES OF SOLAR WATER HEATERS

The simplest system for heating water with the sun is to allow the sun's rays to warm up water directly, in a *passive* water-heating system. If you're a rural American of the pre-plumbing vintage, you may recall bathing in a tub of water that had sat in the sun all Saturday afternoon.

However, most farm applications of hot water require moving the water from the collectors to storage to point of use, which implies some sort of *active* solar system. Figure 6-1 shows typical active and passive systems.

A *drain-back* system, if properly designed, can be the most efficient solar water heater to install. In many of these systems, no heat exchanger is required, since the potable water is circulated through the solar collectors only when temperatures are above freezing and thus no antifreeze is needed. With this system, collectors also can be designed to operate

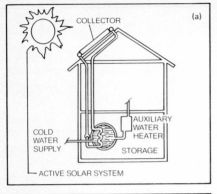

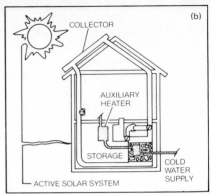

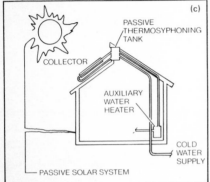

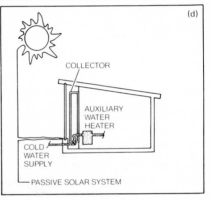

FIG. 6-1 Solar systems designed to preheat water before it enters a conventional water heater can be combined with space heating. *(U.S. Department of Energy.)*

at slightly cooler temperatures, increasing their efficiency. In a drain-back system, all plumbing must be sloped toward the storage tank, away from the collector inlet point. The collectors themselves must be "tilted" to avoid swags and pockets in the plumbing that might trap water when the system automatically drains. See Fig. 6-2.

Key to making a drain-back system work is a differential thermostat that operates both a circulating pump and a solenoid valve. The thermostat senses temperatures in both the water storage tank and the collector. When the collector temperature gets 10 to 15°F above the temperature of water in the tank, the thermostat energizes the circuit to close the solenoid valve and start the pump. (A start-up temperature difference of at least 10°F is necessary to prevent "short-cycling" starts and stops of the pump. As cooler water is admitted to the collector, surfaces cool down slightly. If the temperature spread is too narrow, the controller will shut the system off, the collector will drain then heat up, and the cycle start all over again.)

As the sun gets lower in the sky, collector temperatures drop. When the temperature difference between collector and stored hot water is again 10°F, the thermostat shuts off the pump and the solenoid valve automatically opens to allow water to drain from the collectors and connecting piping. Some drain-back designs require an air vent at the highest point in the system to ensure complete drainage.

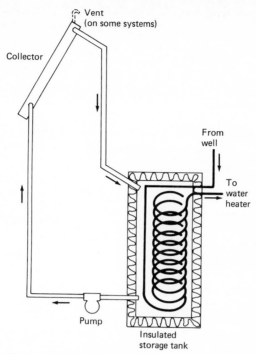

FIG. 6-2   Drain-back water heaters may or may not have a heat-exchanger loop.

A drain-back system can be built without a solenoid valve, thus eliminating one piece of electrical equipment. But not every plumber can understand how to do this. The secret is in the return pipe from the collector. If the return pipe is sized and sloped properly, the system will automatically siphon all water back to the tank when the pump shuts off.

A *closed-loop* water heater is more often used, although it is not as efficient as a well-designed drain-back system. Here, some type of antifreeze is added to the liquid passing through the collector. This "primary" loop of liquid picks up heat from the collector and transfers it to a storage tank through a heat exchanger of some kind—usually copper tubing. This allows the closed loop to utilize a relatively small amount of antifreeze solution to warm up a larger amount of fresh water in the storage tank.

There is a potential hazard with closed-loop systems. The antifreeze solution is toxic, sometimes highly corrosive. Any leaks in the heat exchanger can result in domestic water being contaminated with the antifreeze chemical.

Also, the primary loop—or heat exchanger—must be sized and designed to allow at least 90 percent of the solar heat to be transferred to the fresh water in the storage tank. A certain amount of heat is lost in the process of transferring or exchanging heat from one element to another, which is why most closed-loop systems are less efficient than drain-back methods.

An *air-to-water* heat exchanger can transfer a great deal of solar energy from sun-heated air to water. This is an attractive combination for many farm buildings. For example, a swine farrowing house with a solar-assisted space-heating collector can also preheat water for hot-water piping that is used for zone heat for baby pigs.

In these systems, a blower moves the heated air over pipes or tanks to transfer some solar heat to the water. The air-to-water heat exchanger may be located in the ductwork from the collectors or in a rock heat-storage bin. With proper insulation, this system is freezeproof and requires no antifreeze solution. It's also a feature that can be added with little extra cost, if built in at the time the space-heating system is installed.

THE ECONOMICS OF SOLAR HOT WATER

A solar hot-water system can cost from $15 per square foot of collector area to three or four times that figure. The economics of solar water heating are fairly simple, when measured in terms of the conventional fuel saved. However, the return on the initial investment depends on how much the system costs to install and operate and the price of the fuel that solar energy replaces.

A water-heating system should have a lifetime of at least 20 years, with only minor repairs and replacements. Any economic projection should be based on an estimate of conventional fuel costs over the life of the solar equipment.

In a way, the installation of a solar water-heating system is comparable to buying part of your 20-year fuel supply all at once, since you pay for equipment now that will gather free solar energy for years into the future.

To make the point, let's say a family of four installs a solar water-heating system that will provide half of their domestic hot-water needs for the next 20 years. On the average, each person uses 20 gal of hot water each day.

However, most solar water-heating systems are designed to work in tandem with conventional water heaters. That is, water from the source of supply (well, water district main, etc.) is preheated by solar energy and then piped to a conventional water heater that utilizes gas, electricity, or fuel oil to bring the water up to operating temperatures of 120 to 140°F. During periods when the solar system is not adding much heat to the water, the conventional water heater takes over in normal fashion.

If the water must be heated from a well-pump temperature of 55°F to 120°F, here's how to calculate the Btu's needed to heat 80 gal of water per day:

$$80 \text{ gal} \times 8.3 \text{ lb/gal} \times 65°F = 43,160 \text{ Btu/d}$$

That's 1,294,800 Btu required to heat water in a 30-day month. Heating all the water by LP gas would require about 17.6 gal of LP gas per month. At 60 cents per gallon, that's $10.50 worth of fuel each month, or $126 worth per year.

If the solar system provides half of the Btu's needed to add those 65°F to the water, that's a savings of $63 per year—assuming that LP gas stays at 60 cents per gallon, which is unlikely.

Energy experts predict that LP gas prices will rise from 7 to 21 percent per year between now and 1990. Taking the most conservative forecast of 7 percent per year, the savings in the second year would be $67 and so on, as more expensive fuel is saved each year of the equipment's life. Should LP gas continue to increase in price at an average 7 percent for the first 10 years the solar water heater is in operation, the savings per year would be as shown in Fig. 6-3.

A rule of thumb calls for the installation of 1 ft² of collector for each 1 gal of hot water needed per day, with about 2 gal of hot-water storage capacity for each 1 ft² of collector. For a household that uses 80 gal of hot water each day, figure on 72 to 120 ft² of collector surface area.

Such a system can be built and installed for about $2000—less, if the owner does a good deal of the work. That makes solar water heating a fairly slow-paying proposition, even allowing for a 7 percent boost each year in the price of conventional fuels. However,

First year...........................	$ 63.00
Second year	67.41
Third year.......................	72.13
Fourth year......................	77.18
Fifth year	82.83
Sixth year	88.63
Seventh year...................	94.83
Eighth year	101.47
Ninth year	107.57
Tenth year	116.17
Total	$872.22

FIG. 6-3 Amortization schedule for solar water heater.

tax credits will let you pass on to Uncle Sam up to 30 percent of the bill for domestic solar water-heating equipment. Chapter 23 will have more on this and on the investment and energy tax credits for "business use" of solar power.

At best, however, a solar water-heating system will not let you recover the costs of installation in the form of fuel savings in a couple of years. Economically, it's a long-range decision.

COMBINATION HOT-WATER SYSTEMS

One advantage of a solar hot-water system is that it is—or can be—compatible with a great many other systems. We've already mentioned "harvesting" some water heat from a solar air-collection system. In animal housing, other possibilities quickly come to mind, and some of them have already been put into practice by energy-conscious farmers.

For example, Gale Snow, a dairyman in Hickory County, Missouri, installed a "hybrid" solar-assisted heat-exchanger system (pictured in Fig. 6-4) for his 85-cow dairy operation. The system employs 120 ft² of solar collector on the south-facing roof of the dairy barn, with heat-exchanger coils of copper pipe to transfer heat from the milk-cooling refrigeration compressor.

"Since we put in this new system, we've had plenty of hot water available when we need it," says Snow. "That wasn't the case before; we often ran short of hot water before we got finished with cleanup chores."

In fact, Snow is planning to run the surplus hot water through insulated underground lines to his dwelling. Here's how the system works.

Hot water (90 to 100°F) is stored in a 1500-gal underground tank. The tank is insulated on the sides and bottom with Styrofoam®, and panels of the material float on top of the water in the tank to prevent heat loss through the tank's top. Two conventional LP-gas-fired water heaters are set to boost the water temperature to 110 and 170°F, respectively. The cooler water is used to wash cows' udders; the 170°F water is used to scrub the barn and clean milking equipment.

A circulating pump, controlled by a differential thermostat, moves water through black-painted finned copper tubes in the solar collector. The controller starts the pump when the temperature in the collector gets 15°F above the temperature of water in the underground storage tank.

When the collector temperature drops back within the 15° spread, the controller shuts off the pump. A solenoid valve in the same electrical circuit as the pump drops open to allow water in the collector to flow back to the tank.

FIG. 6-4 Solar collectors are teamed with cooling compressor heat exchangers to provide ample hot water for this dairy parlor. *(Photo: Farm Building News.)*

"We estimate that the collector picks up about 35 percent of the potential heat from the sun," says Snow. "We're actually gaining more heat from the cooling-tank compressor. The entire system not only saves money on LP gas needed to heat water, but saves a lot of time by providing ample hot water for cleanup chores."

Surplus heat from the cooling-tank compressor is "harvested" by circulating the heated freon refrigerant from the cooling coils of the unit through 120 ft of copper tubing submerged in the underground tank. This is a variation of the heat-exchanger system described earlier and shown in Fig. 6-5.

Snow's Holsteins average 50 lb of milk each day; that's about 4300 lb of milk that must be cooled from 95 to 38°F for storage. Snow has the potential of 300,000 Btu of heat to recover from each day's milk production—the heating equivalent of burning 4.5 gal of LP gas. The system is not 100 percent efficient, but, coupled with the solar unit, it heats fresh water from about 55 to 95°F.

Well water, circulating through a 60-ft-long copper loop in the tank, picks up heat from the water storage. The fresh water then is piped to the conventional water heaters. The two gas-fired heaters only have to heat 90°F water to 100°F, rather than heat to 100°F water that comes directly from the deep well at 55 to 60°F.

Snow's system cost just over $5000 to install—much of that for copper tubing that makes up the two heater loops.

"I'm not going to recover that investment very quickly, even with propane costs going up steadily," guesses Snow. "But the beauty of it is the abundance of hot water that lets us get cleanup done in a much shorter time so we can get on to other jobs around the farm."

VERSATILITY OF HOT WATER SYSTEMS

In many livestock housing applications, a hot-water heating system can provide a lot of flexibility. The water can be heated by any number of fuels—solar, LP gas, fuel oil, electricity, wood, coal. A typical farrowing house system is shown in Fig. 6-6.

For example, zone heating systems for pig areas in swine farrowing and nursery build-

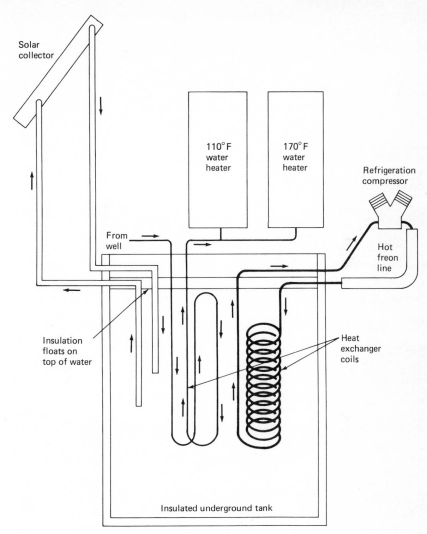

FIG. 6-5 Cross-sectional view of combination hot-water storage system.

ings can minimize pig heat loss by providing an 85 to 90°F area in which pigs can sleep and rest. Because of its versatility, hot water is being chosen for zone heating by more and more swine producers.

All hot-water heating systems have common components:

1. A thermostatically controlled heat source to provide water at a steady temperature

2. An expansion tank filled with air to cushion the expansion and contraction of water as it gains or loses heat

3. A heat-transfer system—radiator, pipes, etc.

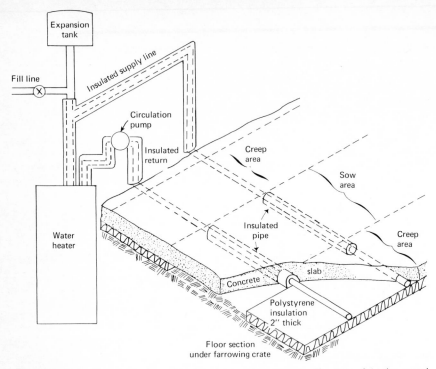

FIG. 6-6 Typical layout of a hot-water floor heat system. Source of heat may be solar, wood, or conventional fuels.

4. A circulating pump to provide a positive water movement through the system

Insulation is needed to make the most efficient use of the energy in hot-water heating. A 2-in-thick layer of expanded polystyrene (R-8) or equivalent foam insulation normally is placed under slab floors to be heated, which cuts costs by about 25 percent.

In areas where the floor does not need to be heated, the transmission piping should be insulated. Insulation for pipe placed in concrete should be waterproof, closed-cell-type foam of either rubber or plastic. This cuts heat loss by 80 percent and can save $1 or more per day in fuel costs in a 20-sow farrowing house.

The total amount of heat needed will depend on the length of the pipe, the shape of the area, and the temperature difference between the heated water and the room. The Btu's per hour required per foot of pipe using 140°F water and 60°F room temperature is:

Component	Btu/h per foot of pipe
Pipe in 5- by 5-in slat	55
Pipe in 5- by 12-in slat	90
Pipes 12 in measured center-to-center in 4-in-thick slab	35
Pipes 18 in measured center-to-center in 4-in-thick slab	50
Pipe with 1-in insulation in concrete slab	10

Residential-type water heaters, fired by either LP gas or electricity, are the most commonly used "boilers" for providing and controlling water temperature.

In some solar-assisted hot-water heating systems, the conventional water heater takes any "fill" or makeup water needed from the solar storage tank. In others, the controller operates valves and pumps to take water from the solar collector when that unit is providing water of the right temperature, then switches to the conventional water heater as a source when the solar unit can no longer supply the temperature needed.

In a closed-loop hot-water system, relatively small electric centrifugal pumps can be used. Pump motors typically are $1/4$ to $1/20$ hp; the smaller of them can be wired direct to the controller without switching relays. A circulating pump should have a capacity to move 1 gpm for each 5000 Btu/h of heater capacity.

Since the system is closed, the only working head the pump has is friction losses in the plumbing system. For good operation, friction loss should be maintained below 5 ft of head. With many heating-pipe loops, this may require a step up or two in pipe size.

The pump normally is located in the return line, near the water heater. This allows the pump to operate at the coolest point in the loop. Pumps with magnetic-drive systems are gaining popularity in hot-water heating systems, as they allow the use of sealed impeller cases and eliminate the need for shaft seals that must withstand high operating temperatures.

An expansion tank allows for expansion and contraction of water in the system, as the liquid is alternately heated and cooled. The expansion tank should be located in a "dead-end" standpipe, above the water heater. The tank should not be connected in the heating loop.

Piping often used is copper or black iron, although plastic pipe rated for hot-water service may be used. Plastic and copper have less friction loss than iron pipe of the same inside diameter. Pipes in concrete-slab floors should be installed directly over the insulation, with the concrete placed over both the insulation and the pipe. This gives more even temperature distribution at the surface of the floor. Pipes in slats should be located near the center of the slat. Figures 6-7 and 6-8 show other options as applied to dairy operations.

IN SEARCH OF QUALIFIED INSTALLERS

The problems with finding contractors who understand solar energy and can install the hardware to harvest it efficiently were discussed somewhat in the previous chapter. That dilemma is even greater with solar water-heating systems.

Without a thorough understanding of what is supposed to happen where and when in a system, even journeymen plumbers are stymied by some aspects of solar hot-water heating. As a result, systems are installed that do not work, even when the hardware is designed to capture 40 percent of the available sunshine and provide 60 to 70 percent of the hot water needed.

There's some shoddy equipment on the market too, and you can pay out dollars for faddish "additions" that are mostly cosmetic. But, to repeat, most of the problems relating to solar water heaters grow out of poor installation. Pipes leak, water freezes in the collector, valves and pumps are put in backward or upside-down, pipes are not insulated, or controllers are improperly installed and incorrectly wired.

The best insurance against costly disappointment is to understand thoroughly what each component of a system is supposed to do, and to write a performance contract that requires the installer to meet rigid specifications. Then, before you pay for the job, take the system through its paces to make sure it works properly. Reputable manufacturers provide warranties with their products; reputable installers should be willing to guarantee their work.

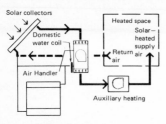

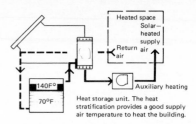

HEATING FROM COLLECTOR Air, the circulating heat transfer medium is drawn through the collector where it is normally heated to about 120-150° F. When the space requires heat, the solar heated air is drawn through the air handling unit in which motorized dampers are automatically opened to direct the hot air to the space. The air then returns to the collector where it is again heated and the cycle repeats itself.

HEATING FROM STORAGE At night or on cloudy days when solar energy is unavailable and when heat is needed in the space, the automatic control system directs the building return air into the bottom of the heat storage unit, up through the pebbles where the air is heated, through the air handling unit and into the space. When the solar heated air does not maintain the space thermostat setting, the automatic control turns on the auxiliary heater to add to the required heat.

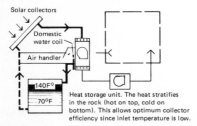

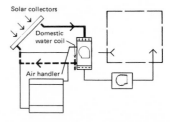

STORING HEAT When the space temperature is satisfied, the automatic control system diverts the air into the heat storage unit where the heat is absorbed by the pebble bed. The air returns to the collector where it is heated and this cycle is repeated.

SUMMER WATER HEATING In the summer, when space heating is not required, air is drawn through the collector where it is heated and then through the water heat exchanger coil. The solar heated air transfers its heat to the water which is being circulated through the coil and the air is then returned back to the collector inlet.

FIG. 6-7 Typical operation of the Solar-Surge air and water heating system designed for dairy operations. *(Source: Babson Brothers Company.)*

FIG. 6-8 Both wash water and milk parlor are heated by the sun in this experimental setup at USDA's Animal Genetics Laboratory, Beltsville, Maryland. *(Photo: U.S. Department of Agriculture.)*

6–10

One safety note: Solar water systems can produce temperatures that will scald human flesh. Mixing valves should be installed on all domestic water faucets to limit temperatures to below 140°F.

BUILD A SOLAR POND?

A small pond of salt water may be the solar-heating system of the future. Scientists at Ohio State University are experimenting with a 160,000-gal pond that effectively collects and stores heat from the sun. See Fig. 6-9.

The pond acts as a heat trap. As solar rays penetrate to the bottom of the pool, water at a 9-ft depth is heated. In freshwater ponds, warm water expands and rises to the top. A solar pond utilizes an absorber of black plastic lining and salt concentrations to store the heat gathered at or near the bottom. Figure 6-10 shows a sectional view. The salt solution naturally stratifies, so denser brine collects at the bottom of the pond, while upper water surfaces are nearly fresh water. The layers of salt concentration act as insulators, so heat is lost slowly.

Israeli engineers have built a large solar pool using naturally saline water from the Dead Sea. The heat collected is used to run a Rankine-cycle engine to generate 150 kW of electricity.

In the Ohio pond, heat is collected by heat-exchanger pipes near the bottom of the pond. Fresh water circulating in the heat exchanger carries heat to buildings, where the heat is distributed by radiators.

Costs of using the solar pond to heat a single building run about the same as using electric heat, say the Ohio researchers. They point out, however, that larger solar pools designed as central heating plants for several buildings would be cheaper to build, per Btu of capacity.

It's important that a solar pond be well-sealed, to prevent the brine from leaching into surrounding soil or seeping into ground water.

FIG. 6-9   Solar pond at the Ohio Agricultural Research and Development Center has a salinity concentration of 20 percent in the bottom to retain heat collected from the sun. *(Photo: U.S. Department of Agriculture.)*

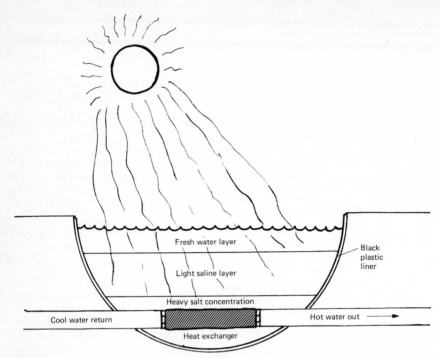

Fresh water layer

Black
plastic
liner

Light saline layer

Heavy salt concentration

Cool water return

Hot water out ⟶

Heat exchanger

FIG. 6-10 A solar pond can absorb and store enough heat from the sun's rays to heat a building of the same cubic feet. Temperatures of up to 190°F have been recorded.

7

Other Heating and Cooling Options

The big trend in livestock and poultry production in the past dozen or so years has been the move to total confinement. The costs of housing animals in completely enclosed, controlled-environment structures has been offset by the advantages:

- One worker can manage many more animal units in confinement.

- Animal production can be a year-round proposition, even in areas with climates unfavorable to raising animals outdoors during part of the year.

- The comfort factor of working in a controlled-environment situation at 2 A.M. on a cold winter night is decidedly in favor of enclosed buildings.

While moving animals indoors has benefited many aspects of livestock and poultry management, it has complicated others—particularly in the area of energy. Confinement housing makes a producer much more dependent on outside sources of energy to heat buildings, provide ventilation, remove wastes, etc. And it makes the livestock more vulnerable to interruptions in the energy supply line.

At the same time, climbing prices for fossil fuels and electricity are rapidly making those production inputs major costs for farmers. Swine producers now spend $1 or more for heat energy for each pig weaned in winter.

Those sources still supply virtually all the energy used in farm buildings. However, farmers are turning to energy alternatives to heat and cool structures. We've already discussed the application of solar energy to livestock buildings. Now, let's look at some optional energy sources that may be useful for many engaged in animal agriculture.

WOOD HEAT

Wood is a plentiful, accessible fuel for many rural Americans. It is relatively clean and comes from renewable sources: forests and farm woodlots.

At first glance, wood heat seems to be more applicable to human dwellings than to animal housing. For one thing, it's more difficult to automate the combustion of wood with the same degree of accuracy that can be used to control heating equipment that uses gas, oil, or electricity as a source. For another, wood is bulkier per Btu of heat potential than most conventional fuels. For a third, not every producer has access to a steady, long-range supply of firewood; if you must haul wood 50 mi or must pay $80 or $90 per cord for it, that fuel may not seem like an attractive alternative.

Still, for some farmers, wood is an abundant, economical source of heating fuel, and modern woodburning equipment makes the use of this source safer and more convenient than it was in the past. Figure 7-1 shows one way to distribute wood-warmed air to a farrowing house.

One pork producer installed a similar system, after spending $1 per pig for LP gas to keep pigs warm in individual farrowing houses during winter. The most abundant on-farm source of heat was firewood from woodlot stands of hardwood trees, so this farmer built a 25- by 150-ft farrowing-nursery building, and virtually weaned it from the gas tank.

At the heart of the heating system is a 4-ft-long firebox, with a thermostatically controlled combustion damper. The firebox is loaded with seasoned oak logs; a thermostat in the farrowing room adjusts the air intake damper to control combustion to maintain the space heat at a preset temperature. The firebox is sheathed with a sheet-metal plenum, much as any other forced-draft furnace is. Warm-air ducts run to each pair of farrowing pens and each pair of nursery pens, with the ducting dropped to within 18 in of the floor to provide an 85 to 90°F zone for baby pigs. An electric blower pushes warm air through the ducting at 3600 cfm.

As a backup system, the furnace is equipped with an auxiliary LP-gas burner (see Fig. 7-2), controlled by a second thermostat, which takes over the heating load if the wood fire dies down.

Firewood is not a *free* source of heat, even when you own the trees. There is cost and labor involved in cutting and hauling the wood, as well as labor to refill the firebox two or three times each day and to remove ashes.

Still, you might be able to do the wood cutting and hauling in odd chunks of time throughout the year and plan the wood supply to have a winter's worth of fuel cut and seasoned before the cool months arrive. The savings of a dollar's worth of LP gas per winter-farrowed pig could pay for the daily chores of stoking the fire and scooping out ashes, tasks that take only minutes each day.

If you have an ample supply of good-quality hardwood on your farm, you may want to apply it to animal housing. How economical firewood is as an alternative source of heat depends partly on where you're located and whether you have a dependable supply of firewood available. It also depends on the cost and availability outlook for more conventional heating sources and whether you can incorporate the extra time required to gather, haul, and use firewood into your management schedule.

For some producers, a better option might be to raise a few more units of livestock or poultry and let those animals pay the extra cost of more expensive conventional fuels. When wood is seasoned after harvesting and burned in proper equipment, it can be a safe and efficient fuel. Whether it is also an economical fuel for a particular farm operation is a matter for the individual farmer to decide.

THE WOOD SUPPLY

The fuel value of wood varies with the type of wood and depends largely on the density and moisture content. Any wood will burn, but the denser (heavier) woods, if properly dried, generally deliver more Btu's per unit of volume. In other words, a pound of wood of any kind, on a dry-weight basis, contains the same number of potential Btu's of heat energy. However, with lighter, less dense woods, a pound makes a bigger pile. We'll get into why that is so important shortly.

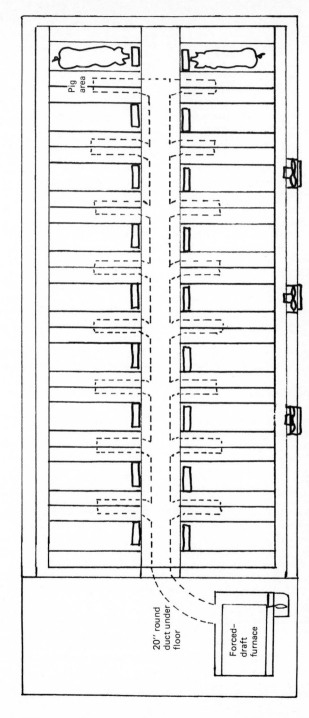

FIG. 7-1 Heated air from a forced-draft wood-burning furnace can be distributed with an underfloor duct system. The laterals provide some floor heat under pig creep areas. With planning, the outlets can be turned to blow air toward pigs in winter and toward the sow in summer—utilizing the same fan for both heating and cooling.

Pig area

20" round duct under floor

Forced-draft furnace

FIG. 7-2 Wood furnaces can be equipped with an auxiliary LP-gas or fuel oil burner, as shown to the right of the firebox door on this unit.

Firewood commonly is measured by the cord, which is the equivalent of a tightly stacked pile of wood that measures 4 by 4 by 8 ft, or a stack of wood occupying 128 ft³ of space, as shown in Fig. 7-3. One *pound* of very dry wood (zero moisture content) of any species has a heating value of about 8600 Btu when burned. Any moisture in the wood reduces the recoverable heat by carrying heat up the chimney during vaporization.

Each pound of water vaporized uses about 1200 Btu. A pound of wood with 20 percent moisture content (the average moisture content of air-dried wood) contains ⅕ lb of water and ⅘ lb of solid wood and has a potential heat value of about 7000 Btu.

Additional Btu's are lost through the formation of volatile liquids and gases during combustion, depending on the efficiency of the equipment in which the wood is burned. In efficient woodburning stoves and furnaces, many of the volatile gases are consumed in the firebox, rather than exhausted up the chimney unburned. This can improve the combustion efficiency of a stove or furnace to as high as 60 to 65 percent—which puts wood in the same efficiency category as gas and oil.

When wood burns, three things happen:

1. Moisture is removed by evaporation.

2. The wood chemically breaks down into charcoal, gas, and volatile liquids, with carbon dioxide and water vapor being the chief end products.

3. The charcoal burns, forming carbon dioxide directly or with an intermediate conversion to carbon monoxide.

Table 7-1 lists the weight per standard cord of various woods, when green (or fresh cut) and when air-dried to 20 percent moisture content, and the potential Btu's of heat in a cord of air-dried wood of each species.

Referring to Table 7-1, what is a cord of air-dried white oak worth? With LP gas priced at 60 cents per gallon, the wood is worth $178 per cord, strictly on the basis of Btu's of available heat.

But there's a catch or two. For one thing, even the more efficient woodburning devices operate at only 55 to 65 percent efficiency. For another, there's no cutting, splitting, and hauling required with the LP gas, and gas-fired equipment is more convenient to operate and easier to regulate than most wood-fired heaters.

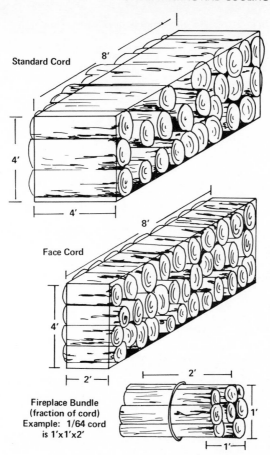

Standard Cord

8'

4'

4'

Face Cord

8'

4'

2'

Fireplace Bundle
(fraction of cord)
Example: 1/64 cord
is 1'x1'x2'

2'

1'

1'

FIG. 7-3 Standard measurement of firewood is the *cord,* defined as the amount of closely stacked wood that will occupy 128 ft³ of space. A *face cord* or *rank* is a fraction of a cord, depending on the length of the logs. *(Forestry Department, University of Missouri.)*

WHAT KIND OF EQUIPMENT?

Today, many wood-fired stoves and furnaces are amazingly efficient, compared with older models. And their operation can be automated to a considerable degree, with thermostat controls to operate dampers and forced-draft blowers or circulating pumps to distribute the heat (see Fig. 7-4). However, selecting the kind and size of heating equipment for a particular heating requirement is not always a simple matter. Manufacturers of wood-fired heating equipment rate their products on the basis of several different criteria, and the claims made can range from reasonable to ridiculous.

Computing the size of equipment needed begins with the same exercise to calculate

TABLE 7-1   Weight and Potential Heat Energy of Selected Woods

Species	lb per cord (green)	lb per cord (20% moisture)	Potential heat per cord, Btu (in millions)
Ash	3940	3370	23.6
Basswood	3360	2100	14.7
Cedar, red	3260	2700	18.9
Cottonwood	3920	2304	16.1
Elm, American	4293	2868	20.1
Elm, red	4480	3056	21.4
Hackberry	4000	3080	21.6
Hickory, shagbark	4980	4160	29.1
Locust, black	4640	4010	28.1
Maple, silver	3783	2970	20.8
Maple, sugar	4386	3577	25.0
Oak, red	4988	3609	25.3
Oak, white	4942	3863	27.0
Osage orange	5480	4380	30.7
Pine, shortleaf	4120	2713	19.0
Sycamore	4160	2956	20.7
Walnut, black	4640	3120	21.8

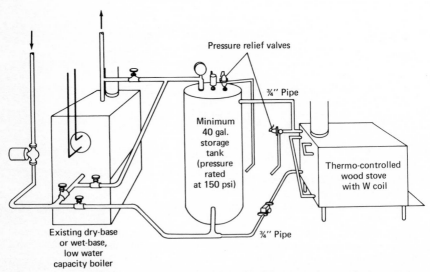

FIG. 7-4   Wood furnaces and stoves can be equipped with hot-water heat exchangers, to serve as auxiliary heating systems along with a conventional boiler or water heater, or both. The loop running to the wood stove should be equipped with pressure-relief valves, located at 9 and 10 in the drawing. *(National Stove Works, Inc.)*

heat losses from a building as explained in Chap. 3. The capacity of the equipment must be adequate to meet the requirements of the coldest day (or night) to be expected, whereas the operating level for the average or normal heating season will be much below that.

As mentioned earlier, burning 1 lb of air-dried wood of any species will release about 7000 Btu of heat energy. However, as can be seen from the comparisons of *weights* of wood in Table 7-1, nearly twice as many Btu's of shagbark hickory can be stored in the same volume as of a lighter wood, such as cottonwood. In other words, burning 1 lb of either hickory or cottonwood releases the same amount of heat, but a firebox filled with hickory will produce nearly twice as much heat as one filled with cottonwood.

Given equal combustion efficiencies, larger fireboxes hold more fuel and therefore more Btu's of heat energy, which means either an increased rate of heat release (more Btu's per hour) or a longer time over which the unit can release a useful amount of heat—or both. The rate of burn is an important consideration if the heating equipment is to be used around the clock as a primary heat source. Firebox size and rate of burn should be matched to allow for a reloading of the firebox at fairly long intervals (8, 10, or 12 hours), without the necessity of relighting the fire.

Rate of burn (and combustion efficiency) is controlled chiefly by the amount of oxygen admitted to the fire. More efficient woodburning units can be dampered tightly to kill the flames. These units hold fires longer and will produce more useful heat than less-efficient stoves and furnaces.

Where the combustion air enters the firebox is also a factor affecting the efficiency of a unit. Stoves and furnaces that admit air directly into or under the fuel bed are not very efficient. For complete combustion of the volatile gases that are formed, the incoming combustion air must be preheated and mixed with the gases at 1100°F. More efficient models pull air into the firebox through a preheating chamber.

Despite advertising claims that often confuse rather than enlighten potential customers, manufacturers of woodburners should be able to provide performance information on their equipment that will let you evaluate how well the unit suits your needs.

Essentially, you will need to know how much heat the unit puts out over a period of time with the kind of wood you will burn. The standard measurement is Btu's per hour. Secondly, you need to know the rate of burn (the time over which this heat output can be maintained with one filling of the firebox).

At present, there are no established industry standards for rating woodburning equipment. However, the average heat output of a stove or furnace is equal to the total amount of energy in a fireboxful of wood times the efficiency of the unit divided by the time to burn a load of wood. This can be roughly computed, given some basic information on the heater and the wood that will be burned in it.

Suppose you will burn white oak wood to heat a swine farrowing house. By referring to Table 7-1, we find that a cord (128 ft³) of air-dried white oak weighs 3863 lb and contains 27 million Btu of available heat energy. One cubic foot of white oak weighs just over 30 lb and contains about 210,000 Btu.

Assuming a woodburning furnace is 60 percent efficient, how many Btu's of useful heat —in the form of air-dried white oak wood—could you load into a firebox that measures 4 by 1 by 1 ft?

$$\text{Firebox volume} = 1 \times 1 \times 4 = 4 \text{ ft}^3$$

$$\text{Firewood density} = 30 \text{ lb/ft}^3$$

$$\text{Heat energy of wood} = 7000 \text{ Btu/lb}$$

$$\text{Furnace efficiency} = 60\%$$

$$\text{Btu} = 4 \times 30 \times 7000 \times 60 = 504,000$$

That figure alone doesn't tell you all you need to know, however.

Will the furnace produce 504,000 Btu of heat for 1 hour? Or will it generate an average 50,400 Btu for 10 hours? You also need to know the "burn time" or rate of burn of the equipment. If the furnace described above is designed to burn a fireboxful of wood over a 10-hour period, you may expect the average heat output to be 50,400 Btu/h.

Heat is heat, regardless of its source. However, a heat distribution system can employ one of several vehicles. Heat generated by burning wood can be used to heat the ventilating air of a building, or it can be used to heat water for underfloor piping or radiators, as discussed in Chap. 6.

With wood-heated water, some extra precautions may be needed in the system. Notice the location of pressure-relief valves in the system shown in Fig. 7-4. A wood fire has a longer "die-down" and "recovery" time than gas burners or electrical heating elements. Larger circulating pumps will probably be needed to move water in a conventional hot-water heating system that is fired by wood, to provide positive circulation and prevent damage to the system.

In addition, a bypass loop should be built into the piping to allow heat to be removed from tubes quickly enough to prevent boiling—which can cause the pump to *cavitate* or lose suction. The selection of materials for the hot-water plumbing is more critical than with other systems, also. Galvanic corrosion can be more of a problem with metals that are subjected to wood flames and the products of combustion. Also, provisions should be made for high heat buildups that would occur if a pump should fail. Pressure-relief valves can be located to handle this kind of emergency, to prevent damage to the plumbing.

In a hot-water heating system, calculating the heat flow from the firebox flames to the water in the heat pickup tubes is necessary, but complicated. The circulating pump should be of a capacity that can provide sufficient flow to keep water temperatures well below boiling.

One way to determine the heat flow is to build the heat-exchanger tubes (some wood-burners are equipped with water-heating elements, as shown in Fig. 7-4), light a hot fire in the firebox, and run water through the system with a garden hose. By varying the rate of flow of the hose, you can change the outlet temperature of the water. Measure the outlet temperature at various rates of flow. When the flow is right for the temperature range you need, measure the flow rate with a bucket and stopwatch. It's also a good idea to install a pressure gauge at the water inlet during this exercise, to determine what the pump head will be.

Another approach to using wood-heated water is to locate the wood stove or furnace "upstream" in the system, to preheat water going to a conventional boiler or water heater. This can be done by circulating the water through a jacket or loop of tubing in contact with the firebox, then piping the water to the conventional heater unit. In this kind of system, the wood stove acts as a water preheater, in much the same manner as solar collectors described in Chap. 6.

Still another version of this "piggyback" approach to using wood heat is to duct a woodburner's heat into the return air plenum of a conventional forced-draft heating device. In this arrangement, two thermostats can be used. One, set at a higher temperature, controls the combustion air on the woodburning unit. The other, set a few degrees lower, starts up the conventional furnace when the woodburner is not producing enough heat to meet the demand. The blower can be wired to move heat from either or both sources.

A WORD ABOUT SAFETY

Wood can be a dangerous fuel to use in homes and farm buildings, but it doesn't have to be. Common sense and a few precautions can prevent costly—and often tragic—fires.

1. The chimney should be properly constructed, kept in good repair, and kept clean of tar, soot, and creosote.
2. The heating unit should be well-designed and sturdily constructed so that burning coals and sparks cannot escape; it must be located on a nonburnable base large enough to catch any coals or sparks that spill out when doors are opened.
3. Walls and ceilings of flammable materials must be protected by adequate distance or heat shields.
4. Burnable materials should be kept well away from the heating unit. Flammable liquids should never be used to start or encourage fires.
5. Enough ventilation should be provided so that oxygen consumed by combustion can be replaced.

ENERGY FROM MOTHER EARTH

Many livestock producers—particularly pork producers—are cutting their heating bills by moving underground. Or at least partially underground. Others duct ventilating air through underground pipes, to pick up some of Mother Earth's residual heat.

Below-grade animal housing is not a new idea. Farmers of two generations ago built "bank barns," with the subgrade level occupied by livestock and the ceiling insulated by a large hay mow on the floor above. The concept still has energy-saving potential, and a few problems.

First, the good points.

The heat-conserving features of below-grade buildings are several. For one, the space is protected from wintry winds, so there's less infiltration of unwanted cold air. For another, the mass of earth absorbs much of the "shock" of rapid temperature changes. If the outside temperature is 10°F and you want to maintain the interior of an above-ground building at 70°F, that's a 60°F difference. If the average temperature of the earth surrounding a below-grade building is 45°F, you only have to maintain a 25°F difference.

Contrary to what some people think, the earth is not a particularly good insulator. But it's a pretty good thermal "shock absorber." The massiveness of earth slows down temperature changes, so daily and seasonal lags occur.

In one recent study, the lowest average outside air temperature for the site was in February, while the lowest average temperature 9 ft below grade was in late May. Conversely, the warmest outside temperature was in early August, while the warmest temperature at the 9-ft depth occurred in November. In other words, summer heat doesn't get through to the lower levels of below-grade structures until November—just when the supplemental heating season begins.

There are limitations to the benefits of building below grade for animal housing, however. With some types of buildings, construction costs for subgrade structures may run 20 to 25 percent higher than for well-built aboveground buildings. Soil type in some areas argues against building below grade, as does the topography of some sites. Subgrade buildings are better bets with south-sloping hills and well-drained soils.

CONTROLLING THE MOISTURE

Perhaps the biggest drawback to building animal housing below grade is that old bugaboo: dampness. As we've mentioned before, animal housing usually needs ventilating to remove excess moisture. While walls in contact with the earth conduct less heat out of a building, outside air drawn into an underground structure needs to be heated, just as if the building were constructed above the ground.

However, there are ways to get rid of excess moisture in a livestock building, without pulling great volumes of fresh air into the building. Using dehumidifiers to control moisture often can reduce the minimum ventilation requirement to just that amount of fresh air needed for the animals' respiration and to remove odors and gases.

SUMMERTIME COOLING

In most sections of the country, the temperature of the earth at a 10-ft depth stays at 55 to 65°F year-round. That stored "coolness" can be used to cool animals in warm weather, just as residual heat from underground can reduce heating bills in winter.

Merril Wernsing, an Illinois farmer, pulls air into his confinement swine building through nine 150-ft lengths of field tile buried 12 ft underground. Figure 7-5 shows a similar system.

Here's how the earth moderates temperatures inside a building in both winter and summer. When outside temperatures range in the mid-90s, the underground cooling system pumps air into the building at 70 to 75°F. A considerable amount of moisture will be condensed out of hot, humid air as it passes through the underground loops of pipe, as much as 5 gallons per hour (gph) on particularly muggy days.

The system also cuts heating bills in winter. With outside temperatures at 25°F, the "earth energy" pipeline admits 45°F air to the furnace inlet.

A more conventional method of using earth energy to cool animals is to spray droplets of cool water directly onto the animals' bodies. Water pumped from a deep well is at a constant 50 to 55°F. Evaporation of the moisture cools the livestock, as the cool water absorbs heat from the animals' skin and surrounding areas. See Fig. 7-6.

With most spray cooling systems, about 0.09 gph per pig is ample, for hogs in semi-

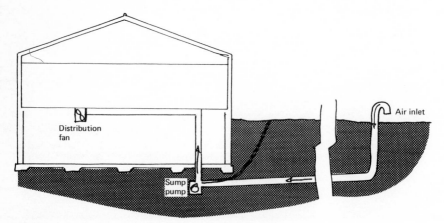

FIG. 7-5 Underground pipes buried at 5 ft or greater depth can cool summer air by 15°F and add a like amount of heat to incoming air in cold weather.

FIG. 7-6 Evaporative cooling for livestock can be accomplished with an intermittent spray of water.

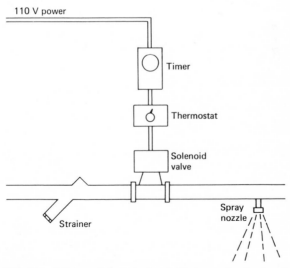

110 V power

Timer

Thermostat

Solenoid valve

Strainer

Spray nozzle

FIG. 7-7 A thermostat can be located between the timer and solenoid valve in a cooling spray system. Set the thermostat at 75 to 80°F and the timer to open the solenoid valve for a total of 2 min out of each hour.

confinement structures, or a spray capacity of 0.045 gpm per pig, when the spray is operated for 2 min out of each hour. The typical control (as diagramed in Fig. 7-7) is a timer set to open a supply valve for 20 s every 10 min, or 30 s every 15 min.

Nozzle design (spray angle and pattern) and the height of the nozzle affect the sprayed area at the floor. The nozzle should cover three-fourths of the pen area with a solid cone of droplets, rather than with a fine mist or fog.

All the heating and cooling options discussed in this chapter—wood and geothermal —involve making initial capital investments with a goal of paying smaller fuel bills in the future. The best time to plan the energy-smart features of a building is in the "pencil-and-paper" stage. Changes can be made fairly economically with an eraser. Once building materials and alternative energy hardware have been bought, it's more expensive to correct mistakes.

8

Energy-Efficient Cropping

Agriculture is the business of capturing solar energy in growing plants and using the live organisms sustained by that energy to produce food and fiber needed by human beings. In the United States, food production has been aided greatly by the use of energy to provide power, fertilizer, and pest-control products—with the result of lower food prices than in countries that rely on human and animal power.

Considering the importance of the industry, agriculture is not a huge consumer of energy. The entire food system, from production to consumption, takes less than 17 percent of the total energy used in this country. Agricultural production, as mentioned earlier, uses only 3 percent of the total—less energy than is used to fuel jet aircraft.

But energy is becoming steadily more expensive. Rising production costs make decisions about which crops to grow more critical to profits. Costs are going up faster than are prices for most farm commodities, and energy leads other inputs in the rate of price increase. Fuel prices in some areas jumped 50 percent between the 1979 and 1980 crop years. That price rise added 25 cents to the cost of growing a bushel of dryland wheat in Kansas and tacked on an extra $15.81 per acre to the cost of growing irrigated corn.

Like most other aspects of alternative energy in agriculture, energy-efficient cropping is a long-range proposition. At least, it is for most commercial farmers. For example, plant breeders are optimistic about developing crop varieties that are resistant to specific insects and diseases. This could be an economical way to conserve the energy now used to produce and apply petrochemicals to control these pests. But you won't find seed corn that is resistant to *all* insects—not this year, or next.

American agriculture today is highly dependent on fossil fuels and electricity generated from the burning of fossil fuels, as indicated by Table 8-1. That dependence grew over a number of years, as farmers adapted energy-intensive technology to cropping operations. The 300-hp, four-wheel-drive tractor did not come off the assembly line a few months after the first tractor was invented. It took time, and it will take time to adapt new energy-conserving technologies to fit total production systems.

There are short-term ways to conserve energy in crop production, irrigation, harvesting, and crop processing. But, for the most part, they involve trade-offs. Either production is decreased for all or part of the acreage or a substantial investment in alternative energy systems is required to maintain production currently achieved by inputs of purchased energy.

The key is to make cropping more energy-efficient without taking undue risks or trimming production output too severely. Let's look at some of the possibilities, taking the most energy-intensive crop inputs first.

TABLE 8-1   Percentage of Energy Derived from Fossil Fuels in Various Farm Operations

Operation	Percent of total
Fertilizer	30–38
Crop drying	15–17
Field operations	10–15
Irrigation	7–13
Pesticides	2– 5
Other	15–20

FERTILIZER

Nitrogen is most often the limiting nutrient in crop production. The lion's share of energy used to manufacture fertilizer is consumed in making nitrogen, most of which is manufactured by combining atmospheric nitrogen with hydrogen from natural gas to form ammonia. Something like 40,000 ft³ of natural gas is used to produce 1 ton of ammonia.

Some day, cheap sources of electricity may permit the manufacture of nitrogen with other processes. The University of Nebraska now is experimenting with power from photovoltaic cells used to create a lightning bolt to "fracture" atmospheric nitrogen out of air, in much the same kind of process that occurs in nature. Figure 8-1 shows how this system works. Similar experimental plants in Hawaii and Nevada use wind and water power to generate the electricity.

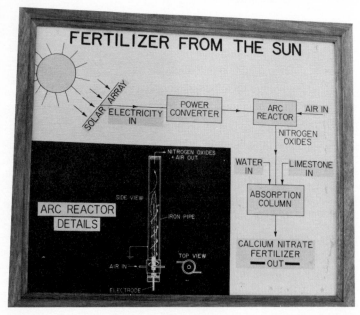

FIG. 8-1   Sun-powered fertilizer plant at the University of Nebraska Field Laboratory uses electricity from photovoltaic cells to make nitrogen from air. The nitrogen is carried in limestone particles in solution in water, to form calcium nitrate.

However, costs are still far out of reach for commercial application of these nitrogen-producing techniques. For the near future, the principal alternative to commercial nitrogen fertilizer made from fossil fuels is a crop rotation that includes one or more legumes to fix substantial amounts of atmospheric nitrogen in the soil for a following nonlegume crop.

Alternatives to growing commercially fertilized nonlegume cash crops successively include rotating corn and soybeans or other valuable legume crops, or growing winter legumes as nitrogen-fixing crops in regions with longer growing seasons.

A crop such as red clover can be worked into a four-crop rotation, such as corn, soybeans, wheat, and clover, to reduce both fertilizer and fuel needs. For example, in the first year of the rotation, a cash crop of corn is grown. The second year, soybeans are planted, followed by a winter wheat crop seeded after soybeans are harvested. In late winter or early spring, red clover is overseeded in the wheat. After wheat harvest in summer, the clover usually will make enough growth for a hay crop. The following year, a hay crop can be made from the clover in early summer, perhaps followed by a clover seed crop later on. Then, the clover regrowth is plowed down to provide nutrients for the corn crop, as the rotation starts over again.

In this kind of rotation planting, red clover can supply up to 110 lb of nitrogen from the atmosphere to help meet the fertility requirements of the following crops. Other legumes fix varying amounts of atmospheric nitrogen:

Legume	Alfalfa	Sweet clover	Ladino clover	Crimson clover	Vetch
Nitrogen, lb/ac	194	119	177	94	80

When coupled with minimum tillage practices, this kind of rotation involves deep plowing once in the rotation—when the clover is turned under. The soybean crop provides much—perhaps all—of the nitrogen needed by the wheat crop that follows. The plowed-under clover feeds the following corn crop a good share of the needed nitrogen.

A bonus with this kind of rotation can be better long-term weed control. The two successive years of row crops (corn and soybeans) allow cultivation and chemical control of perennial weeds. The clover crop, mowed two or three times for hay, effectively controls annual weeds.

Some farmers now are experimenting with legume intercrops grown between rows of nonlegume cash crops. This may prove to be another workable alternative to buying all the nitrogen required.

Organic wastes, such as animal manures and sewage sludge (where these materials are available), can often be substitutes for manufactured nitrogen. If full use were made of the "collectible" animal manure now produced, this source could fertilize 20 to 25 percent of the current U.S. cropland. Available sewage sludge that might be used to fertilize crops has not been accurately inventoried, but probably would be sufficient for only 5 percent of the current needs, or less.

CROP DRYING

Crop drying is the second biggest user of energy in crop production. Chapters 9 to 11 are devoted to ways to save energy in this operation.

Some people suggest that corn growers go back to ear-corn pickers that allow more natural drying in the field. Bruce McKenzie, agricultural engineer at Purdue University, doubts that this will happen on a big scale.

"By going back to ear corn, we'd double the amount of material that has to be handled," he says. "And we'd have to deal with a non-free-flowing product, which would mean major changes in grain-handling equipment."

However, McKenzie notes that the 1000 lb or so of corncobs in each acre represents a lot of fuel potential. In fact, there are enough Btu's of heat energy in the cob to dry the corn that comes off it.

We'll go into ways to dry grain with crop "leftovers" in Chap. 11.

FIELD OPERATIONS

Field operations, including tillage and harvesting, account for about 15 percent of the energy used in crop production. Ideas for saving fuel in field operations range from reducing tillage to substituting animal power for tractors.

A note in passing about substituting *real* horsepower for the internal combustion kind: Few people seriously pose this as a solution to our energy dilemma, although the idea may be a workable option for a few individual farmers. To produce today's crops with animal motive power would require some 16 million horses and mules. To breed up this number from the 5 million or fewer now available as seedstock would take at least 15 years—and it would require half the current U.S. crop acreage to feed them.

In the short run, better maintenance and more careful operation of field machinery probably will do more to save fuel than any other fieldwork alternative. Longer-term, farm-produced fuels will begin to replace fossil fuels. But for the next season or two, a review of tillage practices may be the best way to save tractor fuel on most crop farms.

"Farmers can make big fuel savings and get better weed control by easing back some on tillage," says Maurice Gebhardt, U.S. Department of Agriculture agricultural engineer. "Farmers who plow in the fall shouldn't be going any deeper than 3½ to 4 inches with discs and field cultivators in the spring. If you till deeper than that, you reduce the concentration of the herbicides near the surface, and lose effective weed control."

In fact, conventional tillage practices often seem to be at odds with both energy conservation and erosion control. For instance, many farmers moldboard-plow to bury surface materials, to control weeds and insects, and to "make the field look better." Yet erosion is often much less when material is left on the surface as a mulch, as it is with other methods of deep tillage such as chisel plowing—which consumes much less fuel per acre than moldboard plowing.

As Gebhardt notes, farmers often overdo secondary or "shallow" tillage to provide a well-worked bed for crop seeds. However, water infiltration is better and erosion less when the soil surface is rougher.

Reduction of tillage operations can have a triple benefit: reduced consumption of expensive fuel, fewer trips over the field (less soil compaction), and—in many cases—reduced soil erosion.

Estimated tractor fuel requirements for different tillage implements are listed below, based on information from James C. Frisby, agricultural engineer, University of Missouri:

Implement	Moldboard plow	Chisel plow	Tandem disc harrow	Field cultivator	No-till row planter
Depth, in	8	8	4	8	2
Fuel, gal/ac	4.31	1.64	1.44	1.14	1.53

"The influence of continued use of reduced tillage on insect and weed populations (in all regions and for all soil types) is not fully understood," says Frisby. "However, it is generally accepted that good water infiltration and surface mulch reduce soil erosion. It is also apparent that eliminating primary tillage operations—especially moldboard plowing—reduces fuel consumption."

While the biggest benefit from reduced tillage appears to be soil and water conservation, this cropping technique lends itself to fuel conservation. However, research in Illinois, Minnesota, and Wisconsin indicates that land under no-tillage systems does not produce sustained crop yields that are equal to those from land under conventional tillage. Perhaps the best compromise on some farms is to alternate reduced tillage with conventional tillage.

Matching tractors to the implements they pull is becoming more important as fuel costs rise. Tractors have changed over the past several years. Diesels have replaced gasoline engines in most larger-size tractors. Horsepower has increased rapidly. Tractor weight has increased, too, but not as rapidly as engine horsepower. As a result, the weight-to-horsepower ratio has decreased by about 25 percent in the past five or six years.

These changes dictate changes in the way equipment is operated, if fuel economy is a primary goal. For example, modern tractors, with low weight-to-horsepower ratios, cannot develop enough traction to carry full engine power in lower gear ranges.

This loss of traction, or *wheelslip*, is critical to a tractor's efficiency. For two-wheel-drive tractors, wheelslip should be about 10 percent in uncultivated ground and 12 percent in cultivated soils. Four-wheel-drive tractors should slip slightly less. If the wheelslip is too high, weight can be added to the tractor, or implements can be pulled at shallower depth.

Modern diesel tractor engines are designed to run at high speed. The old two-cylinder John Deere could be overloaded, or *lugged,* below the rated engine speed all day long with little damage to the engine. A diesel engine should not be lugged below rated speed for any length of time.

However, that does not necessarily mean that a diesel tractor is more fuel-efficient at high over-the-ground speeds. Low-speed operation with wider implements pulled in lower gears may require less fuel per acre than high forward speeds with narrower implements.

IRRIGATION

Irrigation is vital in areas where crops cannot be grown successfully without supplemental moisture. Improved water management, better efficiency of irrigation systems, and the building development of solar-powered irrigation are topics discussed in Chap. 12.

PESTICIDES

Despite the widespread use of petroleum-based chemicals to control pests, a recent survey pins the loss of U.S. crops to insects, weeds, and diseases at nearly a third of the annual production. In addition, another 10 percent of the stored foods fall prey to rodents, insects, and microorganisms.

Insects and diseases cause the loss of about a fourth of the total U.S. crop production. While only a small amount of energy is used to produce and apply chemicals to control these pests, the cost of these chemicals is becoming a major crop budget item on some farms.

The principal long-range alternative to chemical control of insect and disease damage appears to be genetic resistance bred into crop varieties. Bred-in resistance has low energy cost, but it is technically difficult and time-consuming to achieve, and often impossible to maintain over time because of pest mutations. The resistant corn rootworm is a prime example of how pests can also become less controllable with chemicals as generations of the insect evolve.

Crop rotation is often a highly effective means of pest control. However, this technique may cut production, particularly if rotation crops of relatively low economic value are needed in the rotation.

Biological control of insects and diseases has met with limited success (see Fig. 8-2). This technique employs the natural enemies of crop pests (predators or disease organisms) to control the damaging insect or disease. One drawback with biological controls is their cost. Another is that the damage is often well underway before the biological agent brings the culprit under control.

Weeds have been described as "plants out of place." These out-of-place plants steal about 8 percent of the annual U.S. crop production. The three primary ways of controlling weeds are mechanical cultivation, chemicals, and hand labor. Minnesota experiments with weed control in corn show that energy input is greatest with mechanical cultivation, less with herbicides, and least of all with hand labor. However, the net profit from the three methods was greatest with chemicals, was somewhat less with mechanical cultivation, and showed a negative return to hand labor.

So-called "pest management" techniques show some potential for reducing both costs and energy. These basically involve judging the need for pesticides by frequent observation of field conditions and then applying the proper amount of a selective chemical to control the pest discovered. This method is somewhat like testing soils for fertilizer needs and then applying only what is needed for the crop.

Near term, the prospects for dramatically reducing the energy used in crop production appear dim. However, there are a couple of bright spots. Two cropping operations that readily permit more immediate energy-conserving practices without cutting into production are grain drying and irrigation. We discuss several approaches to both in the chapters ahead.

FIG. 8-2   Biological control of crop pests has met with only limited success to date. Shown is a beneficial predator, *Podisus,* attacking a cabbage looper larva. *(Photo: U.S. Department of Agriculture.)*

9

Low-Temperature Grain Drying

Natural-air grain drying has been around for a while—at least since the late 1940s, when artificial drying really got underway. The advent of field shelling and the attendant concern over field losses of grain boosted the idea.

"It soon became obvious that advantages were offered by low amounts of heat to speed up drying," recalls Bob George, agricultural engineer, University of Missouri. "The rat race started. Instead of buying additional fan horsepower to push air through more bushels, farmers kept fan horsepower the same and added higher amounts of heat. At that time, it was cheaper."

Since the oil embargo of 1973 and higher energy prices that followed, crop growers are taking a new look at the costs of drying grain, and many of them are turning back to natural-air systems or drying with air that is heated only 2 to 10°F. In Chap. 10, we discuss ways to use solar energy to provide supplemental drying heat.

OVERDRYING IS COSTLY

Actually, higher drying temperatures in batch and continuous-flow dryers brought on some management problems in addition to higher fuel consumption per bushel dried. Grain often is overdried or overheated in these systems.

Overdrying can be a double whammy. It not only uses extra fuel, it also means fewer pounds of grain to sell. Each percentage point of moisture removed from corn below the 15.5 percent prescribed for Number 2 corn costs 2 percent of the market price. For example, corn dried to 14.5 percent moisture content and sold for $2.50 per bushel on a weight basis loses 5 cents per bushel. While elevators and processors discount grain and soybeans below trade standards (wetter than the industry minimum), they seldom pay premiums for grain that is drier.

Overheated grain rapidly loses quality. Its value for both livestock feed and human food is reduced. In Nebraska trials, with corn harvested 52 days after pollination, swine performed better when the grain was dried at 104°F than did swine fed grain dried at higher temperatures. Corn that was harvested 80 days postpollination and thus had dried more in the field produced the best pig grains when the grain was dried at 140°F. However, many high-speed, high-temperature dryers heat the grain to 180°F or more.

USE DRYING ENERGY EFFICIENTLY

The most common measurement of efficiency in the use of energy for drying grain is Btu's per pound of water evaporated. That figure is helpful when it comes to deciding on the most economical systems for given conditions, or how to combine systems most efficiently.

Table 9-1 shows that high-temperature drying requires more Btu's per pound of water removed than do natural air systems.

Natural-air or low-temperature drying is successful mainly because grain releases its moisture to moving air. This is what happens when corn is left on the stalk in the field. In a grain bin, however, the air must be forced through the grain so that moisture can be evaporated, as illustrated in Fig. 9-1.

TABLE 9-1   Energy Consumption of Various Drying Systems When Grain Is Dried to 13 Percent Moisture Content

Drying system	Btu per lb of water removed
Batch (180°F)	3800
Continuous-flow (180°F)	3200
Concurrent-flow (300°F)	1650
Circu-flow (160° rise)	1600
Batch-in-bin (40° rise)	1580
Natural air	300–600 (electric equivalent)

SOURCE: Based on data from Fred Bakker-Arkema, agricultural engineer, Michigan State University.

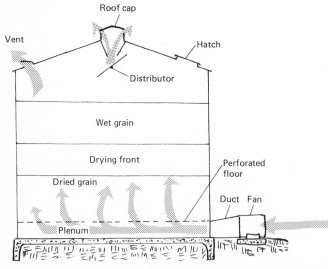

FIG. 9-1   A typical natural-air drying bin. Grain dries in zones, or *fronts,* from bottom to top. A distributor at the auger inlet of the bin will spread fines through the grain, rather than letting them collect under the auger, where they can restrict airflow.

A grain-drying bin ordinarily has a perforated floor. Air is forced into the space under the floor and up through the grain.

Most engineers recommend a fan capacity that will supply 1 cfm of air for each bushel of grain in the bin. Exhaust vents equal to 1 ft² for each 1000 cfm of fan capacity are needed to vent off moisture-laden air. Usually some sort of distributor is used to prevent the *fines* (small particles and broken kernels) from being deposited directly under the bin-loading auger.

The information that follows relates primarily to shelled corn, but the principles apply to all grain.

Molds are the main cause of spoilage in stored grain. The growth of these organisms is a function of moisture and temperature. When grain is kept cooler, moisture content can be higher without molds causing damage. In warmer temperatures, the safe storage moisture content is lower. Table 9-2 shows safe storage moistures for various grains.

For each combination of temperature and relative humidity of the drying air moving through stored shelled corn, for example, there is a related moisture content for the grain. If the air is constant, the grain eventually reaches an *equilibrium* moisture content. How quickly the grain reaches this equilibrium moisture content depends on the rate of airflow through the bin, as drying rate is roughly proportional to airflow. The moisture contents that can be achieved using only natural-air circulation are approximately as listed in Table 9-3.

By controlling the moisture content or temperature—or both—the growth of molds that cause spoilage can be inhibited in stored grain. Combinations of moisture content and temperature determine how long grain can be safely stored. Safe storage time decreases

TABLE 9-2   Safe Storage Moistures for Grains

Grain	Maximum safe moisture content, %
Shelled corn, grain sorghum:	
To be sold as No. 2 grain within 6 months	15.5
To be stored up to 1 year	14.0
To be stored more than 1 year	13.0
Soybeans:	
To be sold by spring	14.0
To be stored up to 1 year	12.0
Small grains (wheat, oats, etc.)	13.0

TABLE 9-3   Equilibrium Moisture Content of Grain

Temperature, °F	Relative humidity							
	10%	20%	30%	40%	50%	60%	70%	80%
20	9.4	11.1	12.4	13.6	14.8	16.1	17.6	19.4
30	8.3	10.1	11.4	12.7	13.9	15.2	16.7	18.6
40	7.4	9.2	10.6	11.9	13.1	14.5	16.0	17.9
50	6.7	8.5	9.9	11.2	12.5	13.8	15.4	17.3
60	6.0	7.9	9.3	10.6	11.9	13.3	14.8	16.8
70	5.4	7.3	8.7	10.0	11.4	12.7	14.3	16.3
80	4.9	6.7	8.2	9.6	10.9	12.3	13.9	15.9

SOURCE: Based on equation by Chung-Pfost, ASAE 76-3520.

by about 50 percent with each 10°F rise in temperature or with each 2 percent rise in moisture content. Table 9-4 gives estimated safe storage times of shelled corn at various combinations of temperature and moisture content.

If you begin combining corn at 26 percent moisture when the grain is at 60°F, you can store the corn in that condition for only 8 days before moldy kernels begin to reduce the market grade of the corn. If the grain is harvested at the same moisture content, but at 50°F, safe storage time increases to 18 days.

Harvest usually begins early in the season, when grain is wetter and temperatures are higher. Natural-air drying alone may not be sufficient to control temperature and moisture quickly enough to avoid some quality losses—particularly on farms that harvest 30,000 bu or more each season.

COMBINATION DRYING

A combination drying system often is the best compromise between speed and energy economy. If grain is discharged hot from a high-temperature dryer at about 20 percent moisture and drying is completed with natural air, the Btu's per pound of water removed is half that required if the grain is dried to 15 percent moisture in the high-temperature dryer alone.

Research at the University of Minnesota backs up the idea of using a combination drying method, particularly for the early part of harvest. A high-temperature, high-speed dryer, fueled with LP gas or natural gas, can be used to dry corn quickly to about 20 percent moisture. Then, the corn is augered into a storage bin, where it is cooled. Drying is completed by circulating natural air through the grain in the bin.

In this system, the first stage of drying is usually completed within 24 hours or less of harvest. The second stage, utilizing only ambient air, takes place over a period of several weeks.

However, by this time, the corn is cooled down, so safe storage time is extended. As harvest continues on into fall, the temperature cools down more and corn in the field dries naturally to a lower moisture content. This reduces the urgency to dry corn quickly, so natural-air drying can be used entirely. Energy is still being used, of course, to operate the fans that circulate air through the grain. But the *total* energy requirement is less than that needed by conventional high-temperature drying alone.

"These combination systems take advantage of the speed of high-temperature drying down to about 20 percent moisture, then allow more economical drying with natural air down to 15.5 percent moisture," says Bob George.

The engineer offers some cautions on managing the natural-air phase of combination drying, or natural-air drying from harvest to dry grain.

TABLE 9-4   Safe Storage Time of Corn at Various Moisture Levels

Corn temp., °F	Days of safe storage at moisture levels of:						
	18%	20%	22%	24%	26%	28%	30%
30	648	321	190	127	94	74	61
40	288	142	84	56	41	32	27
50	128	63	37	25	18	14	12
60	56	28	17	11	8	7	5
70	31	16	9	6	5	4	3

1. Grain in the lower levels of the bin dries first. As drying progresses, the drying zone or *front* should be checked regularly to see if drying is on schedule. If you check the level of the drying front (how far through the bin drying has progressed) when it is supposed to be one-half dried and find that the drying front has progressed only one-third of the depth of the wet grain, the bin has been overloaded for the conditions that exist. Figure 9-2 illustrates how the drying front progresses.

2. Grain harvested early, when moisture content normally is higher—say 25 percent—and dried quickly to 18 percent in a high-temperature system will be subjected to above-average temperatures in the natural phase of drying. In other words, the grain was harvested earlier than normal for 18 percent moisture, and outside air is higher. Molds can develop in less time than normal. In these situations, the bin should be loaded at shallower depths than normal for corn harvested at 18 percent moisture. This will ensure good airflow and speed up drying.

3. The natural-air phase of a large combination drying system may require several large-horsepower fans. It's a good idea to let the power supplier know ahead of time what your horsepower requirements will be so that adequate service will be available.

GRAIN DEPTH IN BINS

The moisture content of grain governs how deep it can be layered in a natural-air drying bin. With corn harvested at 24 percent moisture and put into an empty bin, the bin should be filled only to a 6-ft depth. Then, fans should be run to dry the corn down to 16 percent moisture or less before any more "wet" corn is added.

With corn at 16 percent moisture, a bin can be filled to a maximum of 18 ft. For other moisture contents, Table 9-5 shows initial fill depths, moisture levels, and depth of additional wet grain that can be safely added to the bin.

Table 9-5 assumes that fan horsepower and bin diameter are properly matched to get a fixed airflow rate of more than 1 cfm/bu through the combined depth of the grain. These pairings of fan horsepower with bin diameter are recommended:

Fan horsepower	5	7.5	10
Bin diameter, ft	24	27	30

Depending on fan design, 1 hp generally will provide a minimum airflow for 1000 bu of grain, filled to a depth of 16 to 18 ft. If fan horsepower is one size bigger than shown in the table above, bin filling depth can be increased by 1 percent. If the fan is a size smaller, reduce filling depth by 1 percent.

If a grain-handling and -drying system incorporates several bins of the right size, with properly sized fans, it isn't a long wait until more grain can be added to the first bin. Such a system provides a way to get grain dried in a reasonable time and at less energy cost—without any spoilage during drying.

These revivals of old drying systems have their peculiar handling requirements and limitations, however. Their smooth operation depends on careful planning of the layout of new grain-handling systems and expansion of many existing ones.

The kind of drying system selected and the drying rate in bushels per day should be compatible with the number of bushels harvested and dried each season. In-bin drying systems usually can handle crops of up to 30,000 bu/year. Volumes greater than 30,000 bu/year may require high-temperature drying equipment for at least part of the harvest.

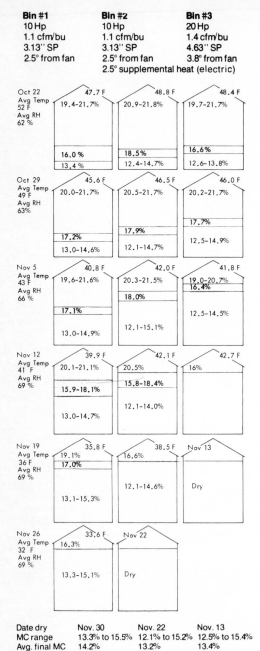

	Bin #1	Bin #2	Bin #3
	10 Hp	10 Hp	20 Hp
	1.1 cfm/bu	1.1 cfm/bu	1.4 cfm/bu
	3.13" SP	3.13" SP	4.63" SP
	2.5° from fan	2.5° from fan	3.8° from fan
		2.5° supplemental heat (electric)	

Oct 22 Avg Temp 52 F Avg RH 62 %
Bin #1: 47.7 F / 19.4-21.7% / 16.0 % / 13.4 %
Bin #2: 48.8 F / 20.9-21.8% / 18.5% / 12.4-14.7%
Bin #3: 48.4 F / 19.7-21.7% / 16.6% / 12.6-13.8%

Oct 29 Avg Temp 49 F Avg RH 63%
Bin #1: 45.6 F / 20.0-21.7% / 17.2% / 13.0-14.6%
Bin #2: 46.5 F / 20.5-21.7% / 17.9% / 12.1-14.7%
Bin #3: 46.0 F / 20.2-21.7% / 17.7% / 12.5-14.9%

Nov 5 Avg Temp 43 F Avg RH 66 %
Bin #1: 40.8 F / 19.6-21.6% / 17.1% / 13.0-14.9%
Bin #2: 42.0 F / 20.3-21.5% / 18.0% / 12.1-15.1%
Bin #3: 41.8 F / 19.0-20.7% / 16.4% / 12.5-14.5%

Nov 12 Avg Temp 41 F Avg RH 69 %
Bin #1: 39.9 F / 20.1-21.1% / 15.9-18.1% / 13.0-14.7%
Bin #2: 42.1 F / 20.5% / 15.8-18.4% / 12.1-14.0%
Bin #3: 42.7 F / 16%

Nov 19 Avg Temp 36 F Avg RH 69 %
Bin #1: 35.8 F / 19.1% / 17.0% / 13.1-15.3%
Bin #2: 38.5 F / 16.6% / 12.1-14.6%
Bin #3: Nov 13 / Dry

Nov 26 Avg Temp 32 F Avg RH 69 %
Bin #1: 33.6 F / 16.3% / 13.3-15.1%
Bin #2: Nov 22 / Dry

Date dry	Nov. 30	Nov. 22	Nov. 13
MC range	13.3% to 15.5%	12.1% to 15.2%	12.5% to 15.4%
Avg. final MC	14.2%	13.2%	13.4%
Temp. range	34.5 F to 34 F	41 F to 39.2 F	44.8 F to 42.7 F
Fan hours, drying	1104	912	696
Kilowatt hrs. used	9,685	16,027	12,212

FIG. 9-2 How the drying front progresses from bottom to top of a bin. Airflow is the principal element in low-temperature drying, even where small amounts of heat are added. The above example is based on 28 years' average weather data for Des Moines, Iowa. *(Midwest Plans Service.)*

9–6

TABLE 9-5   Maximum Depth of Wet Grain To Add to a Bin of Dried Corn (16 Percent Moisture Content or Less)

Depth of dried corn in bin, ft	Depth of fill (in feet) when moisture content of added corn is:					
	18%	20%	22%	24%	26%	28%
Empty	Full	9.4	7.3	6.0	5.2	4.4
2	12	8.8	6.8	5.7	5.0	4.1
4	11	8.1	6.3	5.3	4.6	3.8
6	10	7.5	5.8	4.8	4.2	3.5
8	9	6.8	5.2	4.4	3.8	3.2
10	8.2	6.2	4.8	4.0	3.4	2.8

Combined high-temperature and natural-air drying can be used to good advantage with higher grain volumes.

"The handling requirement of grain should be considered also," says Bob George. "A sketch of the proposed layout will help you analyze the handling system. Mark arrows on the layout sketch to show the flow of grain, and decide on the flow rates needed to make the system work without having bottlenecks that slow down the process."

Table 9-6 shows estimated bushels of capacity for bins of various sizes.

Here are points to be considered in determining flow rates and selecting the size and kind of drying and handling equipment.

1. **Harvesting rate** should at least match the drying capacity. The minimum harvesting rate in many larger grain operations should be twice the designed drying rate.

2. **Drying rate** in bushels each 24 hours is the key to a successful handling system. Size the drying equipment to dry 5 to 7 percent of the total bushels produced, so that the moisture content is reduced by 10 percentage points in a 24-hour period. In other words, if you harvest 40,000 bu of corn at 28 percent moisture content, the drying equipment should have the capacity to dry about 2400 bu of corn down to 18 percent moisture in a 24-hour period.

3. **Wet-corn handling rate** calls for augers or bucket elevators to be sized to handle four to five times the harvesting rate for a farmer who feeds the corn, and five to six times the harvesting rate for a cash-grain operation.

4. **Dryer loading/unloading rate** is often the limiting factor in the flow rate of an entire grain-harvesting and -handling system. Other elements are keyed to the loading and unloading rates of the grain-drying system.

TABLE 9-6   Estimated Bushels of Capacity for Cylindrical Grain Storage Structures

Diameter, ft	Height, ft						
	10	15	20	25	30	35	40
14	1,200	1,800	2,500	3,100	3,700	4,300	4,900
16	1,600	2,400	3,200	4,000	4,800	5,600	6,400
18	2,000	3,000	4,100	5,100	6,100	7,100	8,100
20	2,500	3,800	5,000	6,300	7,500	8,800	10,000
22	3,000	4,600	6,100	7,600	9,100	10,600	12,200
24	3,600	5,400	7,200	9,000	10,900	12,700	14,500
26	4,200	6,400	8,500	10,600	12,700	14,800	17,000
28	4,900	7,400	9,900	12,300	14,800	17,300	19,700
30	5,700	8,500	11,300	14,200	17,000	19,800	22,600

SOURCE: Based on University of Illinois data.

For example, in a natural-air bin layer system, the same equipment is used for wet-corn handling and dryer loading and needs to be sized so as not to be a bottleneck in harvesting. Unloading is controlled by end uses. For feed processing, it could be slow, into an automatic feed mill. If grain is sold for cash, unloading bins into trucks might be fast.

On the other hand, with a high-termperature, continuous-flow drying system, wet-grain holding bins should be sized to allow corn to be handled fast enough not to interfere with harvesting. In this system, the dryer loading rate is equal to the capacity of the dryer—as is the dryer discharge rate.

The main point to keep in mind is that no other part of the grain-harvesting and -handling system should force the combine to wait. That's why high-speed drying equipment—although it uses more fuel—often is a necessary element in large-volume grain operations. Especially in the early stages of harvest, high-speed drying lets a volume of grain be dried down to a moisture content that can safely be handled in slower, natural-air or low-temperature drying systems.

Still, the process of drying grain with natural air can save a great deal of fossil fuels even in bigger cropping operations. If properly managed, these methods of drying grain can result in a high-quality product, as well.

Solar Grain Drying

Each year, U.S. farmers produce about 200 million tons of feed grains, another 70 million tons of food grains, and 2 billion bu of soybeans. Most of this crop must be dried for safe storage. The fuel needed to dry this annual crop is equivalent to about 640 million gal of LP gas.*

At 60 cents per gallon, that's an annual crop-drying bill of $384 million. Energy for crop drying represents a large portion of the total energy used in agriculture.

The motive for crop drying is simple: to reduce crop moisture and temperature for safe storage and handling. The actual process of drying is relatively complex and depends on a number of variable factors. At the risk of repeating some of the information contained in Chap. 9, here's what happens when grain dries.

Air serves two important functions in the grain-drying process. First, it carries heat required for moisture evaporation. Second, it carries away the moisture evaporated from the grain. During crop drying, air is forced through damp grain, picking up water through the process of evaporation.

As moisture is evaporated, heat is drawn from both the grain and the air. The temperature drops until a natural limit is reached. However, warm air can carry more moisture than cooler air. Air at 60°F can hold twice as much moisture at saturation (100% relative humidity) as can air at 40°F. By adding heat to the air, more moisture can be evaporated from the grain.

So, the two ways to increase grain drying are (1) to heat up the air to increase evaporation and (2) to increase the amount of air used to carry away evaporated moisture. Of the two methods, increasing airflow is generally more energy-efficient than adding large amounts of heat. In other words, the drying ability in a natural-air drying system in a grain bin comes principally from the air, not from added heat, as is the case with continuous-flow and batch-type dryers usually located outside the bin. If enough airflow is provided, corn in most regions of the Midwest will dry to a safe moisture content and cool to a low enough temperature during the fall to keep in winter storage. However, more drying often is needed in the spring, as weather again warms up.

Adding supplemental heat does not decrease the need for adequate airflow through the grain. Adding heat in a bin-drying system reduces the grain moisture content in the lower layers of a bin, but does not reduce moisture content in the top of the bin very much.

---

*Based on information by Chau, Baird, and Bagnall, ASAE 78-3014, American Society of Agricultural Engineers, St. Joseph, Michigan.

Often, some heat is needed, however (such as when the natural air cannot dry the grain because the relative humidity is high).

THE SOLAR OPTION

As only a small amount of heat is needed in most cases, a solar collector can often provide enough supplemental heat for low-temperature drying. The slower drying rate—thus longer drying period—allows more time for collecting solar energy and utilizes the investment in solar equipment over more days, weeks, and months. Simple, relatively inexpensive solar collectors can provide the small amount of heat needed. See Fig. 10-1.

Solar collectors used for low-temperature grain drying typically produce temperature rises of 5 to 15°F, at the high airflow rates needed. When solar energy is used to replace or augment conventional sources of energy in a high-temperature drying system, lower airflow rates are often used, to gain more temperature rise through the collector.

The economy of solar grain drying, as fossil fuels increase rapidly in price, is attracting more attention. Research and on-farm experience around the United States show that solar energy can reduce the *operating* costs of low-temperature drying systems. However, to be realistic about the costs and returns of any alternative energy system, you should include the costs of construction, installation, and annual maintenance in the drying costs per bushel.

FIG. 10-1   Solar grain drying at Ohio Agricultural Research and Development Center uses two polyvinyl heat collectors. Heated air is moved through the collectors by fans, which keep the plastic inflated. *(Photo: U.S. Department of Agriculture.)*

WILL SOLAR PAY?

In 1980, Walter G. Heid, Jr., agricultural economist with the U.S. Department of Agriculture, studied the application of solar energy to a combination grain-drying system, and compared it with the costs of using a conventional high-temperature drying setup. The term *combination drying,* you'll recall, describes the technique of teaming up first the use of a high-temperature system, to dry corn quickly to about 20 percent moisture content, then transferral of the warm grain to a drying bin equipped with a fan (and, in the case of Heid's analysis, a solar collector), to complete drying to about 15 percent moisture content.

Heid studied a test case involving 300 acres of corn with an average yield of 100 bu/ac, for a total harvesting and drying requirement of 30,000 bu. Spreading the harvest over 15 working days means that 2000 bu/day would be harvested and dried long enough to remove 5 percentage points of moisture. Based on a 10-hour harvest day, a dryer capacity of 200 bu/h would be needed.

The system employs three 10,000-bu storage bins; each is 22 ft high and 27 ft in diameter, with a floor area of 513 ft². The wide bin diameter-to-height ratio is particularly well-suited to a low-temperature, solar-assisted drying method, since the daily filling depth can be kept to a minimum.

When corn is dried quickly to 20 percent moisture and then placed in the bin, the low heat produced by a solar collector should be sufficient in most cases to finish the drying process without spoilage.

A high-temperature batch-type dryer with the capacity to remove 5 percentage points of moisture from 200 bu of corn per hour represents a capital investment of about $9000 (in terms of 1979 dollars). This unit would be used at near full capacity at the beginning of harvest, when corn was at about 25 percent moisture, and at a lower capacity as harvest progressed and corn dried naturally in the field. A similar dryer, sized to remove 10 percentage points of moisture from 200 bu/h in a conventional high-temperature system, would cost about $16,500.

To study the solar-assisted part of the process, Heid uses a portable solar collector— or collectors—similar to the 12- by 24-ft Illinois design shown in Fig. 10-1.

"It's important to construct or purchase collectors that may be related to other uses when they are not being used to dry grain," says Heid, and notes that two portable collectors might be hooked up "in tandem" for some chores.

In his study, Heid outlines four levels of collector use for grain drying: 100, 75, 50, and 25 percent. A 75 percent rate means the collector is used three-fourths of the time for grain drying and one-fourth of the time for something else.

When the collector is not used full-time for drying grain, the fixed costs of the unit are prorated among the various uses. Further, Heid assumes that a homemade flat-plate collector covered with fiberglass will have a 10-year life expectancy and will cost $4 per square foot of collector area to build.

Heid also uses three different ratios of collector-to-bin floor area: 1:1, ¾:1 and ½:1. Collector surface requirements vary with climatic region; generally, more collector area is needed in humid regions. More collector area is also usually needed if solar energy is to play a larger role in total crop drying than it does in a combination system.

For each collector regardless of size, a constant extra investment of $1800 is needed for fans (7.5-hp centrifugal models) plus related transition ducting and equipment. For the three 10,000-bu bins and the three collectors to serve them, this means an additional investment of $5400 compared with the conventional high-temperature drying setup. The cost of the bins is not included in either system.

With electricity priced at 5 cents per kilowatthour and LP gas at 35 cents per gallon (it's higher than that in most areas), here's the way the two systems compare on breakeven costs.

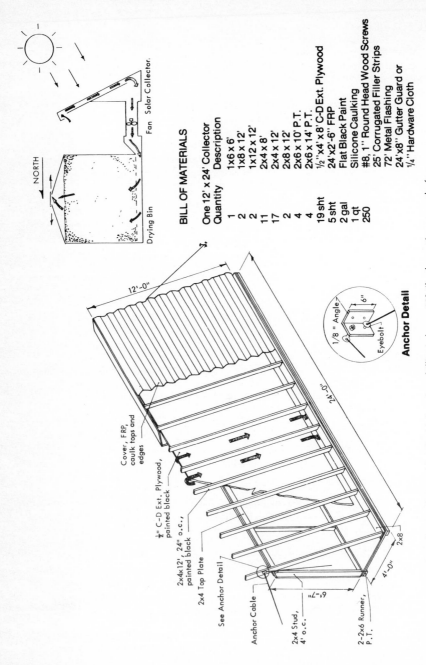

BILL OF MATERIALS

One 12' x 24' Collector

Quantity	Description
1	1x6 x 6'
2	1x8 x 12'
2	1x12 x 12'
11	2x4 x 8'
17	2x4 x 12'
2	2x8 x 12'
4	2x6 x 10' P.T.
4	2x6 x 14' P.T.
19 sht	½" x4' x8' C-D Ext. Plywood
5 sht	24'x2'-6" FRP
2 gal	Flat Black Paint
1 qt	Silicone Caulking
250	#8, 1" Round Head Wood Screws
	25' Corrugated Filler Strips
	72' Metal Flashing
	24'x8" Gutter Guard or
	¼" Hardware Cloth

Anchor Detail

FIG. 10-2 Portable collector designed by the University of Illinois has a 63° tilt, the optimum angle for fall grain drying in northern Illinois. The collector should be anchored against high winds. *(University of Illinois.)*

With 30,000 bu of corn dried from 25 percent moisture down to 15 percent, assume the following annual costs:

$$0.15 \text{ gal LP gas} \times 30,000 \text{ bu} = 4500 \text{ gal} \times 35¢ = \$1575$$

$$0.14 \text{ kWh electricity} \times 30,000 \text{ bu} = 4200 \text{ kWh} \times 5¢ = 210$$

$$\text{Total energy costs} = \$1785$$

$$\$16,500 \text{ initial cost of dryer} \times 10\% \text{ annual interest} = \$1650$$

$$\text{Total annual costs} = \$3435$$

$$\text{Drying cost per bushel} = 11.5¢$$

Those 30,000 bu of corn dried from 25 percent moisture to only 20 percent would require a smaller high-temperature dryer, with the following result on energy and ownership costs:

$$0.07 \text{ gal LP gas} \times 30,000 \text{ bu} = 2100 \text{ gal} \times 35¢ = \$735$$

$$0.07 \text{ kWh electricity} \times 30,000 \text{ bu} = 2100 \text{ kWh} \times 5¢ = 105$$

$$\text{Energy costs for initial drying phase} = \$840$$

The use of solar collectors and related fans and other equipment to dry the corn from 20 percent moisture to 15 percent would require an additional energy outlay:

$$0.5 \text{ kWh electricity} \times 30,000 \text{ bu} = 15,000 \text{ kWh} \times 5¢ = \$750$$

Capital (ownership) costs for the solar equipment would depend on the collector-to-bin floor ratio and also on the percentage of use of the collector that is charged to the grain-drying operation. With 1 ft² of collector for each 1 ft² of bin floor (1:1 ratio), the collector cost would be about $6156. At a ratio of ¾:1 the cost would be $4617, and at a ratio of ½:1 the collector investment would be $3078.

When the standard $5400 investment in fans and other solar-related equipment is added, here's the way the capital costs look, with the investment in the smaller high-temperature dryer added in:

Equipment	Collector-to-bin ratio		
	1:1	¾:1	½:1
Solar equipment	$11,556	$10,017	$8,478
High-temperature dryer	9,000	9,000	9,000
Total capital costs	$20,556	$19,017	$17,478

Charging 10 percent interest on the capital investment would have this effect on total annual costs of the combination high-temperature–solar-assisted system:

	1:1 ratio	¾:1	½:1
Capital costs (10% interest)	$2,056	$1,902	$1,748
Energy costs (from above)	$1,590	$1,590	$1,590
Total annual costs	$3,646	$3,492	$3,338
Cost per bushel	12.2¢	11.6¢	11.1¢

At this rate, only the collector-to-bin floor ratio of ½:1 is competitive in costs per bushel of grain dried with the conventional high-temperature system. As shown earlier, the high-temperature drying from 25 percent down to 15 percent moisture incurred total annual costs of $3435, or about 11.5¢ per bushel.

However, that's where having alternative uses for the solar collector helps pay the bill. If only half of the initial investment in the collector is charged to grain drying, the economics of going solar look much better.

Also, farmers in many parts of the United States would quarrel with Heid's assumption that LP gas costs only 35 cents per gallon. If the cost of LP gas to run both high-temperature dryers is put at a more current figure—say 60 cents per gallon—grain-drying costs with both systems increase, but costs for the high-temperature-only system go up much faster.

With LP gas plugged in at 60 cents per gallon, here's how the options compare:

	High-temperature dryer	1:1 ratio	¾:1 ratio	½:1 ratio
Total annual costs	$4560	$4171	$4017	$3863
Cost per bushel	15.2¢	13.9¢	13.4¢	12.9¢

SOLAR FOR DRYING ONLY?

Heid's study featured portable solar collectors and assumes that other on-farm uses can be made of the heat-gathering equipment when it is not needed for grain drying. Chuck Harley, of Jackson County, Ohio, built the 12- by 24-ft portable collector shown in Fig. 10-3 and paid for the structure in LP gas savings the first year.

"I bought it all at the local cash-and-carry hardware," says Harley. "The total cost was

FIG. 10-3 This 12- by 24-ft solar grain dryer was built by Ohio farmer Chuck Harley for less than $600 worth of materials. Harley spent more than $600 for LP gas to dry grain the year before; this unit repaid materials costs in the first year. *(Photo: U.S. Department of Agriculture)*

just a little under $600. Since I spent a bit more than $600 last year for the propane gas I used to dry my corn, I figure this dryer will pay for itself pretty soon."

Harley lets the solar collector provide all the supplemental heat needed to dry corn in a 5000-bu bin. An electric fan draws warm air off the collector through a duct, into the bottom of the bin.

Harley built his dryer of plywood, greenhouse-type clear fiberglass sheets, 2- by 4-in studs, hardware cloth, and black paint. The dryer is mounted on skids so that it can be used at other farm structures, although the primary purpose of the unit is to dry grain.

INTEGRATED COLLECTORS

Another way to put multiple-use collectors to work is to build the solar collector on a nearby building—machine shed, shop, livestock building, etc.—that can make use of the solar energy in winter, after grain drying is completed. Figure 10-4 shows a 48- by 96-ft solar machine shed and shop that is adaptable to other uses. Engineers put the day-long average efficiency of the $4/12$ pitch roof collector at about 40 percent, the wall collector at nearly 50 percent.

One caution when building a structure with a solar attic: Provisions should be made to vent the attic in summer, when the collected heat is not needed. One square foot of ridge vent for each 100 ft² of collector area, plus a 4- by 8-ft screened opening in each gable, should be ample. But even so, the building may be hotter in summer than one without the solar feature.

Also, a big drying fan creates a lot of suction—enough to collapse lightweight ductwork in some cases. The ducting and transition fittings should be designed to handle the pressure differentials to be expected.

Some farmers have built topless grain bins and covered the row of grain structures with a pole barn covered with clear fiberglass roof and walls. This greenhouse-type collector traps solar energy in the building. Fans pull the warmed air *down* through the topless bins and exhaust the "used" moist air outside.

TO STIR OR NOT TO STIR?

Although a bin-stirring machine is not always a paying piece of equipment for a well-managed natural-air drying system, there are some advantages to stirring grain in a low-temperature drying setup. For one, a stirrer mixes wet and dry grain in the bin. This, in effect, gets rid of the drying front and lets you avoid overdrying grain at the bottom of the bin while waiting for the grain at the top to drop to a safe moisture level.

Figure 10-5 shows one type of bin-stirring device.

Stirring can also increase airflow by loosening grain and distributing fines throughout the bin. Increasing airflow cuts drying time, which means less expense to run fans. Any heat added—by solar collectors or other sources—also reduces the time the fans must be run by speeding up drying.

The question of whether to add heat to soybeans in bin-drying systems has been debated across farming country. There's pretty general agreement that only small amounts of heat can be added; beans dry quickly.

In many areas, air is all that is needed to dry soybeans from a harvest moisture content of 16 percent or so down to safe storage levels of 13 percent. However, there's always the hazard of overdrying soybeans and of throwing away test weight when beans are below standard moisture limits for sale and storage. The problem of overdrying can often be sidestepped by using small amounts of heat and a mechanical stirring device.

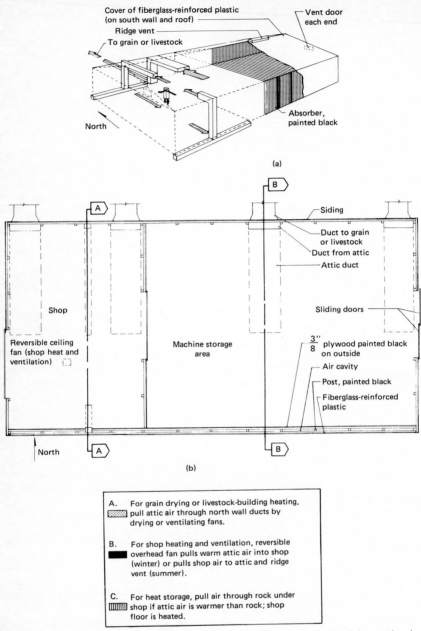

FIG. 10-4 *(a)* Schematic and *(b)* floor plan for 48- by 96-ft machine shed and shop with solar attic. Full plans are available from Midwest Plans Service, Iowa State University, Ames, IA 50011. Ask for MWPS Plan 81901.

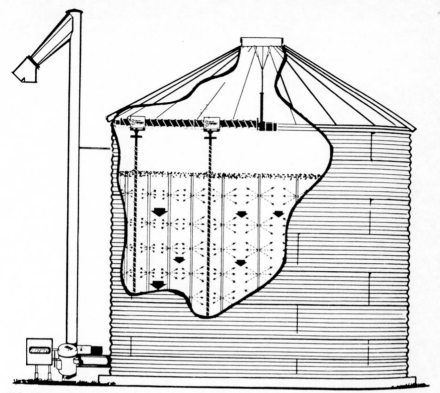

FIG. 10-5   When using heated air (solar or other sources), a bin-stirring machine and perforated wall tubes can enhance drying and prevent overdrying of bottom layers of grain. *(Sukup Manufacturing Company.)*

A stirrer adds to the capital cost of drying. The equipment typically costs about 20 percent of the cost of erecting the bin, for a 10,000- to 12,000-bu structure. The device also takes up part of the bin space—more than a foot of vertical space with most models. These disadvantages will have to be weighed against the benefits of quicker drying and prevention of costly overdrying.

There are times, however, when overdrying can be used to save fuel without running the risk of selling too-dry grain. Suppose you have old-crop corn that is still in the bin in late summer. A solar collector can be used to *over*dry the stored grain down to 8 or 10 percent moisture before harvest begins. Then, the super-dry grain can be mixed with new corn at 25 to 27 percent moisture, to produce a mixture of grain at an average of about 20 percent moisture content.

This is often called a *dessicant* drying technique. Grain elevators use this method of blending to have grain at precisely the industry trade standard when they sell it to a processor, terminal elevator, or exporter. On the farm, it can be a way to get part of the drying chores done before harvest even begins, and it's another way to get more mileage out of an investment in solar collectors.

Table 10-1 gives optimum mixtures of super-dry and wet grain for this technique.

TABLE 10-1 How to Mix Wet and Overdried Corn To Achieve a 20 Percent Moisture Blend

Moisture content of new corn, %	Bushels of wet corn to add when moisture content of overdried corn is:								
	5%	6%	7%	8%	9%	10%	11%	12%	13%
21	12.5	11.8	11.0	10.3	9.5	8.8	8.0	7.2	6.4
22	6.1	5.8	5.4	5.1	4.7	4.3	3.9	3.5	3.1
23	4.1	3.8	3.6	3.4	3.1	2.9	2.6	2.3	2.1
24	3.0	2.8	2.7	2.5	2.3	2.1	1.9	1.7	1.5
25	2.4	2.2	2.1	2.0	1.8	1.7	1.5	1.4	1.2
26	2.0	1.8	1.7	1.6	1.5	1.4	1.3	1.1	1.0
27	1.7	1.6	1.5	1.4	1.3	1.2	1.1	1.0	0.8
28	1.4	1.3	1.2	1.2	1.1	1.0	0.9	0.8	0.7

SOURCE: Bern, Anderson, and Wilcke, Iowa State University.

11

Drying Grain with Crop Wastes

That all-encompassing term *biomass* takes an interesting meaning where grain drying is concerned. Renewable crop residues are among the cheapest, most readily available energy sources on crop farms. Some 330 million tons of corn stalks, soybean stems, wheat straw, and other crop aftermath are grown in the United States each year, and that tonnage represents several billion Btu of potential energy.

CORN STALKS

A 95 bu/ac corn crop yields 5500 lb of stalks, leaves, and cobs, which contain the energy equivalent of 320 gal of LP gas. (See Fig. 11-1.) If half of the available residue is harvested for use in drying grain and the balance left on the soil, that's the equivalent of 160 gal of LP gas. At 50 cents per gallon, that's an energy value of $80 per acre.

At least one manufacturer of drying bins and equipment, Stormor, Inc., markets a furnace that attaches to a drying bin and utilizes corn residue as a source of grain-drying heat. In fact, the furnace accepts almost any combustible material—stalks, cobs, straw, hay, etc. The device is shown in Fig. 11-2.

During a normal year, less than 10 percent of the crop residue left after corn harvest could be burned in such a furnace to provide enough energy to dry the grain grown on that acre. The prototype Stormor unit, used in conjunction with a 24-ft-diameter Stormor Ezee-Dry bin, dries 1000 bu of corn in about 7 hours, removing 10 percentage points of moisture and consuming 1 ton of corn stalks.

The furnace is designed with a vertical heat exchanger that picks up heat from the firebox, to heat air that is blown through grain in the bin. Stormor defines its crop-residue burner as an "alternate or assisting energy source" for grain drying. The heat chamber, or firebox, measures 10 by 10 by 10 ft and accommodates stacks up to 7 by 8 by 8 ft or round bales up to 7 ft in diameter.

The burner operates at drying air temperatures of 180°F, and the quality of dried grain is equal to that dried in gas-fired units. Stormor technicians describe the operation of the furnace as follows:

1. After the crop is harvested, the stalks are cut and chopped with a rotary scythe or similar equipment. Stalks should be left in the field for two or three days to allow them to dry.

Propane   Corn stalks   5-9 bu. corn

FIG. 11-1   Eighteen pounds of corn stalks and leaves at 25 percent moisture content can replace 1 gal of LP gas, to dry 5 to 9 bu of corn from 25 percent down to 15 percent moisture. A ton of corn crop residue can produce enough heat to dry 1000 bu of corn.

FIG. 11-2   Biomass grain dryer features wide ground-level doors through which big bales or stacks of crop refuse can be loaded into the firebox. *(Photo: StorMor Corporation.)*

2. Stalks and leaves are harvested with a mechanical stacker or large round baler. Then the big packages of crop residue are placed into the furnace, one at a time, through two large ground-level doors at the front of the unit.

3. The doors are closed and the material is lighted through an exterior port. Two small, fractional-horsepower blowers provide combustion air to the flame. Larger (10- to 15-hp) fans blow air between the walls of the burner and an outer jacket, picking up heat, which is carried through ductwork to the drying bin.

4. Temperature of the drying air is controlled by a thermostat in the dryer plenum, wired to the two small fans that control the combustion air.

5. The residue burns fairly completely, leaving only a small amount of ashes that need not be cleaned after each filling of the furnace.

Stormor points out that the furnace has other potential on-farm uses, such as heating animal confinement buildings. However, the company has not fully tested uses other than grain drying at this writing.

CORNCOBS

A pound of dry corncobs has a potential heat energy of about 7000 Btu, more than enough to dry 10 percentage points of moisture from the kernels of corn that are shelled off the cobs, if the material is burned in an efficient furnace.

For most crop farm applications, a system to burn cobs directly probably would be more practical, but a *gasifier* can be even more efficient at gathering heat from cobs. Pioneer Hi-Bred International, a major seed corn company, has developed a system to burn cobs with limited oxygen, to produce a mixture of combustible gases.

The gases produced are drawn off and ignited in a sort of "afterburner" chamber, to heat air that is blown through the grain. Pioneer engineers and Crom Campbell, a Des Moines, Iowa, heating consultant who helped design the system, say that the cobs from 1 bu of corn will provide enough heat to dry 2 bu of grain.

Quality control is more critical with drying seed corn than with regular field corn, and Pioneer spent five years and three-quarters of a million dollars to perfect the automated cob burner. In 1979, the company put the units through trial runs in Iowa, Indiana, and North Carolina, with satisfying results. More of these gasification furnaces are planned, and the company expects the equipment to make its crop-drying operations energy-self-sufficient. The technique is applicable to farm drying of commercial corn as well; however, capital costs of the equipment probably will make it practical only for larger crop farms.

As seed corn is picked in the ear, then carefully shelled, Pioneer has a ready supply of cobs close at hand. For the time being, crop farmers who adopt the idea of drying grain with corncob heat probably will need to harvest part of the crop as ear corn and then shell it to have the cobs available as grain-drying fuel.

Steve Marley, an agricultural engineer at Iowa State University, believes combine designs can be altered rather easily to separate and collect cobs during a single harvesting trip. In fact, three of Marley's students at Iowa State—Dave Bengston, John Sundberg, and Patrick Weiler—designed a workable "cob catcher," shown in Fig. 11-3, that can be attached to combines. The unit separates cobs and pieces of cobs from chaff with a blower and collects the cobs in the combine. When this equipment is perfected, it will let a farmer harvest shelled corn and the fuel to dry it at the same time.

HOW MUCH RESIDUE SHOULD BE USED?

It's a mistake to consider crop residues as simply material that "goes to waste" in the field. Stalks, leaves, stubble, straw, and cobs have some value—as livestock forage and bedding or as organic material to be returned to the soil.

"Crop residues provide our best, cheapest and easiest soil erosion control, protecting the soil from wind and water erosion," says William E. Larson, soil scientist with USDA's Agricultural Research Service.

Also, the residues contain a great deal of fertilizer that can be returned to the soil. Residues from an acre of land that produces 150 bu of corn will contain about 93 lb of

FIG. 11-3 "Cob-catcher" attachment for combines is explained by inventors John Sundberg, left; David Bengston; and Patrick Weiler—all agricultural engineering students at Iowa State University. The device uses a blower to separate pieces of cobs from chaff, so that cobs can be collected and used as fuel.

nitrogen, 15 lb of phosphorus, and 112 lb of potassium. If these nutrients are removed, they would need to be replaced by additional fertilizer applications, and energy would be required to produce, transport, and apply the fertilizer to the field.

How much crop residue can safely—and economically—be removed from a field will vary with area, topography, soil type, annual rainfall, and a host of other factors. For example, on sloping wheat land in Oregon, Larson's computer advises leaving all the wheat straw on the land to hold the soil against water and wind erosion.

On the other hand, with fairly flat Corn Belt land that is high in organic matter, perhaps half of the residues could be removed without running much additional risk of soil erosion.

The long-term costs of increased soil erosion may be tough to figure, in terms of annual dollars spent. But that "free" source of energy in the form of crop stalks, straw, leaves, and cobs should be charged with the fertilizer replacement costs involved, as well as the costs of gathering, hauling, and handling the material.

There's abundant potential for energy savings in the "leftover" plant parts that remain after grain is harvested. But it is not an absolutely free source of energy.

12

Irrigation Cost Cutting

The financial risks of crop production have never been greater. Average costs (except land) to grow an acre of corn in the United States were about $210 in 1980, up from $175 the year before, and showing no signs of slackening the climb.

With that kind of money already in the pot, many farmers say they cannot afford to be without the extra insurance that irrigation provides. Indeed, in many areas of the country, crop production depends heavily on pumped water. This is particularly true in more arid regions in the western United States. Nationally, crops grown on irrigated acres total the amount of farm exports to foreign buyers, which help pay for petroleum imported by the United States.

A few years ago, when energy was both plentiful and relatively inexpensive, irrigation often was taken for granted as a normal farm cropping input. Today, it's a major crop budget item on irrigated farms—particularly where water must be pumped from considerable depths. Thus, irrigation is a major candidate for cost- and energy-cutting practices.

With the fast-changing energy supply and price picture, looking ahead with any degree of success is difficult. A major influence on irrigation will be what happens to prices of fuel in relation to farm crop prices. A brief look back doesn't encourage a very bright forecast for the future in this energy-to-crop price ratio. For example, in 1978, diesel fuel prices averaged about 50 cents per gallon in the western Corn Belt. Corn sold for an average $2.40 per bushel, soybeans for $6.20. Two years later, diesel fuel had shot to nearly a dollar per gallon, and crop prices were little changed from 1978 levels.

Other factors enter the picture, too. Population pressure on land and continuous land price inflation would have strong effects. As land becomes higher-priced, more intensive production per acre—via irrigation and other techniques—may be more attractive from a net profits standpoint than expansion through the purchase of more land.

Irrigators are painfully aware of the rapid increases in the cost of all types of energy used. With current prices for diesel, LP gas and natural gas, farmers in Nebraska, Kansas, Texas, and other High Plains states are spending $35 to $85 per acre to get supplemental water to their land. How the energy-irrigation connection affects a particular crop grower depends partly on whether an irrigation system is already installed or is merely in the planning stages.

If you already have a system in place, you're most concerned with variable irrigation costs and the added income realized from applying supplemental water to the crop. If the added yield times price exceeds the variable costs of irrigating, a crop can be watered profitably.

For example, a corn grower in the western Corn Belt might apply 4.5 in of irrigation water per acre and expect to increase yields by 40 bu. If the cost of operating irrigation equipment is $50 per acre, the supplemental water will pay a profitable return only so long as the corn price exceeds $1.25 per bushel (40 bu × $1.25 = $50).

See Table 12-1 for a comparison of costs.

However, a farmer planning to install an irrigation system is faced with a more complicated economic decision. Irrigation will not only need to boost crop production (and income) by enough to pay all extra costs of operating the equipment, but also need to provide enough additional income to pay the fixed capital costs. If an irrigation system will cost $500 per acre to purchase and install, that "willing water" must boost income by *another* $50 per acre—over the extra cost of operating the equipment—to pay the interest on the investment at 10 percent per year.

IRRIGATION METHODS

More ideas on how to irrigate crops have been developed in the past 20 years than in all previous history, and the motive has been largely cheap energy. Methods used to put supplemental water on crop soil fall into three general categories: sprinkling over the top, furrow and flood irrigation on the surface, and drip or trickle irrigation beneath the soil's surface. Each of these methods has peculiar benefits and drawbacks.

To a great extent, the soil type, the amount of water to be applied, the crop to be grown, and the topography of the field dictate the method used. Let's look at some of the systems.

First developed in the mid-1960s when flexible water-supply hose came on the market, early *traveling guns* had LP-gas engine drives. The hose was at first dragged from one position to another, then hose reels were developed for transporting and storing the supply line. In recent years, water-powered drive systems have replaced gas engines.

Currently, traveling-gun sprinklers use hose sizes from $2\frac{1}{2}$ to $4\frac{1}{2}$ in inside diameter and have travel distances of $\frac{1}{8}$ to more than $\frac{1}{2}$ mi. Most gun sprinklers built in the United States have been of the *cable-tow* type, where a cable is anchored at one end of the irrigating run and the sprinkler tows itself along a special travel lane, usually of grass sod.

The *reel-type* traveling gun, a similar piece of equipment developed in Europe, is catching on here. The main difference between the two is that cable-tow travelers use a soft flexible rubber or PVC plastic hose, whereas the reel-type irrigator uses a semirigid polyethylene pipe. Initial cost of the cable-tow type is somewhat less—perhaps 65 percent of the cost of a reel-type machine with the same capacity. Cable-tow rigs are smaller and more portable; reel-types are simpler and require less labor to operate.

Advantages of traveling guns over pivot-type sprinklers are their lower initial cost, their

TABLE 12-1   Irrigation Cost Comparisons

Water source and system	Average cost per ac-in
Center pivot from reservoir	$2.22
Traveling gun from reservoir	$3.18
Center pivot from shallow well	$2.27
Center pivot from deep well	$4.17
Diesel-powered systems (all)	$2.74
LP-gas-powered systems (all)	$3.13
Electric-powered systems (all)	$2.99
Natural-gas-powered systems (all)	$0.89

SOURCE: Norlin Hein, University of Missouri.

mobility, and their adaptation to small and odd-shaped fields. Also, these rigs can be operated around obstructions more handily than can center-pivot sprinklers.

More than 60,000 *center-pivot irrigation* systems are in use in the United States, and nearly half of them have been built by one Nebraska firm: Valmont Industries. These systems are predominant in the Midwest and Great Plains states.

Center-pivot sprinklers rotate or pivot about a central water-supply point. They are typically built in sizes to irrigate a square quarter-section of land (160 acres), although smaller and larger units are in use. These rigs use electric, oil hydraulic, and water drives for their motive power.

Advantages of center-pivot systems include their low labor requirement and reliability of operation. They are among the easier self-propelled irrigation systems to mechanize, and newer models can be operated at medium and low pressures. Where field size, shape, and topography allow their use, center-pivots are the most labor-efficient and among the more energy-efficient sprinkler systems used.

Disadvantages are initial costs and the problem of moving the equipment any distance. Also, the circular path of the pivoting sprinkler inadequately irrigates field corners, although end guns and corner-covering devices help.

Lateral-move irrigation systems now are being developed that can give many of the advantages of center-pivots to rectangular fields. These sprinklers move sideways across a field, rather than pivot about a central point. One problem with these rigs is the requirement for a movable water supply. On flatter land, the water supply for lateral-move rigs is often an open channel. Some of these systems make use of a flexible hose, similar to that used with a cable-tow traveling gun.

Surface irrigation generally depends on the land being at a particular grade, to allow gravity either to move water slowly through furrows or to trap water on the soil for periods of the crop-growing season—as is the case with crops such as rice. Land forming is a major cost associated with surface irrigation, which generally limits this method to relatively flat land.

Once the initial cost of land forming is repaid, surface irrigation methods are among the more energy-efficient in operation, although some require more labor than sprinkling systems. Energy costs with surface irrigation often depend on the depth or distance of the source of water and the power source used to pump it; operating pressures can be much lower than with sprinklers.

Trickle or drip irrigation costs about $200 more per acre to install than a conventional sprinkler system, but once installed, these systems can save 50 to 90 percent of the energy needed to operate sprinklers and can irrigate crops adequately with 30 to 40 percent less water.

A drip-trickle system distributes water underground or on the surface through plastic pipes. An opening, or *emitter,* is located at each plant, so that water is released only in the plant's root zone and does not wet the area between plants.

These systems operate at very low pressure—typically less than 10 psi. This compares with 40 to 50 psi for low-pressure sprinklers and 80 psi or more for a self-propelled traveling gun. To date, drip-trickle irrigation systems have most often been installed to irrigate perennial crops—such as fruit and nut trees—and some high-risk crops, such as vegetables. In fields that normally are plowed every year or two, the system obviously would need to be located so as not to interfere with tillage operations.

Other disadvantages of drip-trickle systems include (as mentioned) high initial cost and the fact that lines and emitters plug easily. The small orifices in emitters and the low operating pressures make filtering of the irrigation water an absolute necessity. In some areas and with some water supplies, chemical treatments are also needed for special water-quality conditions.

IMPROVING PUMPING-PLANT EFFICIENCY

Some irrigation systems in river basins and in the mountain and plains regions of the west utilize gravity to move water. However, most of the irrigation water applied requires some mechanical pumping, and studies across the irrigated part of the United States show that pumping efficiency could be greatly improved. The operating efficiency of a pumping unit (pump, main shutoff valve, and drive system—electric motor, diesel engine, etc.) is the amount of useful work output per unit of energy input.

In studying the irrigation pumping systems in Idaho, R. C. Stroh, R. D. Wells, and J. R. Busch, of the University of Idaho, found that energy often is wasted in the poor design and mismatching of components and by inefficient operation of irrigation systems. For example, the engineers found that throttling or "valving" of electric pumping plants is a frequent management practice. In one system where a 200-hp electric pump was used to irrigate alfalfa hay and barley, the operator partially closed the main valve but left the pump running while the hay crop was being harvested. The result was that the pumping plant operated for that period at about 8 percent efficiency, requiring 2593 horsepower-hours (hp-h) per acre-foot of water applied. Later, when the valve was opened and the system was fully operating, only 575 hp-h was required per acre-foot of water applied.

"Electric-powered pumps should be selected to operate at high efficiency over the majority of the [irrigating] season, without extensive throttling at the main valve," concluded the Idaho engineers, writing in ASAE paper No. 78-2553. "Variable-speed pumps can be adjusted to apply needed water at the lowest allowable pressure, especially at the beginning and end of the irrigation season when crop water requirements decline."

One way to reduce the amount of horsepower required is to reduce pumping pressures. Pressure in a sprinkler irrigation system pushes water through the pipes and causes the water to spray out in a desirable pattern at each nozzle location. Low pressures can move water through the pipes. However, higher pressures are needed with many systems to spray water on the field at the proper rate. As pressure is lowered to a sprinkler nozzle, spray diameter and distance are reduced. Also, the stream of water has less tendency to break up into small droplets; solid streams of water can damage crops and cause "puddling" and erosion of soils.

In recent years, manufacturers have developed sprinklers to use at lower pressure ranges. The big advantage of going to low-pressure systems is that reducing the horsepower required to pump water lowers pumping costs. In some systems, as much as 50 hp can be trimmed from the power source when pressure at the pump is reduced by 50 psi.

Appendix B outlines how to measure water flow in pipes at various diameters and pressures.

HOW MUCH WATER—AND WHEN?

Timeliness of irrigation is critical. In much of the Corn Belt, irrigation is used mainly as "crop insurance," to supplement normal precipitation. In other areas, irrigation at regular intervals is vital to successful crop production.

Water leaves the soil by evaporation from the soil's surface and by transpiration through the plants' leaves. Losses from both causes—often called *evapotranspiration*—can vary from 0.1 in or less on a cool, cloudy day to as much as 0.5 in per day when hot, dry winds are blowing. A rough rule of thumb says that when the available moisture, or *field capacity* (the amount of water in soil that plants can use), in a crop's root zone drops to 50 percent, it's time to irrigate. (See Table 12-2.)

This is doubly true when the crop is in a stage of growth when lack of moisture is critical. All crops need ample moisture to germinate and make good early growth. After germina-

TABLE 12-2   Moisture Availability Related to Soil Texture

Percent of available moisture in soil	Soil texture		
	Sand to sandy loam	Loam to silty loam	Clay loam to clay
100% field capacity (ideal)	Soil clings; when squeezed, outline of ball is left on hand	Wet outline left on hand when squeezed; sticks to tools	Sticky enough to cling to fingers
50 to 75% (good)	Balls under pressure, but falls apart easily	Somewhat plastic; forms a ball, but will not stick to tools	Forms a ball; will ribbon out between fingers when squeezed
Time to irrigate			
40 to 50% (low)	Loose; does not feel moist	Forms weak ball when squeezed	Pliable, but not slick; will ball under pressure
0% (wilting point for most plants)	Dry, loose; flows through fingers	Powdery; easily crumbled	Hard, cracked; difficult to crumble

tion, the more crucial periods vis-à-vis soil moisture vary with the crop. From early tassel through early dent stages, corn yields can be cut by moisture shortages. If irrigation is delayed during this period, pollination and grain formation on ears may suffer. Grain sorghum can stand watering any time the available moisture falls to 50 percent. However, delaying watering is not as critical with sorghums as with corn. Unless the moisture shortage is severe over a period of time, later maturing of the crop may be the only appreciable result.

From full bloom through pod fill, soybeans need moisture, although they can stand several days of moisture stress. Long delays in watering can affect pod set, and moisture shortages during pod filling can cut the crop yield.

Early boll-set is the most critical time for cotton; delays in irrigation can limit yields. Actually, the size of the crop can be affected by moisture shortages anytime between first bloom and when 75 percent of the bolls are set on.

Forage crops can use needed supplemental moisture anytime during the growing season. Unless serious moisture shortages occur, any delays in irrigation will only decrease the current production. Longer-term moisture stress can damage the stand.

WATER "BANKING"

Irrigation systems in more humid areas of the country should be sized to take into account that some rainfall can be expected during the irrigating season and that the soil reservoir will be replenished by normal fall, winter, and spring precipitation.

Even in more arid regions, the concept of using stored soil moisture as a sort of water "bank" is not new. Often, preseason irrigation can be used to store water in the soil. Donald Sisson, product manager for Ag-Rain, Inc., an irrigation company at Havana, Illinois, outlines this concept:

The irrigation system's designed capacity should be equal to the crop's moisture requirements, minus moisture stored in the soil, minus expected effective rainfall during the crop growing season. Sisson presents this arithmetically as

$$Q = C - S - R$$

where Q = designed system capacity

C = total crop moisture requirements

S = stored soil moisture

R = rainfall during irrigation season

Assuming that the soil moisture reservoir is full at the beginning of the season, this stored moisture can be used up gradually during the irrigation period, which is typically July and August for summer crops. Depending on the climate, crop, and soil type, growing plants will take up an average 0.25 in of moisture per day.

The irrigation interval is the time required to use 50 percent of the moisture stored in the root zone and depends on the soil type and total amount of moisture the soil can hold. Effective rainfall, of course, is the amount of precipitation that can be expected during the growing season. To be on the safe side, figure the amount of rainfall expected in four out of five years (for 80 percent probability).

Table 12-3 shows the amount of water that soils can hold at field capacity.

The capacity of the irrigation system needed to maintain soil moisture will vary from a minimum of 5 gallons per minute per acre (gpm/ac) to 10 gpm/ac or more. For example, at 5 gpm, a capacity of 450 gpm would be required to irrigate 130 acres with a center-pivot sprinkler, if the system were operated day and night.

Table 12-4 shows the effect of stored moisture and rainfall on the designed size of equipment to irrigate 130 acres of corn grown on sandy loam soil.

If we assume that the soil is at field capacity of moisture when irrigation begins and that the four-out-of-five-year rainfall rate is accurate, the capacity of the system can be about 28 percent smaller than if these assumptions are not made. If we further assume that energy consumption is directly related to pumping rate, the smaller irrigation rig would use 28 percent less energy.

Now, let's turn to the use of alternative energy to power irrigation systems.

SOLAR-POWERED IRRIGATION

The old saying "There's nothing new under the sun" appears more and more to be true. More than 350 years ago, a French engineer, Salomon de Caus, built a tiny solar pumping

TABLE 12-3   Time Interval between Irrigations Related to Soil Type

	Soil moisture, in			
	Coarse sandy soil	Sandy loam	Silty loam	Clay
Top 1 ft of soil	¾	1½	2	1¾
Second 1 ft of soil	¾	1½	1¾	1¾
Third 1 ft of soil	¾	1½	1¾	1¾
Total in 3-ft root zone	2¼	4½	5½	5¼
Amount lost before 50% of capacity is reached, in	1⅛	2¼	2¾	2⅝
Days required to lose 50% of field capacity if ¼ in moisture is lost per day	4½	9	11	10½

TABLE 12-4   Two Methods of Sizing Irrigation Equipment

	Allowing for stored moisture and rainfall	Not assuming stored moisture and rainfall
Stored moisture in root zone	4.50 in	4.50 in
Moisture used by crop during irrigation interval (¼ in/d)	2.25 in	2.25 in
Irrigation interval	9 days	9 days
Stored moisture considered	4.50 in	
Moisture depletion rate (4.5 in ÷ 60 days)	0.075 in/d	
Stored moisture depletion each irrigation interval (9 days × depletion rate)	0.68 in	
Minimum rainfall expected each irrigation interval	0.10 in	
Irrigation required each interval	1.48 in	2.25 in
Irrigation rate	0.16 in/d	0.25 in/d
Net operating time per day	24 hours	24 hours
System capacity	4.29 gpm/ac	5.96 gpm/ac
Required pumping rate	558 gpm	775 gpm

device and used it to water his garden. Then, the world mostly lost interest in solar-powered irrigation for three centuries, until 1913, when a 45-kW (60-hp) solar pump was built in Egypt.

More recently, the world's fuel woes are renewing interest in this forgotten technology, and new ways to pump water with sunshine are being developed. Today, there are several sun pumps in various regions of the United States, and more being built all the time.

Solar irrigation systems are of two general types: (1) those in which solar collectors heat fluids that drive heat-cycle engines to run pumps and (2) those in which direct solar-generated electricity powers electric pumps. Both types are shown in Figs. 12-1 to 12-3.

Heat-operated solar pumps use concentrating collectors to heat a fluid that is expanded through a turbine-drive engine to run a pump. The U.S. Department of Energy and New Mexico State University have an operating 19-kw (25-hp) solar-powered pump on a farm near Albuquerque. This system uses curved concentrating collectors to heat an oil-based fluid, which in turn heats freon to drive a turbine.

The Albuquerque setup also includes a storage tank for the heated fluid, so that the irrigation system can operate 24 hours a day in clear weather. There's also a plastic-lined storage reservoir that holds 4.5 ac-ft (196,020 ft³) of water that can be used when the solar pump is not working.

When irrigation water is not needed, the system can provide 20 kW of electrical power or a like amount of heat to be used for other farm applications.

A larger solar-powered irrigation system is being installed by the Battelle Memorial Institute near Gila Bend, Arizona, on land owned by Northwestern Mutual Life Insurance Company. The insurance company owns some 76,000 acres of irrigated cropland, and the fuel bill to operate irrigation pumps runs well over $1 million each year. That's a high motive for going to alternative sources of pumping water.

The first large-scale solar-cell-powered irrigation system is in operation at the University of Nebraska Field Laboratory, near Mead. University engineers are using the electrical

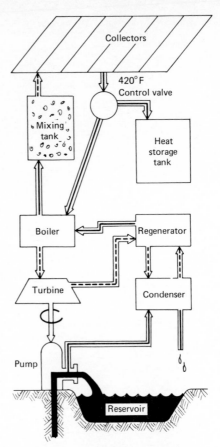

FIG. 12-1 How a solar-powered, heat-engine pumping plant works: (1) Oil-base heat-transfer fluid is pumped to concentrating collectors; (2) fluid heats to 420°F (215°C) in collectors; (3) hot fluid flows to boiler (heat exchanger), where an inert gas (usually freon) is vaporized, or to a thermal storage tank; (4) heated, high-pressure freon drives a turbine to operate a conventional irrigation pump; (5) freon exhausts from the turbine to a regenerator heat exchanger, then to a condenser, where the gas is liquefied; (6) the liquid freon passes through the boiler–heat exchanger, where the working fluid cycle starts over again; (7) the oil-base heat transfer fluid flows from the boiler to a mixing tank; (8) the mixing tank controls the inlet temperature to the collectors.

FIG. 12-2 Nearly 100,000 silicon cells generate electricity to pump irrigation water for 80 acres of corn and soybeans at the University of Nebraska. Brad Rein, agricultural engineer pictured, says the system works well, but costs are far out of line with conventional pumping systems.

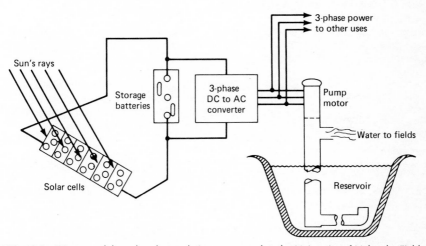

FIG. 12-3 Diagram of the solar photovoltaic system used at the University of Nebraska Field Laboratory.

energy generated by photovoltaic cells to power a redesigned automatic gated-pipe irrigation system that waters 80 acres of corn and soybeans.

About 100,000 individual solar cells mounted on 28 flat panels constitute the generating array, which gives the unit a peak power output of about 25 kW. The cells are interconnected by electrical wiring to lead-acid storage batteries that can store up to 90 kWh of energy for off-peak use, and also to inverters that convert the direct-current (DC) power produced by the cells to alternating current (AC) to power pump motors and other loads.

The solar-cell power operates a 15-hp pump 12 hours a day, to pump 1000 gpm of water from the irrigation reservoir, which also serves as a tailwater reuse pit. At night, the reservoir is recharged from a deep well by pumps operated on utility power. Water is pumped into an automatic gated-pipe irrigation system. Irrigation is scheduled by a computer, which utilizes readings from sensors (electrical resistance blocks) installed in the field. The entire system is closely monitored by engineers at the University of Nebraska.

The major drawback to both solar-powered heat engines and direct electricity generation by solar cells is cost. A kilowatt (about 1⅓ hp) of electricity produced by photovoltaic cells represents a capital investment of about $15,000. At best, solar cells cost $5 to $6 per peak watt of output. Another disadvantage may be the space occupied by the cell array. The solar-powered system at Mead, Nebraska, ordinarily would occupy a third of an acre. Engineers there stacked the cells on the reservoir berm, rather than use up valuable crop land.

Depending on the design of the system, solar heat turbines are not a great deal more economical per unit of energy produced. The Department of Energy now is sponsoring design and development projects for larger solar-powered pumps (100- to 300-hp range). The technology of pumping water with sun power is well-established. The big task now is to get costs more in line with conventional sources of pumping power.

Farmers with long-range plans to put acres under irrigation may want to keep abreast of developments in solar-powered systems.

WIND-POWERED IRRIGATION

In the short run, the kinetic energy in moving air appears to be a more attractive alternative energy source for irrigators. A happy coincidence of geography puts much of the cropland that requires regular irrigation in the plains regions, where the wind moves unimpeded for long distances.

R. N. Clark and A. D. Schneider, agricultural engineers at USDA's Southwestern Great Plains Research Center, Bushland, Texas, have tested ways to use wind energy in irrigation pumping, including direct mechanical drive, compressed-air pumping, and a "piggyback" system that lets wind power assist electric pump motors. Of the methods studied, the hybrid wind-plus-electric systems appear to offer the best combination of reliability and energy economy.

The engineers coupled a 40-kW vertical-axis wind turbine to an existing irrigation pump powered by an electric motor. The power produced by the wind turbine averaged 36 kW and cut energy costs by 65 percent. An overrunning clutch was used to synchronize wind power with the electric motor.* (See Fig. 12-4.)

In the southern Great Plains region, irrigation pumping consumes nearly 50 percent of the total energy used on irrigated farms. That makes irrigation a major cost of crop production. With their "hybrid" irrigation system, Clark and Schneider designed the pump drive to use both a Darrieus wind turbine and an electric motor to power a conventional vertical-turbine pump. The electric motor was sized to operate the pump with no additional power, and ran continuously, even when the wind turbine did the pumping. However, because the pump idled or ran with no load for much of the time, power consumption was greatly reduced.

The wind turbine was coupled to the system through an overriding clutch (see Fig. 12-4) and powered the pump when wind speed was above 6 m/s (about 12 mph). Power used

*For a full report of research results, see ASAE paper No. 78-2549, available from the American Society of Agricultural Engineers.

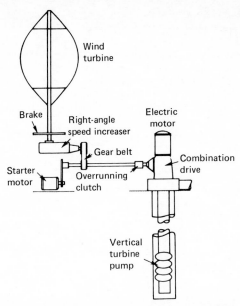

FIG. 12-4 Schematic of a 40-kW wind-assisted irrigation pumping system. The vertical-axis turbine takes over at wind speeds above about 12 mph, with the result that 65 percent of the electrical energy normally required to operate the pump is saved. The wind turbine is not self-starting—note the small electric motor used to start the device, at left in the drawing. *(Courtesy: R. N. Clark, USDA engineer, Bushland, Texas.)*

by the electric motor decreased as wind speed increased, from 51 kW at full load to about 3 kW when the wind blew at 20 m/s.

During the test period, the wind turbine supplied 65 percent of the energy needed to pump water. All the equipment needed—clutch, gear drive and connecting shafts—for assembling the wind-assisted pumping system is available "off-the-shelf" in most irrigation supply firms.

The Texas engineers concluded that wind-assisted pumping systems have the potential for becoming an important source of energy in areas that have strong, steady winds.

A similar system, using a wind-driven induction generator, can assist or supplement the electrical load to the pump motor. This allows more flexibility in locating the wind generator with respect to the well head and pump. Also, power can be generated and fed into the utility grid when water is not being pumped. For more on wind power, see Chap. 19.

A well-designed, well-operated irrigation system can be good insurance, even in cropping regions with 30 in or more of expected annual rainfall. The big reason: The rain doesn't always come when the crop needs it most. The ability to water at will has often meant the difference between a decent crop yield and total disaster.

However, the premium on that kind of insurance goes up with each increase in fossil fuel prices. To get the most out of both the energy and the water used in irrigation calls for knowing the crop's moisture requirements and the ability of the crop to withstand adverse moisture conditions without suffering unprofitable yield losses.

Knowing the soil's ability to hold moisture is a key to the successful scheduling of irrigation, too. A "water log" can help keep an account of the amount of moisture available in the crop's root zone. The account is credited whenever rain falls and debited for hot, rainless days. This way, you can schedule irrigation when the available moisture in the soil falls to 50 percent of field capacity—rather than by days elapsed on the calendar—and apply only enough water to bring the root zone back up to field capacity.

13

Farm-Grown Fuels

Few ideas have grabbed the imagination of American farmers as has the prospect of producing fuel on the farm. This development has brought on a rash of hurry-up technology, as manufacturers of alcohol stills, methane digesters, and other on-farm fuel plants scramble for a piece of the action while the idea is hot.

Much the same can be said of publications. This is not the first book to discuss on-farm fuel production. It undoubtedly will not be the last.

The interest by agricultural producers in providing at least part of their needed energy is understandable. More than 60 percent of the on-farm energy consumption is in the form of liquid fuels—primarily gasoline and diesel. Another 20 percent or so of the directly consumed energy is in the form of natural gas or liquefied-petroleum (LP) gas. Finding a home-grown replacement for these fuels has the potential of saving up to 80 percent of the total outlay for energy.

There are several processes by which farm-raised products can be turned into liquid and gas fuels. With some minor modifications, gasoline engines operate very well on 170- to 180-proof ethanol (grain alcohol). Vegetable oils can operate diesel engines with little or no mechanical modification. Methane from anaerobic digesters can fire furnaces and water heaters, as well as modified internal-combustion engines.

However, the technology to produce these fuels has a price tag—in dollars, time, labor, and altered farm management priorities. With conventional fuels priced as they are as this is written, about the best a farmer can hope for in producing his own fuel is to break even. That is, to recover—over time—the dollars he has invested in energy equipment and to reap virtually as much energy in the form of fuels as was contained in the original biomass material. At that, the breakeven price assumes a relatively cheap source of feedstock for the fuel and full use of the by-products from the fuel-production process.

You'll hear arguments that petroleum, and the fuels made from crude oil, will increase in price in the future and therefore make farm-grown fuels more economically attractive. There's strong consensus that conventional fuels will increase in price, but it does not necessarily follow that alternative, homemade fuels will be relatively less expensive. The cost of most farm inputs is tied to fuel, of whatever source. As petroleum costs climb, so will the cost of fertilizer, ag chemicals, steel, rubber, transportation, and nearly everything else a farmer must buy to stay in business. So, the costs of producing a bushel of corn will increase, whether the corn is marketed for cash, fed to livestock, or distilled into ethanol.

In other words, an alcohol still does not guarantee a perpetual supply of fuel at $1.20 per gallon regardless of what happens to the price of gasoline. The same thing is true of vegetable oil substitutes for diesel fuel and methane alternatives to natural gas and LP gas.

However, price is not always the sole consideration. Fuel *availability* may be of equal —or greater—concern for many farmers. The supply of petroleum from which gasoline, diesel, and LP gas are made is being depleted steadily, and most of the known reserves are held by foreign countries. Any interruption of fuel supply could have a great impact on food and fiber production potential. If a sharp cutoff in foreign oil supplies forced a strict fuel rationing program in the United States, it's likely that farmers would have to share in the shortage.

LIQUID FUELS

Let's look first at liquid fuels, which account for more than half of the on-farm energy consumption. In farm operations, these fuels are primarily gasoline and diesel, fuels that are convenient to handle and use, relatively safe, able to store a lot of energy in a small space, and—until recently—relatively low in cost.

There is considerable research underway into various types of "heat" engines—external-combustion engines that use low-grade sources of energy. But, so far, the designs tested have only limited application for self-propelled vehicles.

Regardless of the engine used and the source of its energy, its purpose is to transform stored energy into mechanical work. It is reasonable to assume that the internal-combustion engine and the well-matched hydrocarbon fuels will be the power for most vehicles into an indefinite future.

Table 13-1 lists the energy-storing ability of various fuels, with the respective values of energy given in Btu's per pound of material. Right now, biomass is the alternative energy source receiving most attention as a basic feedstock for fuels. Briefly, biomass is any organic substance that grows and stores energy—trees, field crops, manure, seaweed, algae, microorganisms. The biomass most in vogue today includes grain for alcohol and manure for methane.

Biomass is the stuff Mother Nature started out with when she began making coal and petroleum eons ago. However, with petroleum, much of the preliminary work in converting organic materials to usable fuels has already been done. In converting crude oil to gasoline or diesel, only about 10 percent of the energy potential is used. In other words, for each 100 gal of petroleum pumped out of the ground, 10 gal must be consumed in the process of changing the other 90 gal into fuels that can be used in internal-combustion engines.

When the timetable is speeded up—as in the case of fermenting grain to make ethanol —more energy is required. More than a third of the potential energy stored in a bushel of grain (or the equivalent) must be consumed to convert the grain into burnable alcohol, and that doesn't take into account the energy that must be used to produce the grain.

TABLE 13-1   Energy-Storing Ability of Various Fuels

Fuel	Energy, Btu/lb
Liquid hydrogen	61,000
Methane	23,900
Premium gasoline (123,000 Btu/gal)	20,300
No. 1 diesel (135,000 Btu/gal)	19,650
Vegetable oil (127,900 Btu/gal)	16,650
Ethanol—grain alcohol (83,000 Btu/gal)	12,700
Methanol—wood alcohol (64,500 Btu/gal)	9,800
Ammonia	8,900

On the average commercial farm, much of the heavier equipment—tractors and combines—is powered with diesel engines. There is also a need for gasoline to power trucks, autos, and other vehicles. To date, no single farm-grown fuel can work equally well in both compression-ignited (diesel) engines and spark-ignited (gasoline) engines. Fuel properties required in the two engine types are greatly different.

Gasoline engines can readily be modified to operate on ethanol. And early work indicates that diesel engines operate very well on vegetable oils, such as oil from sunflowers or soybeans. But the substitution of one fuel for another is a limited proposition.

For example, vegetable oils readily blend with diesel fuel. Anhydrous (water-free) ethanol can be mixed with diesel fuel, also, but there are unresolved problems with injector pump lubrication and uneven vaporization of the fuel.

Devices have been developed that inject a mixture of ethanol and water into the combustion air of diesel engines, but fairly small amounts of ethanol are added by this equipment. Also, there is some question about the damage an ethanol-water mixture might cause to engine blower turbines.

While anhydrous (200-proof) ethanol blends readily with gasoline and Number 2 diesel, any water in ethanol will separate out of the mixture and cause problems Few ethanol plants designed for on-farm use can produce 200-proof alcohol. (The common mixture for Gasohol is 10 percent anhydrous alcohol and 90 percent gasoline.)

Many farmers now are asking: "How do we produce our own fuel?" Perhaps a better first question to ask would be: "*Should* we try to produce our own fuel?" Each farmer will need to do his own homework well before he makes the heavy investment necessary to get into fuel production on any kind of scale.

HOW MUCH FUEL DO YOU NEED?

Earlier, we suggested that you study your annual consumption of all types of energy. How much gasoline and diesel fuel do you use each year? Which fuel has the higher priority on your farm, gasoline or diesel? Which fuel could you produce most economically? If your farm operation consumes many more gallons of gasoline than diesel, you may want to put your first emphasis on producing substitute fuels to replace gasoline, or vice versa.

In the past several years, much of the high-horsepower farm equipment has been shifted from gasoline- to diesel-powered engines. Diesels typically perform more work per gallon of fuel consumed than do gasoline engines. They have a longer, more trouble-free working life, on the average.

On the other hand, gasoline engines are usually less costly to buy initially and are more adaptable to some uses than diesels.

Table 13-2 gives estimated fuel requirements for several field operations performed with both diesel- and gasoline-powered equipment. Also, horsepower requirements are given per foot of machine working width.

For example, a five-bottom, 16-in moldboard plow has 6.67 ft of working width and requires 85.3 hp (6.67 × 12.8) on the average. If you're plowing in heavy clay, the horsepower (and fuel) requirements would be higher. Sandy loam soil would require less horsepower and less fuel per acre to plow.

The fuel requirements shown in Table 13-2 are estimates for actual field operations and do not include trips to and from the field or long idling periods. As you can see, using gasoline-powered equipment to moldboard-plow, disc, harrow, plant, cultivate, and harvest a corn crop requires about 6.5 gal of fuel per acre. The same crop, put in with

TABLE 13-2   Estimated Fuel Requirements of Various Field Operations

Machine	Ground speed, mph	Required horsepower per ft width	Fuel consumption, gal/acre	
			Diesel	Gasoline
Moldboard plow (8″ deep)	4.5	12.8	1.68	2.35
Chisel plow (8″ deep)	5.0	9.3	1.10	1.54
Disc, tandem:				
Cornstalks, stubble	5.0	3.5	0.45	0.63
Chisel-plowed soil	5.0	4.3	0.55	0.77
Moldboard-plowed soil	5.0	5.6	0.65	0.91
Field cultivator	5.5	5.2	0.60	0.84
Harrow, springtooth	5.5	3.5	0.45	0.63
Harrow, spiketooth	6.5	2.5	0.30	0.42
Cultivator, row crop	4.5	3.2	0.45	0.63
Rotary hoe	7.0	2.3	0.25	0.35
Planter, row crop:				
Conventional	4.5	3.5	0.50	0.70
No-till	4.5	2.0	0.35	0.49
Grain drill	4.5	2.5	0.35	0.49
Combine harvester:				
Small grains	3.0	3.0	1.00	1.40
Soybeans	3.0	3.3	1.10	1.54
Corn, grain sorghum	2.0	4.0	1.60	2.24
Corn ear picker	2.5	3.5	1.15	1.61
Mower:				
Cutter bar	6.0	2.3	0.35	0.49
Mower-conditioner	5.0	3.0	0.60	0.84
Rotary (brush-hog)	5.0	5.0	0.80	1.12
Swather	6.0	3.0	0.55	0.77
Rake	5.0	1.0	0.25	0.35
Baler*	2.0	*	0.45	0.63
Sprayer	4.0	1.0	0.10	0.14
Forage harvester:				
Greenchop	3.0	2.0	0.95	1.33
Haylage	3.0	2.5	1.25	1.75
Corn silage	1.5	5.0	3.60	5.04

*Working widths of hay balers are difficult to specify; horsepower requirement is not included.
SOURCE: University of Missouri Agricultural Engineering Department.

diesel-powered equipment, and performing the same operations, requires just over 5 gal/ac.

Each farmer will need to review tillage operations and the type of fuel as they fit his own situation. Using diesel equipment to chisel-plow, field-cultivate, plant, and harvest an acre of corn would require only 3.8 gal/ac—a significant savings over using a moldboard plow and disc. If this tillage method would produce the same level of yields, it is a management option well worth considering.

The type of fuel substitute that can be produced most economically on the farm may also have a bearing on the kind of equipment used and the mix of farming operations performed. However, if the cost of producing vegetable oils as diesel fuel substitutes is within the range of the cost of producing ethanol as an alternative to gasoline, the decision may well rest on the amount of fuel needed to get jobs done. And when it comes to fieldwork performed per gallon of fuel, diesel has the edge.

THE NATIONAL BIOMASS "BANK"

Assuming that U.S. agriculture will remain primarily devoted to producing food and fiber, how much energy could be made available from surplus cereal crops, forests, agricultural by-products and residues, and animal wastes? In other words, if our farms and forests maintain production, how much energy could be produced from surplus biomass not needed for food, feed, and fiber?

Let's take just the prospect of producing ethanol in one major farm state: Nebraska.

Annually, Nebraska uses about 1 billion gal of gasoline and more than 150 million gal of diesel fuel. To replace the crude petroleum fuels with ethanol would require nearly half of all the grain, potatoes, sugar beets, hay, crop residues, and wood products grown in Nebraska each year, plus half of all the whey and urban solid organic waste.

This assumes that all energy for operating ethanol plants comes from coal, electricity, crop residues, solar, etc., and does not require any additional gasoline or diesel for this purpose. If only grains were used for ethanol production, more than 65 percent of Nebraska's yearly production would need to be converted to fuel.

Nationally, the substitution of ethanol from agricultural products for gasoline offers less potential than some forecasts would indicate. To produce enough ethanol for a 10 percent blend with gasoline (to make Gasohol), more than 3.8 billion bu of grain would be needed each year. That's equal to nearly half of the annual U.S. cereal crop production.

So, while farm-produced fuels may in time make U.S. agriculture virtually independent of outside sources, biomass fuels may not make much of an impact on the nation's total needs.

What is the potential for biomass fuels nationally? If the value of crop residues for fertilizer and erosion control are not taken into account, how much energy is available?

The 50 states have about 2.3 billion acres (917 million hectares) of land. In 1980, farmers had some 354.5 million acres in principal crops, out of a total of 460.5 million acres of cropland. (See Fig. 13-1.)

Actually, the share of the total in crops and forest land has changed little since 1900, although acreages devoted to specific crops have changed. However, average production per acre has increased by more than 80 percent in just the past 30 years. U.S. agriculture now produces enough food and fiber for domestic use and, in addition, exports about a third of its production.

Grains are the most abundant raw materials produced as cultivated crops. Representative production of grains is as follows:

Crop	Corn	Wheat	Sorghum	Rice
Tons (in millions)	198	54	21	7

All cereal grains contain starch as a major component—starch that can readily be converted to ethyl alcohol by fermentation. The average composition of major U.S.-produced grains, as a fraction of dry-weight grain, is shown in Table 13-3.

The theoretical yield of ethanol (ethyl alcohol) per pound of starch is 0.568 lb. In actual practice, however, ethanol yields are closer to 90 percent of the theoretical maximum. The maximum potential in a bushel of corn is about 2.6 gal of 200-proof (100 percent) ethanol.

Lower-quality grains, surplus grain, and grain-processing by-products are among the more promising sources of alternative engine fuel for the next five to ten years. However, *top-quality grain* has a great many potential uses for food and livestock feed, which makes it less attrractive as a source of fuel.

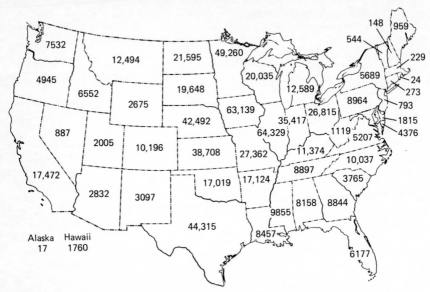

FIG. 13-1 Total U.S. agricultural residues amount to nearly 678 million tons per year. The map shows regional distribution. Animal wastes make up about 186 million tons, corn stover adds another 179 million tons, and small-grains straw accounts for some 139 million tons.

TABLE 13-3 Average Composition of Various Grains

Grain	Starch, %	Protein, %	Oil, %	Fiber, %	Other, %*
Corn	72	10	2	2	14
Hard wheat	64	14	2	2	18
Soft wheat	69	10	2	2	17
Sorghum	71	13	2	2	12

*Includes minerals, vitamins, sugars, and other elements.

Corn is the major U.S. grain crop. One hundred bushels of corn grain at 14 percent moisture content contains the heat energy equivalent of 38.5 million Btu. Converted to ethanol, those 100 bu can produce about 19 million Btu heat equivalent, plus 1930 lb of by-product that can be used for food or livestock feed. We'll go into ethanol production more thoroughly in Chap. 14.

Even more energy is available in the stalks, cobs, leaves, and husks that produced the 100 bu of corn grain. The heat equivalent from the residue is about 50 million Btu. That's a total of about 88 million potential Btu's of heat energy with each acre of 100-bu corn.

With wheat, the comparative energy equivalency is 12 million Btu in the grain and another 30 million Btu in the straw. Overall, there is about twice as much energy in the crop residue as in the grain itself. Potentially, at least.

More than 495 million tons of straw, hulls, husks, shells, cobs, and other crop "leftovers" are produced annually. The heat equivalent of this material is something more than *6000 trillion* Btu. If even half of this residue could economically be collected and used, it could provide well over agriculture's share of the national energy consumption.

When you add in about 185 million tons of animal waste (on a dry-matter basis), the production of energy just from waste materials could total more than 5 percent of the U.S. energy needs. That is considerable energy potential, barely tapped. However, making economic use of that potential is not among the easier problems to solve.

It is not insurmountable. With some crops, the use of residue and by-products as energy is well-established. For example, sugar cane stalks (called *bagasse*) are often used for boiler fuel in sugar mills. As a result, sugar production has a low "makeup" energy requirement.

BIOGAS PRODUCTION

Every farm is a potential gas well. Agricultural residues, wood wastes, and animal manures can be converted into gases through fermentation or through pyrolysis of the material.

Bacterial fermentation of a ton of dry crop residue can yield about 10,000 ft³ of biogas, having a heating value of some 500 Btu/ft³. If raw materials are charged at $20 per ton, this would be equivalent to $2 per 1000 ft³ of gas that has a heating value of 500,000 Btu. Fermentation of the total 680 million tons of residue produced each year could yield some 3400 trillion Btu worth of gas.

Pyrolysis of a ton of residue with heat and pressure could produce 6800 ft³ of mixed gases, with a net heating value of about 150 Btu/ft³.

At present, neither fermentation nor pyrolysis techniques are very competitive economically with the production of natural gas and LP gas, but that is changing quickly. Also, some waste materials, such as livestock manure, must be managed for pollution control. If part of the cost of producing, say, methane in an anaerobic digester is charged to manure handling, the procedure may look more profitable.

We'll discuss biogas produced from waste materials more thoroughly in Chap. 16.

VEGETABLE OILS

There's growing interest in the use of vegetable oils as a substitute for diesel fuel. A great many unknowns still exist—such as how long-term use of these alternative fuels may affect engine wear and lubrication oil contamination—but early results are encouraging. Diesel engines operate at near full power on vegetable oils, and these oils can be blended with diesel fuel at any ratio.

The economics of producing vegetable oil as a substitute for diesel fuel may need more work, however. For example, an acre of sunflowers that yields 1500 lb of seed can produce 65 to 70 gal of oil. A gallon of sunflower oil weighs 7.7 lb, but most oil extraction processes leave some of the oil in the pressed meal or cake. For each gallon of oil, some 12 lb of sunflower meal is produced, which can substitute for soybean meal or other protein supplements in most livestock rations.

At that rate, those 1500 lb of sunflower seed per acre will yield about 70 gal of fuel substitute and 900 lb of protein supplement. At current prices for Number 2 diesel and soybean meal, the fuel and feed value of an acre of sunflowers comes to about $165. If the selling price for sunflower seed on the market is 11 cents per pound or less and you assume that the oil could be pressed out for about the same energy cost as the crop could be hauled to market, growing sunflowers for fuel and protein supplement is a paying proposition.

On the other hand, if the market price for sunflower seed is more than 11 cents per pound—as is often the case—a farmer would be money ahead to sell the crop for cash, then buy diesel fuel and protein supplement.

We'll have more on vegetable oils for diesel fuel substitutes in Chap. 15.

IT'S NOT ALL GRAVY

Earlier in this chapter, we estimated that it takes about 6.5 gal of gasoline to produce an acre of corn with conventional tillage methods. If that acre produces 150 bu of corn, that is potentially 300 or more gallons of ethanol and something more than a ton of distillers grain by-product that can be used as livestock feed.

If this same fuel-production ratio is applied to 100 acres of corn, the potential is 30,000 gal of ethanol and 130 tons of distillers grains. Let's assume that crop residues provide the heat needed to manufacture ethanol and that 2000 gal of fuel equivalent is used in the form of fertilizer, herbicides, hauling, and other indirect energy uses. Added to the 650 gal of gasoline consumed to grow the crop (at 6.5 gal/ac), that is 2650 gal of gasoline equivalent.

At this point, the budget looks pretty good—at least from the standpoint of Btu's of energy. With an investment of some 2650 gal of fuel, about 30,000 gal of ethanol are produced.

But the picture changes when studied more closely. To complicate things, the quality of ethanol produced by most farm-sized stills cannot be substituted gallon-for-gallon for gasoline. Even a gallon of 200-proof ethanol contains only 68 percent of the Btu's in a gallon of gasoline. If the ethanol made is only 175 proof (about 12.5 percent water), the Btu equivalent is only about 60 percent of gasoline. This cuts the 30,000 gal of ethanol down to 18,000 gal on a gasoline-equivalency basis.

And there remains that 130 tons of distillers grains to dispose of—that's 130 tons on a *dry-matter* basis. The stuff contains 9 lb of water for each pound of distillers grains, which means about 1300 tons of material must be handled somehow. Unless you have livestock close at hand to consume this material as is, it will take the equivalent of 25 percent of the heat energy in the ethanol to dry, transport, and store this feedstuff. That reduces the gasoline-equivalency value of the ethanol to 13,500 gal.

If the ethanol replaces gasoline that costs $1.20 per gallon, the value of the 100 acres of corn (at 150 bu/ac) for fuel is $16,200. Add $19,500 for the distillers grains as a substitute for protein supplement, and the total comes to about $35,700 income for the 100 acres. That makes the 15,000 bu of corn worth only $2.38 per bushel.

Admittedly, this is a rather crude budget. It assumes no cost for grinding the grain or interest on the investment in ethanol-making equipment. Even so, the economics of producing fuel on the farm are fraught with hazard.

Much the same is true of sunflowers grown for diesel fuel substitutes, as pointed out earlier. Here is one problem: There is an established market for corn, wheat, soybeans, sunflowers, and other farm grains and oil seeds that can be used to produce fuel. As yet, there is no market for distillers grains (especially not if they're wet), sunflower oil cake, or other by-products of on-farm fuel production.

For another thing, widespread production of oil seeds for diesel fuel and concurrent growth in the production of grains for ethanol could radically change crop-growing practices in the United States. Here's why: Each gallon of sunflower oil extracted leaves 12 lb of high-protein meal that can be substituted for soybean meal in livestock rations. Similarly, each gallon of ethanol produced by fermentation of cereal grains leaves about 6.5 lb of high-protein by-products that can also be substituted for soybean meal or other protein supplements in livestock rations.

This trend could see 1 bu of soybeans replaced by 1 bu of sunflowers, or 3 bu of corn grown for fuel. Of course, soybeans are a high-oil crop, too, and may well develop as an important diesel fuel substitute. However, with the present demand for vegetable oils in human foods, and the attendant demand for plant protein (such as soybean meal, a by-product of soy oil extraction), a fairly delicate balance has been worked out. Anything that reduces the demand (or increases the supply) of high-protein feedstuffs will have the effect of reducing the market price for soybeans—or of increasing the price of soybean oil for human consumption.

Fuel production also takes extra time and labor. The operation of even a semiautomated ethanol plant requires at least a couple of hours each day, to grind a new batch of grain and transfer materials among the various vessels. The spent sludge from a methane digester generally must be stored and handled to some extent. The value of by-product materials as feeds or fertilizer may pay for all or most of the "downstream" labor associated with a farm fuel factory, and typically extra time and labor demands are involved.

New skills are needed, also. Many farmers are ingenious, mechanically. Not so many are skilled in chemistry and microbiology—talents needed to manage fermentation organisms successfully. It's possible—even likely—that more farmers are qualified to build an ethanol still or methane digester than are qualified to operate the plant after it is constructed.

The point to keep in mind is this: There are no *cheap* sources of fuel; there are no *super-simple* ways to convert farm products and residues into usable liquid and gas fuels.

But don't be too discouraged. There's terrific energy potential in American farms and forests. U.S. farmers have proved themselves to be unequaled in the efficient production of food and fiber. That same productive genius can—and will—successfully be applied to energy production and conservation.

BEWARE OF HUCKSTERS

There are always people ready to cash in on the enthusiasm of others or to make a fast buck from an uncertain situation. The farm fuel situation is no exception.

Be leery of claims for equipment that sound too "pie in the sky." There is no equipment or material now known that will let you make "fuel for 20 cents per gallon," as one advertisement for an ethanol still claims—not if you realistically charge all costs involved in producing the fuel. And there is no practically sized hardware or technique that will produce a year's supply of fuel with only a few weeks' work during the slow season.

The fact is, there's a certain amount of faddishness in the more glamorous alternative energy systems: solar, ethanol, wind power, etc. And farmers are no less susceptible than other people to the allure of attractive gadgetry. There's no doubt that new developments and new processes will make biomass fuel production simpler and more efficient in the future. But *cheap* fuel is a thing of the past.

Unfortunately, few standards have developed as yet for fuel-making hardware and operating techniques. For that matter, there is considerable lack of hard information on the fuel properties of ethanol, vegetable oils, and other farm-produced fuels and how engines and other equipment tolerate them on a long-term basis. Where the technology of biomass is concerned, we are now at the point where a lot of Henry Fords are busy inventing their own version of the Model T.

Even so, before you buy any equipment, plans, or components, have the design checked out by a competent engineer. Make sure the equipment operates as advertised and that you understand fully how to operate it. If chemical or microbiological processes are involved, as with fermentation alcohol plants or methane digesters, you may also want to have a chemical engineer or microbiologist pass judgment on the design and operation of the system.

Now, let's look at various farm-produced fuels—how to make them and how to use them.

14

Ethanol

Among the more popular topics in farm conversations these days is on-farm production of alcohol—"do-it-yourself" fuel made from grains and other materials.

There are several reasons for enthusiasm about grain alcohol as a substitute for gasoline in internal-combustion engines. For one thing, the source is renewable. For another, the technology is within reach of most farmers who seriously study the lessons. Thirdly, despite rather gloomy economic analyses by some forecasters, a well-designed, well-operated alcohol plant can produce fuel that is reasonably competitive in price with gasoline.

Listening to all the talk about alcohol developments, it's easy to get the impression that this fuel is brand-new in America. That's not the case at all. Alcohol fuels have a long but checkered history.

In fact, alcohol came into wide use long before refined petroleum fuels were developed. In the mid-1800s, alcohol replaced whale oil as lamp fuel. It was clean, odorless, and fairly widely available. In 1876, Nikolaus Otto, the recognized inventor of the four-stroke-cycle internal-combustion engine, advocated alcohol as the fuel for his new engines.

Henry Ford was an early champion of alcohol for fuel. His earliest automobile, the *quadricycle,* was designed to run on alcohol. Later, the venerable Model T was built with an adjustable carburetor that could be set to burn pure alcohol in the engine. In 1922, Alexander Graham Bell (perhaps foreseeing today's energy crisis) suggested alcohol as an alternative to petroleum-base fuels, as "a beautifully clean and efficient fuel that can be produced from vegetative matter of almost any kind, even the garbage of our cities."

After World War I, France and Germany mixed gasoline and alcohol, in ratios as high as 50:50. During the Depression in the United States, low prices for farm products kept alive interest in alcohol. In 1938, an alcohol fuel plant was built at Atchison, Kansas, and many older farmers in the Midwest may remember burning Agrol, the trade name given to the mixture of gasoline and alcohol produced there.

When allied forces cut off Hitler's fuel supply lines in World War II, most of Germany's military machinery—including virtually all the Nazi aircraft—shifted over to alcohol. In the United States, wartime shortages spurred production of alcohol as a component of synthetic rubber and as fuel for submarines and torpedoes. In 1944, the United States produced nearly 600 million gal of alcohol—100 million gal more than the U.S. Department of Energy's target for the end of 1980.

Through the past 100 years, people in a fuel pinch have turned to alcohol. Between emergencies, the relatively inexpensive price for gasoline has made alcohol fuel uneconomical to produce on large scale. Now, we're in a fuel crisis that appears to be permanent—at least where petroleum-derived fuels are concerned—and alcohol is again assuming the spotlight. Only this time, it's a genuine "grass-roots" kind of movement, with individual farmers, farm cooperatives, and other farm-related businesses getting into the game.

However, when you look at the nation's consumption of gasoline, ethanol made from grain is at best a stopgap fuel. Americans consume about 100 billion gal of gasoline each year. If *all* the corn, wheat, sorghum, and other cereal grains produced were converted to ethanol, that would supply only about 20 percent of the annual total. In fact, it would take nearly half of all U.S.-grown grains just to produce enough anhydrous (200-proof) alcohol to blend with gasoline to make Gasohol—which is 10 percent ethanol and 90 percent gasoline.

Still, ethanol is the only alternative fuel available now and the only one likely to be available in any quantity before 1985 or later. On a regional basis—particularly in major agricultural regions—alcohol fuels no doubt will become important sources of energy.

WHAT IS FERMENTATION ETHANOL?

Ethanol (also called *ethyl alcohol* and *grain alcohol*) is a colorless, flammable liquid with a chemical formula of C_2H_5OH. It's the alcohol product of fermenting the sugars in natural raw materials with yeast, and it's the stuff that makes alcoholic beverages intoxicating.

Actually, ethanol is produced by two different processes: (1) fermentation of sugars and (2) chemical synthesis of petroleum or natural gas. Most industrial ethyl alcohol is made by the latter method; all beverage alcohol is produced by fermentation. Chemically, the two types of ethanol are the same, but we're primarily interested in fermentation ethanol.

The conversion of sugars to ethanol through fermentation is accomplished by the action of microorganisms, or yeasts. These yeasts consume the sugar and give off waste products in the form of alcohol and carbon dioxide. Most living plants contain sugar in one form or another, but the yeasts can only utilize simple sugars, called *monosaccharides.* This means that complex sugars, such as the starch in grains, must be converted to simple sugar before fermentation can take place. The processes can be shown chemically as

$$(C_6H_{10}O_5)_n + nH_2O \underset{\text{enzymes}}{\overset{\text{heat}}{=}} nC_6H_{12}O_6$$
$$\text{(Starch)} + \text{(water)} \qquad\qquad \text{(dextrose)}$$

High-sugar raw materials, such as sugarcane and sweet sorghum, can be mixed with water and fermented directly, without the preliminary conversion of starch to sugar:

$$C_{12}H_{22}O_{11} + H_2O = 2\ C_6H_{12}O_6$$
$$\text{(Sucrose)} + \text{(water)} = \text{(dextrose)}$$

The fermentation process converts simple sugar to equal parts of ethanol and carbon dioxide:

$$C_6H_{12}O_6 = 2\ C_2H_5OH + 2\ CO_2$$
$$\text{(Dextrose)} = \text{(ethanol)} + \text{(carbon dioxide)}$$

Grinding and cooking grain to release the starch is common to virtually all fermentation processes (see Fig. 14-1). There are two major ways of converting the ground, cooked starch to sugar.

1. Depending on the grain used, part of the whole grain can be sprouted to release enzymes that convert the grain's starch to simple sugars. This is called *malting,* and malt from barley was a major ingredient (perhaps is yet, in some parts of the country) in a moonshiner's brew.

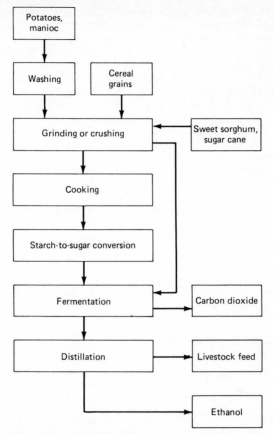

FIG. 14-1 Processing schematic shows the steps required to turn various feedstocks into fuel alcohol by fermentation and distillation.

2. The more common method of converting starch to sugar is to use commercial enzymes, catalytic proteins that are produced by living microorganisms to promote the chemical change in the grain.

The mixture of grain, enzymes (or malt), and water is called *mash.*

There are some standard rules of thumb for computing how much fermentable sugar can be converted from starch, but generally, 1 lb of starch will convert to 1 lb of sugar. That pound of sugar then can be fermented to produce ½ lb of ethanol and ½ lb of carbon dioxide. Corn is about 72 percent starch. The theoretical yield of alcohol from converted starch is 0.568 lb per pound of starch, but in actual practice, yields are generally closer to 90 percent of theoretical. Thus, 100 lb of corn should break down into 33 lb of ethanol, 33 lb of CO_2 and about 34 lb of distillers' grain.

Good fermentation is a basic requirement of successful ethanol production. Those yeasts that convert sugar to ethanol and CO_2 are living plants (although they do not contain chlorophyll) and require a fairly specific environment to grow and multiply. Under good conditions, the yeast population doubles about every 2 hours. These single-celled members of the genus *Saccharomyces* function best at temperatures between 80 and 90°F. At temperatures much below 75°F, the yeast go dormant. At temperatures much above 95°F, they begin to die.

Fermentation yeasts are also sensitive to pH. The pH of the mash before yeast is added should be about 4.0 to 4.5; the activity of the yeast itself will change the pH somewhat. Newer yeasts are being developed that are tolerant of a wider pH range, but fermentation activity is more vigorous in the moderately acid range.

The fermented mash, or *beer* is typically about 90 percent water and solids and only 10 percent ethanol—a mixture that is not very useful as a fuel. The distillation unit is that part of the ethanol apparatus that separates ethanol from the water mixture in fermented beer. There are several ways of doing this, but most small-scale stills utilize *thermal* separation.

The physical laws governing the temperatures at which various liquids vaporize, or boil, allow the ethanol to be vaporized out of the water, then recondensed into liquid as concentrated ethanol. For water at atmospheric pressure, the vaporization temperature, or boiling point, is 212°F. For ethanol, the boiling point is much lower—about 173°F. The technique of distillation heats the beer to the point where ethanol vaporizes and rises as steam, while the water portion stays in liquid form. The alcohol vapor is drawn off, cooled, and condensed back to a liquid for storage and use.

(There's a complication inherent in this method of distilling ethanol, because some water vaporizes at virtually all temperatures above freezing. This prevents distilling ethanol to concentrations higher than about 96 percent, or about 191 proof. We'll get into the why of this shortly.)

The distillate that is boiled off the ethanol-water mixture is fermentation alcohol. Table 14-1 compares the properties of 200-proof ethanol and unleaded gasoline.

PROS AND CONS OF ON-FARM ETHANOL PRODUCTION

It isn't hard to get into an argument on the subject of a farmer producing his own fuel. As with any issue, there are two sides to the question. Depending on which side of the debate a speaker happens to represent, you can hear farm-made alcohol described as either a fool's dream or the best thing to happen to rural America since hybrid corn.

It's wise to weigh both pros and cons before making a decision. On the plus side, ethanol offers these advantages:

1. It's a good, clean-burning fuel for gasoline engines, either in a Gasohol blend or as a complete engine fuel. The major air pollutants from burning gasoline are oxides of nitrogen and lead. With ethanol, nitrous oxide emissions are reduced by about 50 percent, and the fuel contains no lead.

2. On-farm production could provide an assured supply of fuel for farming and transportation needs. With higher prices and uncertain supplies of gasoline, this is a major consideration with many farmers who decide to buy or build ethanol stills.

3. The fuel would create a new market for surplus farm crops, which would help bolster prices to some extent. Also, as technology develops to make ethanol from waste products, this fuel can add new uses for crop residues and other "waste" materials.

4. An on-farm ethanol still can be operated during the slack seasons; the use of spare-time labor to make alcohol would help even out the year-round work load on some farms.

TABLE 14-1   Comparison of Unleaded Gasoline and 200-Proof Ethanol

Property	Gasoline	Ethanol
Chemical formula	C_4 to C_{12}	CH_2CH_3OH
Molecular weight	Varies	46.07
Carbon/hydrogen ratio	5.25:1	4.00:1
Carbon, % by weight	84	52.1
Hydrogen, % by weight	16	13.1
Oxygen, % by weight	0	34.8
Boiling point,* °F	90 to 410	173.3
Freezing point,* °F	−65 to −161	−174.6
Vapor pressure,* psi at 100°F	9 to 12	2.5
Latent heat of vaporization, Btu/lb at 77°F	147.30	395.00
Heating value, 77°F:		
Low value, Btu/lb	19,000	11,550
High value, Btu/lb	20,250	12,780
Low value, Btu/gal	110,960	76,230
High value, Btu/gal	118,250	84,350
Weight per gallon, lb	5.84	6.6
Flash point,† °F	−45	56
Ignition temperature, °F	720	685
Optimum air/fuel ratio	15:1	9:1
Compression ratio (spark-ignited)	8.5:1	15:1
Energy of optimum fuel/air mixture, Btu/ft³	94.8	94.7

*Properties given are at 1 atmosphere.
†Flash point is the temperature at which a combustible liquid will ignite when a flame is introduced.

5. High-protein distillers' by-products can be a source of livestock feed supplement.

6. Several alternative fuels can be used to provide heat for the alcohol-making operation: corn stalks and cobs, timber wastes, sawdust, methane from livestock manures, even a solar collector or the ethanol itself.

7. With a well-designed plant, the surplus heat can be recovered to help heat buildings or water.

8. Ethanol as a fuel for farm vehicles may be a money-saving proposition, compared with buying petroleum-derived fuels. There are several ways in which a fuel-making operation can be made more energy-efficient.

Now, for the gloomy side of the picture:

1. Perhaps as important as any other argument against an on-farm still right now is the fact that no one has yet come up with the ideal design for a plant. There's a lot of talk, and several farmers have installed plants—workable ones. But, in the course of conducting research for this book, the author visited more than a dozen farm ethanol stills and found only two of them working at the time. The rest were in some phase of modification to make them more efficient or easier to operate.

2. Also, not every farmer has the skills to operate a still—at least, not right off. You're breaking new ground and have few successful patterns to copy, few failures to study and avoid. The process of converting grain to useful engine fuel is complicated and

involves skills that are new to even most "jack-of-all-trades" farmers: chemistry, plumbing, microbiology, steam engineering.

3. The cost of an ethanol plant, even if most of the components are built on the farm, makes demands on capital. At this stage of the game, costs for turnkey installations vary all over the map.

4. At best, the economy of producing your own fuel hinges closely on the market price of feedstocks. With a grain-to-alcohol still, it takes something more than ⅓ bu of corn to produce 1 gal of alcohol—which makes the cost of the fuel very sensitive to the market price for grain.

5. No major engineering firms or machinery manufacturers have developed long-term test results on engine wear, deposits, and other possibly harmful effects of using "wet" ethanol in internal-combustion engines. You'd be traveling at your own risk, to a great extent, when using homemade fuel.

6. While the procedures for obtaining licenses and permits have been simplified to some degree, the paperwork connected with producing ethanol is still a formidable record-keeping task. (We'll go into the legal aspects of fuel production in Chap. 20.)

7. Both public and private lending institutions are "loosening up" somewhat on lending money for small-scale distilling plants; however, not every lender is willing to lend money on equipment with no proven performance standard and no guaranteed payback.

8. A big part of making an on-farm still pay is making economical use of the by-products. Generally, this means some way to make profitable use of 17 or 18 lb (dry-matter basis) of distiller "leftovers" that remain after ethanol is distilled out. This material typically is about 90 percent water, which means it must either be used as feed immediately or dried and handled with an expenditure of energy.

SHOULD YOU PRODUCE ETHANOL?

Even after weighing some of the arguments for and against on-farm production of ethanol, you should make a careful study of your own situation before you decide to join the growing number of farmers producing fuel.

Many of us are "tinkerers," devoted to the idea of trying something just to see if it will work. And a confirmed tinkerer often needs no greater motive than that. But for farmers who are serious about producing some or all of their farm-used fuel, the standard practices of any business investment should be brought to bear on the decision. This generally means finding out as much as you can about new processes and equipment, and how they will fit your farm operation, before you start writing checks or borrowing money.

Happily, the fact that the technology of making ethanol fits about any scale can allow an operator to gain some skills with small, inexpensive setups.

As you consider the fuel-making potential of your farm, it might be the better side of wisdom to ask yourself some basic questions:

1. What is the value of having my own fuel source? Do I want to modify engines to operate on a fuel that is not in great supply?

2. Am I willing to devote the time, talent, and energy necessary to become an expert in producing ethanol?

3. If I produce more ethanol than I can use on my farm, what do I do with the surplus?

4. Is this the best use I can make of several thousands of dollars of capital? If the investment

in ethanol equipment limits the amount of capital for other enterprises, which would have the higher rate of return?

5. Where can I get financing?

6. Will I need additional storage for grain or fuel?

7. How will I use the spent stillage left over from ethanol production?

8. What permits, regulations, and licenses are required? How about liability, insurance, and safety measures that should be considered?

At best, the on-farm production of ethanol is a complicated, time-consuming process. It adds another, totally new enterprise to the mix of farm jobs. The successful ethanol producers will be those who apply the same zeal to producing fuel they would apply in trying to grow 175 bu of corn per acre or maintain a weaned-litter average of 10 pigs.

Once enough information has been gathered to answer the above questions, there are decisions yet to make regarding the amount and kind of feedstocks (grains, usually) that can be produced without interrupting the normal farming operation or requiring additional farming equipment. For example, 1 ton of sugar beets can produce about 20 gal of ethanol, but if a farmer is not now growing sugar beets, shifting part of the acreage to this crop would involve investments in specialized equipment to harvest and process the raw material.

You'll need to make accurate estimates of the time and labor involved, too. Managing an ethanol plant—even one that is moderately automated—requires daily attention. How much time during normal farming routines can you devote to running a still? If you must hire additional labor, the economics of producing your own fuel change rapidly.

With preliminary homework done, and the enthusiasm for producing alcohol undampened, the next step is to learn how to make fuel-grade alcohol in the most efficient way possible. We'll cover the basics shortly, but the operation of an ethanol plant is nearly as much art as science, and not all the techniques can be learned from a book—neither this one nor any other being published. A certain amount of "hands-on" experience is necessary to sidestep potential trouble.

Land-grant universities can provide a lot of valuable information on the chemical and microbiological processes involved. More than 40 community colleges and technical schools now are offering short courses and workshops on the safe production of ethanol and its use as motor fuel.

Here's a partial listing, with telephone numbers to call for more information:

Alabama	Talladega College, (205) 362-8800
Alaska	University of Alaska, (907) 479-7631
Arkansas	Mississippi County Community College, (501) 762-1020
California	Modesto Junior College, (209) 526-2000 College of Siskiyous, (916) 938-4463
Colorado	Lamar Community College, (303) 336-2248
Delaware	Delaware Technical and Community College, (302) 678-5416
Florida	Brevard Community College, (305) 632-1111
Georgia	Georgia Tech, (404) 894-2400 Rural Development Center, (404) 542-2154
Idaho	College of Southern Idaho, (208) 733-9554
Illinois	Kankakee Community College, (815) 933-0345 Lakeland Community College, (217) 235-3131 Lincoln Land Community College, (217) 786-2200

Indiana	Vincennes University, (812) 882-3350
Iowa	East Iowa Community College, (319) 242-6841 Iowa Central Community College, (515) 576-3103 Corn Promotion Board, (515) 225-9242
Kansas	Colby Community College, (913) 462-3984
Kentucky	Paducah Community College, (502) 442-6131
Louisiana	Nicholls State University, (504) 446-8111
Maryland	Cecil Community College, (301) 287-6060
Massachusetts	Springfield Technical College, (413) 781-6470 Clark University, (617) 793-7711 University of Lowell, (617) 452-5000
Michigan	Mott Community College, (313) 762-0237
Minnesota	Southwest State University, (800) 533-5333
Mississippi	Northwest Mississippi Junior College, (601) 562-5262
Missouri	State Fair Community College, (816) 826-7100
Nebraska	Southeast Community College, (402) 761-2131
New Mexico	Navajo Community College, (505) 368-5291
New York	Onondaga County College, (315) 469-7741
North Carolina	Pitt Community College, (919) 756-3130 Carteret Technical College, (919) 726-1171
North Dakota	North Dakota State School of Science, (701) 671-2221
Oklahoma	Panhandle State University, (405) 742-2121
Oregon	East Oregon State College, (503) 963-2171
Pennsylvania	Lehigh County Community College, (215) 799-1141
South Dakota	South Dakota State University, (605) 688-4111
Texas	Texas Technical University, (806) 742-2121 Navarro Junior College, (214) 874-6501
Vermont	University of Vermont, (802) 656-2990
Washington	Washington State University, (206) 748-9121
Wyoming	Eastern Wyoming College, (307) 532-7111

FEEDSTOCKS

Ethanol *can* be manufactured from almost any material that contains sugar in one of its many forms—which includes virtually all living plant tissue. As a practical matter, however, the ethanol feedstocks used by any individual farmer will likely be those crops that can be grown, harvested, and stored most handily. From the simple to the complex, those six-carbon compounds that we call sugar take three forms.

Sugar crops, such as sugarcane, sugar beets, and sweet sorghum, plus many fruits and berries, contain *monosaccharide* forms of sugar, such as glucose, which can readily be fermented into alcohol by yeast with no intermediate processing, other than that necessary to make the sugar available to the microorganisms.

Starch crops, such as grains and Irish potatoes, contain sugar units that are tied together in long chains. Yeast cannot use these *disaccharide* forms of sugar until the starch chains are converted into individual six-carbon groups. This conversion process can be done fairly simply by the use of cooking and enzymes or by mild acid treatment.

Cellulose crops, such as corn stalks, wood wastes, and other fibrous plant material, contain chemicals called *polysaccharides*— long chains of sugars that are bonded together and coated with lignins. These chains must be broken down to release the sugar. Breaking the chemical bonds of cellulose is more complicated—and to date, more expensive—than breaking down starch to simple sugar. The conversion typically is done with extreme heat and pressure or with sulfuric acid hydrolysis, although research with special enzymes looks promising.

The choice of raw material to use in making ethanol plays a big role in plant design, cost, and method of conversion. On the surface, it would seem that the best feedstock to use would be a commodity high in simple sugar—sweet sorghum, sugarcane or beets, vegetables such as pumpkins, or fruits such as apples or grapes. This would let you eliminate the mash step of the process, where the starch in grains is converted to sugar.

Unfortunately, the "keepability" of most sugar-type products is limited—at least in the raw state. And it's seldom practical to build a plant with the capacity to distill, say, 100 acres of pumpkins before they spoil.

In addition to the yeast that forms alcohol, a host of damaging microorganisms find sugar products delicious. With any length of storage, some kind of preliminary conversion step is necessary—as by converting sweet sorghum to molasses—to prevent spoilage. In many cases, these processes are more involved and costly than "mashing" grains.

In practical terms, for the short run, the choice of feedstocks for ethanol production will depend on the type of crops grown and will vary from farm to farm and from region to region. In the Pacific Northwest, wheat and potatoes are major crops that can be used. In the northern Great Plains, wheat and sugar beets are potential feedstocks; in southern Great Plains states, corn and grain sorghum. The traditional Corn Belt, of course, produces more corn than any other potential ethanol source. Hawaii and some Gulf Coast regions grow an ideal ethanol crop: sugarcane.

A big consideration of feedstocks for an on-farm ethanol plant is the economics of producing the raw material. If your region, soil type, climate, equipment, and skills are best suited to growing corn, this grain will probably be your first choice.

Table 14-2 gives estimated crop yields, gallons of ethanol per acre, average crop prices for the 15 years from 1963 to 1977 (in terms of 1979 dollars), and an estimated cost per gallon of 190-proof ethanol produced.

Some economists have pointed out that the cost of making fermentation alcohol is very sensitive to the market price of feedstocks being used. An increase of 50 cents per bushel in the market price for corn, for example, adds 18 to 20 cents to the cost of a gallon of alcohol. These are *opportunity costs,* or the price the corn would have brought if sold on the market, rather than converted to fuel.

However, ethanol can be produced from low-quality products that would not command top market price: overheated or drought-damaged grain, overripe or insect-damaged fruit, etc. In some years, it may be possible to sell high-quality grain on the cash market, then buy lower-quality grain at a reduced price to use as ethanol feedstock.

One additional note on feedstocks: Ethanol can be made from a number of food-processing wastes, such as citrus pulp or whey from cheese-making plants. Farmers in close proximity to a processing plant may be able to contract these residues for ethanol feedstocks at lower cost than they can produce grain or other crops. However, in these situations, the cost of handling and hauling the material needs to be computed carefully and included in the ethanol budget.

TABLE 14-2   Estimated Yields and Costs of Ethanol From Various Sources

Feedstock	Typical crop yield	Ethanol produced, gal/ac	Average crop price	Ethanol cost/gal
Corn	91 bu/ac	214	$2.69/bu	$1.14
Grain sorghum (milo)	56 bu/ac	125	$2.40/bu	$1.08
Wheat	31 bu/ac	79	$3.46/bu	$1.36
Rye	25 bu/ac	54	$2.36/bu	$1.07
Barley	44 bu/ac	83	$2.35/bu	$1.24
Oats	56 bu/ac	57	$1.46/bu	$1.43
Rice	44 cwt/ac	175	$11.44/cwt	$2.88
Potatoes, white	261 cwt/ac	299	$4.98/cwt	$4.35
Potatoes, sweet	111 cwt/ac	190	$9.78/cwt	$5.72
Sugar beets	21 tons/ac	412	$31.58/ton	$1.43
Sugarcane	37 tons/ac	555	$23.66/ton	$1.56
Sweet sorghum	44 tons/ac	500	*	$0.44†

*Market prices for sweet sorghum are not established.
†Estimated cost is based on crop production cost of $200 per acre.
SOURCE: USDA and Development Planning Associates.

BASIC ETHANOL PRODUCTION

Essentially, the steps involved in producing ethanol from grain are (1) harvest, preparation, and storage of the feedstocks; (2) cooking and converting starches to sugars; (3) fermenting; (4) distilling; (5) solids separation, storage, and use; (6) ethanol denaturing; and (7) fuel storage and use. See Fig. 14-2. We'll get into more detail on each of these procedures shortly.

Earlier, we mentioned that it is possible to produce only about 191-proof (96 percent) ethanol with conventional distillation equipment and techniques. The theoretical maximum concentration is 194.4 proof—or 97.2 percent ethanol—but the practical limit with most farm stills is closer to 191 proof.

Even a theoretically perfect still cannot produce ethanol with less than about 2.8 percent water. The reason is that ethanol and water form what is called a *minimum-boiling azeotrope.* This mixture boils at 172.6°F (about half a degree lower than the boiling point of pure ethanol). Boiling the azeotrope longer would never concentrate the ethanol further.

However, to be blended with gasoline successfully, ethanol must contain less than 1 percent water. Otherwise, the water and ethanol separate from the gasoline. Producing ethanol that is less than 1 percent water (anhydrous alcohol) is an involved commercial process called *entrainment distillation* with benzene. These processes are beyond the scope of this book, and beyond the capability of farm-scale ethanol plants.

There is encouraging research that alcohol may be "dried" to 1 percent water or less with selective filters, and we'll touch on these developments later. But, with the basic distillation system, you'll have to settle for something less than "pure" ethanol, and here's how that product is made from field corn:

The grain should be milled or ground to a fine meal; about ⅛- or ¹⁄₁₆-in screens in a hammer mill produce a fine enough grind. This reduces particle size so that water, enzymes, and yeast can come in contact with most of the grain. Water is mixed with the cornmeal (usually at about 25 gal/bu of grain), although some operators prefer to add part of the water later, to help raise or lower temperature.

The proper amount of water to mix with the grain will depend in part on the expected

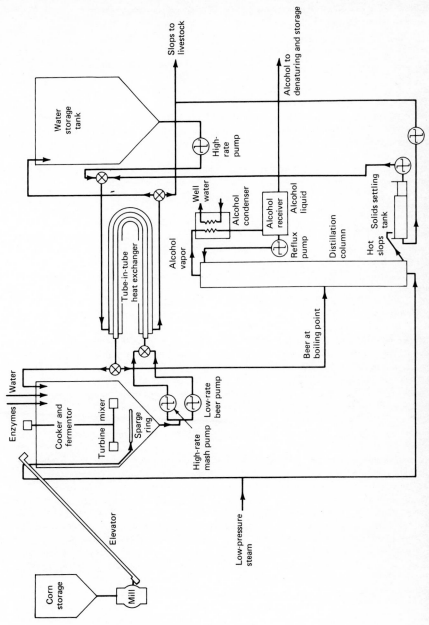

FIG. 14-2 Operation of an experimental ethanol plant at Iowa State University, Ames, Iowa. The plant produces about 10 gph of 180-proof ethanol. *(Iowa State University.)*

14–11

yield of ethanol, and whether or not the mash will be *pumped* from the cooking container into a fermentation vat. Mix the water and ground grain well.

Adjust the pH level of the grain-water mixture—now called mash—to a 6.0 to 7.0 range. (*Note:* Some newer enzymes are tolerant of wider pH ranges, but most work best at the pH range indicated.) Use sodium hydroxide or ammonium hydroxide to make the mash more alkaline, or raise the pH level; add dilute sulfuric acid to make the mash more acid, or lower the pH. Heat the mash to 120°F, and add *alpha amylase* enzymes at about 2 oz of enzyme powder for each 5 bu of grain.

Continue heating the mash, stirring or agitating continuously, to the temperature range recommended by the manufacturer of the particular enzyme used. The alpha amylase enzyme converts the starch in grain to dextrin, in what is called the *liquefying* step. Without this process, as the cell walls of the starch rupture when heated, the mash would turn to a thick, puddinglike consistency rather than a liquid. Different enzymes require different temperatures and different periods of "holding" time to perform their function—usually about an hour, although some work in 20 to 30 min. Generally, the higher the temperature, the shorter the time.

Continue stirring the mash during the cooking and holding time. Then, cool the mash to 85 to 90°F by dilution with cold water or by circulating cold water in a heat exchanger in the cooking tank. In more efficient operations, the fermented mash that is ready to be distilled is used as the cooling medium. Adjust the pH level to 4.5 to 5.0 and add the second enzyme, *glucoamylase.* The glucoamylase enzyme further breaks down the dextrins to glucose, the form of sugar that can be used by fermentation yeast. This is the *saccharification* step.

The temperature of the mash should be held in the range recommended by the enzyme manufacturer. Keep stirring.

Fortunately, this second enzyme can work at the same time the yeast are converting the sugar to ethanol. Most operators add the yeast at the same time as the enzyme, or shortly afterward. This saves time.

With the temperature of the mash at 85 to 90°F, inoculate with live yeast culture, at the rate of 2 to 4 oz of granulated or cake yeast per bushel of grain. Bakers' yeast can be used, but many operators prefer to use special distillers' yeast cultures. Use the freshest yeast possible. The more live organisms in the culture, the faster and more efficient the fermentation. Some ethanol producers use a "seed" from an actively fermenting batch of mash to inoculate the new mash.

How much yeast to add for good fermentation depends on a number of variables, but generally, the more live organisms added, the better. Yeast organisms grow rapidly in the right environment; however, the length of time required to convert sugar to ethanol is greatly dependent on the number and vitality of yeast cells added per quantity of sugar. The concentration of sugar in the mash can affect yeast activity, also. Yeast are killed off by their own waste, in the form of ethanol and carbon dioxide. If the concentration of ethanol in the solution gets too high before all the sugar is consumed, some of the sugar is wasted. The ideal is to use enough yeast inoculation to start vigorous fermentation right away and to keep a healthy yeast population until the solution reaches 8 to 12 percent ethanol by volume.

Fermentation is a critical part of the ethanol process and one that often gives operators of farm-scale stills the most problem. Ethanol content of the fermented mash—called beer—governs the yield of ethanol per bushel of grain or other unit of feedstock.

Fermentation takes 48 to 72 hours, depending on the temperature of the mash, yeast activity, and dilution of the mixture. The liquefied mash must be stirred until the yeast are well mixed. Some directions call for the stirring or agitation to stop at this point, probably so the yeast can work in an anaerobic (without oxygen) environment. However, in sealed fermenting tanks with enclosed stirrers or agitators, mixing can continue through most of the fermentation process.

The activity of the yeast generates heat and can cause the temperature of the mash to rise during fermentation. Some kind of temperature control (water jacket, heat exchanger, etc.) is needed to keep the mash temperature below 95°F.

At the end of fermentation, the beer will contain from 8 to 12 percent ethanol—a specific gravity of 1.04 to 1.07 on a hydrometer. Once the beer reaches 12 percent ethanol, the yeast begin to die off and fermentation stops. Some operators, as noted above, remove 5 percent or so of the beer volume before fermentation ends—as on the second day—and use this to inoculate the next batch of mash. Timing is important, so that live, active yeast are recovered.

When all the sugar possible has been converted to ethanol, the beer is ready for distilling—the process of separating ethanol from the ethanol-water mixture. In some stills, the solids need to be separated from the liquid in the beer, by screening, filtering, or decanting. In other plants, the entire mixture of solids and liquid is passed through the still. The wet grain contains some of the ethanol, so the latter method can recover more of the alcohol in the beer.

Here are typical distilling operations, with two types of alcohol recovery:

Process one requires the separation of grains from the liquid beer. The beer is heated to near its boiling point (about 170°F) and passed through a single rectifying column still. The heated beer is fed into the column near the top, while steam circulates through a heating coil near the bottom of the column.

In this kind of plant, the column either contains a series of perforated sievelike plates or trays or is filled with some kind of packing—often glass marbles. As the still operates, the top of the column is at 173°F—the boiling point of ethanol; the bottom plates will be near 212°F—the boiling point of water.

Vapors leaving the upper end of the column are rich in ethanol. The *thin stillage,* which is mostly water, flows out the bottom of the column. Column size and the flow rates of beer entering the column are critical design factors. The diamater of the column governs the distilling capacity; the height of the column governs the proof, or ethanol percentage. Ethanol vapor is condensed by passing it through a condenser, usually water-cooled.

A well-designed still of this type can produce 160-proof (80 percent) ethanol in one pass, if the equipment is operated properly.

Process two begins with heating the total beer—solids and liquid—in a container, or "pot" still. Boiling continues until the condensate contains very little ethanol. The product of this first stage will test about 100-proof (50 percent) ethanol.

In the second stage of this process, the 100-proof ethanol is passed through a rectifying column, as described in process one, above. The output from this second distillation should be about 190 proof, when operating conditions are correct.

Most modern stills designed to make higher-proof ethanol incorporate both processes in a double column, but the idea is the same: Redistilling concentrates the ethanol percentage.

Federal regulations require that fuel ethanol be *denatured* to make it unfit for human consumption. The "revenuers" want to make sure that none of your production is bootlegged as tax-free booze. Approved denaturing formulas are provided by the Bureau of Alcohol, Tobacco, and Firearms as part of their permit procedure and are discussed further in Chap. 22. Generally, denaturing is accomplished by adding specific amounts of methyl (wood) alcohol, gasoline, or other substances.

DESIGNING THE PLANT

Ethanol stills—both homemade and commercial models—are getting better all the time. However, price, capacity, and relative ease of operation vary all over the board, from about $5000 for a 15-gal batch unit to upward of $100,000 for a continuous-flow plant that turns out 40 or more gallons of ethanol per hour.

Some manufacturers are building good, serviceable equipment; others are turning out rigs destined for an early visit to the scrap-heap. Some plants are marketed by salespeople with sound technical information; others are peddled by "snake-oil" salesmen who will promise anything to make a sale.

The best way to tell which is which is to have a good rudimentary understanding of what an ethanol plant does and how it does it. You might also want to contact the National Alcohol Fuel Still Certification Board, a sort of "Underwriters Laboratories" for the alcohol fuel business, at Box 7013, Overland Park, Kansas, 66207. This is a nonprofit reviewing agency that evaluates the design of stills and other alcohol equipment on the market.

You'll find a listing of people who offer plans, kits, and complete components in App. D. However, the inclusion of firms and individuals on the list in no way guarantees their products or services.

Whether you buy a complete turnkey package or buy components and build your own plant, make sure the equipment is covered with a full warranty. If possible, make only partial payment for the hardware until it is installed and operating properly. Spell out in the warranty or service contract the performance specifications you expect, and who will make repairs or adjustments to the equipment during the shake-down period. A dealer or manufacturer who is selling reliable equipment should not object to being tied down to a warranty.

As with any rapidly changing technology, there's always the hazard that new developments in alcohol plants next month or next year will make even the most efficient of today's stills obsolete. It's the same technology march farmers faced when they went from mules to tractors and from ear-corn pickers to combines, and you have to review the economics as they apply to your particular operation now. Try to build or buy as much flexibility and efficiency into the alcohol plant as possible, but realize that the "best" ethanol still probably has not yet been invented.

Farm-scale stills range from single-pot moonshiner models to 24-in-diameter columns that can produce 40 or more gallons of ethanol per hour. To a great extent, plant design will depend on the volume required, the feedstocks used, the source of heating fuel, and the purity of ethanol needed.

Perhaps equally important for most farmers is the amount of capital that can be invested and the labor that will be available to operate, clean, and maintain the equipment. If you will need 15,000 gal of ethanol each year to replace gasoline now used, that supply could be manufactured in 50 days with a 300 gal/d plant or in 300 days with a 50 gal/d rig. The fixed capital cost would be higher with the former setup; labor costs would be higher with the latter. Most farmers, not having unlimited capital *or* labor, probably will choose a plant size that makes the best compromise between initial capital outlay and operating time.

The volume of ethanol produced and the available labor supply will influence the extent of automation of the equipment, on most farms. The successful operation of an ethanol plant requires rather precise controls over volumes, pressures, and temperatures—all functions that lend themselves to automation, but at a price. Controlling these elements in a distillation plant is somewhat more complicated than controlling them in an automatic clothes washer, although the principles are much the same.

Heat source and comparative costs of heating fuels also will dictate some features of plant design and operation. Steam is the most convenient and flexible vehicle for carrying heat where it is needed in the ethanol process. A boiler can be fired with electricity, natural or LP gas, fuel oil, diesel fuel, wood, crop residues, coal—even the ethanol produced. Also, with steam, the heat source can be located some distance from the still—a decided safety factor, should leaks occur.

Another important consideration in plant size decisions will be the amount of distillery by-product that can be used. The spent grains contain from 24 to 30 percent protein, on a dry-matter basis. However, as the stuff comes out of the still, a calf would need to consume 10 to 12 lb—of mostly water—to get 1 lb of distillers' grains.

The moisture content of the still by-product can be lowered by straining, screening, or centrifuging (spin-drying) to about 70 percent. Assuming a feeding rate of 2 lb per head per day, eight or nine cattle would be needed to consume the spent grains from each 2.5 gal of ethanol produced each day. The distillers' grains can be dried and stored, of course, but this cuts into the energy efficiency of the operation.

The following are calculation steps to help plan a still capacity and operation schedule:

1. Gallons of ethanol needed to replace gasoline used. (On a Btu basis, 1 gal of ethanol contains about 83,000 Btu, compared with 120,000 Btu for 1 gal of gasoline.) To get an accurate estimate of how much 170- to 180-proof ethanol will be needed, multiply gasoline gallons by 1.5: _____.

2. Number of days (or work hours) that can be devoted to ethanol production: _____.

3. Bushels of grain to be fermented to yield needed gallons of ethanol. Under good production conditions, figure 1 bu of corn will yield 2.5 gal of ethanol: _____.

Example:
    2000 gal of gasoline × 1.5 = 3000 gal of ethanol
    3000 ÷ 2.5   = 1200 bu
    100 days of operation   = 12 bu/d

4. The mixing-cooking tank should hold 12 bu of grain, plus 240 gal of water (at a 1:20 ratio), plus about 33 percent reserve: 448 gal, in this case. This allows for a final grain-water ratio of 1:32 with 10 percent head space.

5. Fermentation tanks can be the same size as the mixing-cooking tank, or slightly larger. In fact, the mixing-cooking tank can also serve as a fermentation tank. This eliminates some plumbing and pumping, but if a batch of ethanol is to be run every day, you'd need to be able to mix, cook, and ferment in at least three tanks. If a batch is fermented every 48 hours, two fermenting tanks will be needed for daily production; if a batch is fermented every 72 hours, three tanks would be needed.

6. A beer holding tank will be needed with most operations, to store beer before distillation.

7. The column (rectifying-column still) would need to be sized for the number of gallons to be produced per hour, at the quality desired. On the average, a batch system takes about 4 hours for the distilling process, with each batch of beer. Capacity is a function of column diameter:

Diameter, in	12	15	24
Capacity, gal/h	22	32	90

8. Enough storage tank capacity must be available for the ethanol produced. Federal

regulations require that storage be "secure," which usually means locked storage tanks or tanks inside a locked structure.

9. Heat and water demands for the various processes must be accurately estimated.

10. Estimate the capital investment required.

Example:
6000-gal plant

Cost of materials	$ 20,900
Assembly labor	20,600
Building	20,000
Total	$ 61,500

40,000-gal plant

Cost of materials	$ 48,000
Assembly labor	31,000
Building	25,000
Total	$104,000

Based on an annual interest rate of 10 percent and a 10-year useful life (depreciation period), the capital costs of ethanol produced would be $1.54 per gallon for the 6000-gal still, and about 39 cents per gallon for the 40,000-gal plant.

However, this example assumes that all materials and labor are purchased at current (1980) market price averages and that a new insulated building would be constructed to house the alcohol plant. It's possible—even likely—that most farms would already have a building that could be economically insulated and equipped to house the 6000-gal plant. Some farms may have existing quarters for the larger plant.

And farmers with welding and plumbing skills could save many of the labor dollars charged by doing much of the installation work. On the other hand, the ethanol operation should be charged with capital costs for additional grain-storage and -handling equipment required strictly by the fuel-making enterprise.

Economies of scale apply to alcohol production. However, the cost of hauling raw materials and disposing of distillers' grains often skews the economics in favor of small- and medium-sized operations.

There are nearly as many variations of ethanol plant details as there are individuals building stills. A 22 gal/h design pioneered by Paul Middaugh, a microbiologist at South Dakota State University, has been copied on several farms in the Midwest, often with alterations to fit the individual farm. Middaugh's still features two 12-in-diameter sieve-tray columns, each 16 ft high. See Fig. 14-3.

The first column (the *stripping,* or beer, column) is fitted with 18 steel plates, with 8 percent of the surface area of each plate perforated in a pattern of $7/16$-in holes. Each of these plates has a $1\frac{1}{2}$-in-diameter downcomer pipe attached, which extends from about 1 in above the plate surface, through the plate, and to within 1 in of the next lower plate. These plates are designed to conduct water down the downcomer pipes and to allow ethanol to vaporize and rise to the top of the column.

The fermented mash—beer—is pumped into the column near the top, above the third plate from the top, by a positive-displacement gear pump at 3 lb pressure. Eight to 10 pounds of steam per minute is admitted to the bottom of the column. The solids in the beer and half of the water flow through the downcomers, under the force of gravity. The steam, carrying ethanol vapors, is exhausted out the top of the column and into the bottom of the second column—called the *absorption* or *rectifying* column. At this point, the vapor is half water, half ethanol—or 100-proof ethanol.

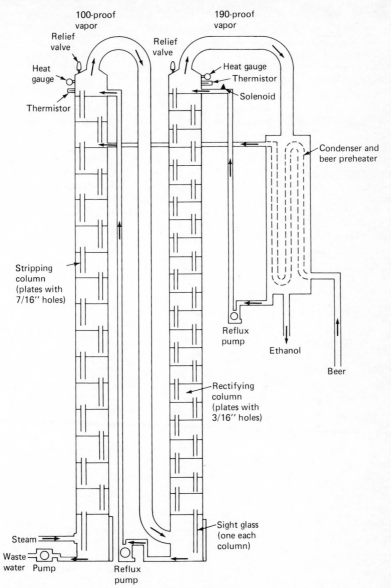

FIG. 14-3 Sectional view of a sieve-plant still, similar to a design developed by Paul Middaugh, microbiologist at South Dakota State University. The two columns are each 16 ft high, 12 in in diameter.

The second column contains 24 steel plates, similar to those in the beer column, except that the holes are smaller—$3/16$ in in diameter. In both columns, the downcomer pipes on the bottommost plates extend to within about 2 in of the bottom of the column. This submerges the lower downcomer below the surface of the liquid that gathers in the bottom of the column and ensures that steam entering the column will rise through the plates, rather than through the downcomer tubes.

As the vapor rises in the rectifying column, it is rapidly condensed and revaporized many times as it comes in contact with the perforated plates. The water, with lower vapor pressure, falls through the downcomer tubes and is pumped back to the top of the first column. (This redistilled water is often referred to as the *reflux*.) The enriched ethanol vapor continues on up the rectifying column and is drawn off the top at 180 to 190 proof.

From the top of the rectifying column, the vapor goes into a condenser (water-cooled heat exchanger) and is cooled enough to liquefy the ethanol. In the Middaugh still, the condensed ethanol is passed through a volume-measuring meter and an automatic denaturing step (where methyl isobutyl ketone and kerosene are added) and is stored in a locked tank.

This type of still (shown in Fig. 14-3) is much more efficient at separating alcohol from water than the old moonshiner-type "pot" still that merely boils mash in a pressure-cooker sort of vessel and recovers the vaporized mixture. There are other column designs that

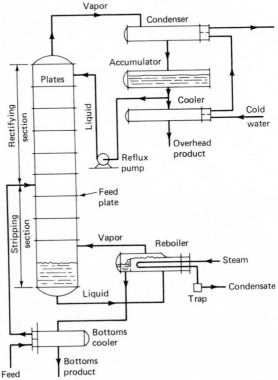

FIG. 14-4   Some plants combine stripping and rectifying functions in a single column, as does this design at Iowa State University.

make the process even more efficient, although a sieve-tray column is probably the best choice if both the liquids and solids in the beer are pumped into the first column. Even then, the holes in the trays will plug up as a gradual coating of protein builds up on the plates. For this reason, it's a good idea to plan access to the plates—such as stacking them on rods so the entire plate assembly can be removed for inspection and cleaning. The option is to pump a caustic cleaning solution through the column periodically.

The rectifying column does not handle the high solids content of the beer column, thus protein buildup is not so much of a problem. A *packed* column offers some advantages over plates. The packing can be ceramic, copper, glass, fiberglass, or other noncorroding material. This packing provides for greater surface area of contact, for the continuous vaporization and condensation inside the still. The downflow of water and the upflow of ethanol vapors take place in much the same way as in a sieve-tray column.

Ethanol is acidic, and hot. It is extremely corrosive to many materials. Generally, copper and stainless steel are the better choices of materials for columns and internal fittings. However, these materials are high in cost, compared with mild steel, and do not adapt as well to fabrication in a typical farm shop. No one seems to know the life expectancy of steel columns and components, but the lower initial cost (compared with using more corrosion-resistant metals) makes steel a popular choice.

In time, the solvency of ethanol will affect rubber hoses and some types of plastic pipe. Where rubber hoses are used, as in the delivery hose from tank to vehicle, the hose should be drained after each use to protect its life. Alcohol also will dissolve some plastic containers. Tanks made of plexiglass or ordinary fiberglass are not suitable for ethanol storage. Some plastic tanks can be used, but make sure you know the composition of the material before you buy a tank.

IMPROVING PLANT EFFICIENCY

Large quantities of raw materials, water, and heat are used in ethanol production. Anything that conserves any of these elements adds to plant efficiency.

There are several ways in which fuel-making operations can be made more energy-efficient:

1. High-moisture grain can be used, eliminating the cost of drying after harvest.

2. Costs of transporting the grain to market and hauling in petroleum fuels can be eliminated when you use farm-grown raw material in the ethanol plant.

3. Poor-quality grain can be used—overheated, cracked, and other low-grade grains that would be severely discounted on the market can still be used to make ethanol, so long as the usual starch content is retained in the grain.

4. Strategic use of insulation (see Fig. 14-5) and heat exchangers to conserve and recycle heat make a plant more energy-efficient. For example, the new beer (on its way to the beer column) often is piped through a heat-exchanger–condenser. The hot ethanol vapors coming off the rectifying column give up heat to the beer, which serves to condense the vapors to a liquid and also preheats the beer before it enters the still.

A variation of this idea is to circulate beer through cooling coils in the cooking-mixing vessel, after the mash has been cooked and must be cooled to about 90°F. The beer picks up heat from the mash, then is routed to the first column of the still.

Depending on the design, the heating and cooling systems of a still may not require fresh water all the time. Water from creeks, ponds, stock tanks, and other sources may be used

FIG. 14-5 Insulation is critical to the thermal efficiency of the alcohol-making process, particularly when—as here—the plant is located in an unheated building.

in heat exchangers, although the feed water for boilers and water mixed with grain should be essentially pure.

One Iowa farmer runs the hot spent beer through pipes in winter, to heat a nearby farrowing house and pig nursery, replacing about $2000 worth of LP gas each season. He also burns corn cobs to fuel the 12 gal/h ethanol plant, and figures 100 acres of corn will produce enough cobs to provide the heat needed to make 14,000 gal of ethanol.

To date, most solar-powered ethanol stills have been disappointing, generally. However, solar collectors could no doubt be used to provide some of the heat needed in the cooking and mashing operations and to preheat beer going into the first column of a conventional still.

The fermentation process is a limiting efficiency factor in many on-farm ethanol operations. Poor fermentation impinges on efficiency in two ways: (1) Conversion efficiency is reduced because less of the available sugar in the feedstock is converted to alcohol and (2) energy efficiency is reduced because the same amount of heat energy is required to produce a lower yield of ethanol. Improved yeast strains and better design of fermentation vessels are helping the situation, but much is yet to be learned about consistently efficient fermentation.

HOW MUCH CAN YOU AUTOMATE?

In some cases, the yield of ethanol per bushel of grain may be less important than the smoothness of the entire operation. More small-scale plants are being automated to some degree, and automatic controls are typically easier to apply to flow systems than to start-and-stop batch operations. (See Fig. 14-6.)

"We have succeeded in completely automating the distilling end of the process, as well as part of the earlier functions," says Duane Wessels, farm builder, Cedar Falls, Iowa. "We have the time requirement down to two hours per day to run 60 gallons of consistent 180-proof alcohol."

Wessels and his son, Larry, are testing the prototype of a plant they will put on the market. Most of the operation of the plant is controlled by a solid-state controller that regulates pressures and temperatures, operates solenoid valves, and monitors production at several points.

"The operator spends about two hours per day, to grind grain for the new batch of mash, supervise the initial cooking and add enzymes and yeast," says Larry Wessels. "Automation is the key to farm ethanol plants. Most farmers don't have the time to spend six or eight hours a day watching a still."

The Wessels' cooker uses 30 bu of ground corn per day, mixed with 25 gal of water per bushel. The mash is cooked with steam from an LP-gas-fired boiler. After the cooking time has elapsed, valves open and pumps start to circulate the hot mash through heat exchangers to preheat new beer, then into one of two fermentation tanks.

The mash ferments for 48 hours. Then, solids and liquid are separated, and the liquid is transferred to the beer tank, where it is preheated by the next batch of hot mash; the

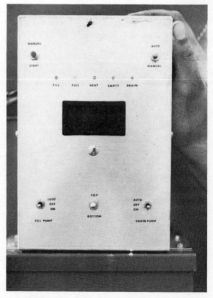

FIG. 14-6 Solid-state controllers perform many alcohol-making functions for Larry Wessels, Cedar Falls, Iowa. *(a)* Wessels monitors the operation of automated controls on his 60 gal/d still. *(b)* A close-up of the still controller. *(Source: Farm Building News.)*

beer is preheated more—to about 150°F—by passing through the cooling coils of the ethanol condenser.

The controller starts a pump to run the 150°F beer into a marble-packed column, and circulates cooling water at the top of the column to hold temperature at a constant 174°F for consistent 180-proof alcohol.

"We've designed this plant as a modular unit, so that any component of the still can be replaced at any time," says Larry. "I'm sure that more efficient cookers, fermenters and other elements will be developed as we go along. As they are, we can up-date these components on any still we have manufactured."

So far, the Wessels are sticking to batch methods for cooking and fermentation. However, other operators are trying different approaches to continuous cooking and fermentation.

A continuous cooker can be smaller than the fermenting tank, and this process adapts well to automated, unattended operation. The material can be cooked more quickly, often with less energy, than with a batch-cooking system. With heat exchangers to heat mixing water while cooling the cooked mash, a boiler can be operated with a more level load.

One problem with continuous-cooking methods is the addition of the proper amount of enzyme at the right time and temperature. However, recent research indicates that the enzymes can be attached to an inert bed or substrate, and the mash passed over them. This way, the enzymes do not go into solution in the mash, but may be reused several times.

For another minor headache, with a continuous-fermentation system, solids must be separated from the liquid in the mash before the yeast is added. As much as 20 percent of the sugars (converted from starch by cooking and enzyme activity) can cling to the solid material. One way to recover much of this sugar is to wash the solids with water that is being pumped into new mash.

Still another potential problem is the contamination or buildup of undesirable microbes in the continuous fermentation system. These microorganisms consume sugar and inhibit the activity of the ethanol-making yeast. In a continuous-flow operation, a considerable amount of mash might be ruined before the contamination was detected.

A new design for fermentation vessels, developed by Emil Wick, food technologist at USDA's Western Regional Research Center, in California, appears promising as a way to avoid some of the problems with continuous fermentation. The tank has a slanted, hopper-type bottom, and new liquid is continuously pumped in at the narrow bottom, while fermented liquid overflows at the shallow side of the top. (See Fig. 14-7.)

The fermentation process is started in the tank, and once it is going actively, fresh liquid is steadily pumped into the vessel. Wick operated the fermenter for 52 days and measured the yeast cells in the tank. Yeast volume was 15 percent of the total volume, much higher than yeast populations in typical batch fermentation. At the end of 52 days, there was no evidence of infecting organisms.

From his tests with the continuous fermenter, Wick concludes that a fermentation rate of one vessel volume each 4 hours can be maintained almost indefinitely.

ON THE DRAWING BOARD

New developments now being tested promise to make the production of ethanol faster, easier, and more efficient. Some of this new technology is fairly adaptable to farm-scale fuel alcohol plants, although with many the economics at this point are uncertain.

Among the more promising is a process developed by Iowa State University researchers that allows the extraction of ethanol from fermented grain mash without using large

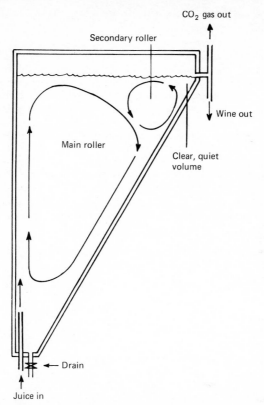

CO$_2$ gas out

Secondary roller

Wine out

Main roller

Clear, quiet
volume

Drain

Juice in

FIG. 14-7 Sectional view of experimental fermenter,
showing flow patterns maintained during continuous
fermentation.

amounts of heat. In effect, this process does away with the need for conventional distillation, and can potentially produce 200-proof alcohol.

The method uses a special silica filter to trap the alcohol fraction of the beer, but let
the water molecules pass on through. The alcohol can then be removed from this "molecular sieve" and the filter reused again and again.

As the beer soaks down through the silica in a column, the filtering agent adsorbs* the
alcohol. One cubic foot of the special filter material will take up 1 to 1.5 gal of alcohol.
If the method proves out as hoped, it could be a real boon to producing alcohol with a
continuous-flow system; it could produce ethanol of a purity (nearly 200-proof) that could
be mixed with gasoline, potentially with much less energy than is required with a conventional distilling process.

Concentrating or "drying" ethanol to 199- or 200-proof (99.5 percent or higher) so that
it can be mixed with gasoline has been a problem with small-scale alcohol production, as

---

*The term is *adsorb,* which means to gather a material on the surface in a condensed layer, rather than
absorb, which means to swallow or drink up, as a sponge absorbs water.

we mentioned earlier. Now, it appears that alcohol can be dehydrated fairly simply on the farm, using a common farm material—cornmeal.

Purdue University scientist Michael Ladisch passes ethanol-water vapor through cornmeal that has been dried to about 2 percent moisture content (see Fig. 14-8).

"We obtain anhydrous [waterless] ethanol with this process, with only 10 percent of the energy required to produce anhydrous ethanol when using conventional multiple distillation with benzene in an intermediate plant," says Ladisch.

Ladisch starts with an ethanol-water mixture at the azeotrope concentration, mentioned earlier. This is where water has been distilled out of ethanol until it reaches 97.2 percent ethanol and 2.8 percent water, at which point further distillation does not concentrate the ethanol.

Ladisch and his colleagues found that fewer than 1000 Btu of heat is needed to redry the slightly dampened cornmeal after it had adsorbed the last 2.8 percent of water from 1 gal of ethanol. Other selective vapor adsorbents probably could perform the same

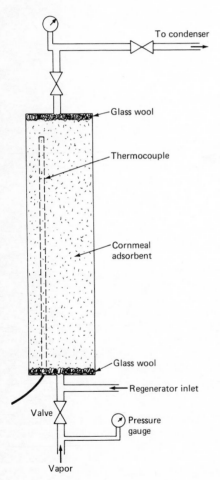

FIG. 14-8 Schematic of Purdue University's cornmeal adsorption unit, used to "dry" ethanol from 190-proof to 200-proof. Temperature measured near the top of the filter shows when water has been adsorbed through the unit. The cornmeal is regenerated by passing nitrogen over it at 120°C.

function, say the Purdue researchers. However, cornmeal was chosen because it is a popular feedstock used to make grain alcohol.

In the two-column distilling plant described earlier in this chapter, the cornmeal filter would be installed at the top of the rectifying column. Some means would be needed to measure the amount of water adsorbed by the cornmeal. Ladisch uses a thermowell to measure the temperature in the top layers of the cornmeal. When the temperature indicates that water has been collected through the adsorption unit, the heat is shut off. In operation, the system would probably use two or more filters, with the ethanol-water vapor diverted to a duplicate unit when one filter had adsorbed its capacity of water.

Presently, producing alcohol from cellulose materials, such as crop residues, is a complex, costly process not very well suited to on-farm fuel plants. But that may change. With discoveries by Donald T. Wicklow, a USDA microbiologist at the Northern Regional Research Center in Peoria, Illinois, farmers may before long start turning straw and cornstalks into ethanol—with only one or two additional microbiological steps involved. (See Fig. 14-9.)

Wicklow has discovered a fungus, found growing on cow manure in a Michigan pasture, that breaks down the lignin bonds that plaster cellulose fibers together in plant tissue. When the lignin covering is removed, the cellulose inside is exposed as long-chain starches that can be converted to sugars by enzymes. In tests, the fungus—*Cyanthus stercoreus,* more commonly called "bird nest" fungus—digested 45 percent of the lignin in wheat straw and exposed 61 percent of the cellulose.

Another technique to break down the lignin in plant material is being tested by Purdue scientists. Called the *Tsao process,* this method involves subjecting crop residues to a series of acid hydrolysis treatments and then fermenting the converted sugars to ethanol.

So far, both the microbiological and acid processes release only part of the potential sugar in such materials as straw, leaves, and stalks. But the relatively low cost of the feedstock makes crop residues potentially more important as a fuel source than the grain. At current costs, cornstalks can be harvested, packaged (as with a big round baler), and hauled up to 30 mi at a cost of about $30 per dry-weight ton. Corn (grain) costs $100 per ton or more.

FIG. 14-9 Chopped wheat straw inoculated with a fungus from aged cow dung is studied by Donald Wicklow, microbiologist at USDA's Northern Regional Research Center.

STORING, HANDLING, AND USING ETHANOL

There isn't much point in making a fuel on the farm unless the product can conveniently be put to useful work. The potential uses of fuel ethanol include powering stationary engines (for irrigation pumps, electricity generators, etc.), drying grain, and firing boilers and furnaces.

But the most practical use, particularly in the short run, is as fuel for self-propelled vehicles. We'll get into the "how" of using ethanol shortly; first, let's look at a few potential problems with handling and storing the stuff.

We mentioned earlier that ethanol is a fairly effective solvent that can deteriorate some materials, such as rubber and many plastics. If part of the storage and handling system is to be made of material that might be damaged by contact with ethanol, it's a good idea to find out about it beforehand. The same caution holds for equipment that will utilize the fuel. Flexible fuel lines and carburetor parts may need to be replaced with more durable materials.

Federal regulations, as mentioned earlier, require that fuel ethanol be denatured to make it unfit for drinking. The Bureau of Alcohol, Tobacco, and Firearms approves several denaturing ingredients and formulas, depending on the proof of the ethanol. The Bureau also requires locked, secure fuel storage.

Also, ethanol over time will absorb moisture from the air, which causes the fuel to lose proof. Storing the product in a pressurized tank may help prevent this, but you may want to get the advice of a registered chemist or petroleum engineer on this point.

Alcohol fuels are inherently no more dangerous to handle and use than their gasoline counterparts—except in one respect. Ethanol and mixtures of alcohol with gasoline have higher vapor pressure than straight gasoline. This can cause pressure to build up in fuel tanks —particularly on tractors with mid-mounted tanks that gain a lot of heat from the engine.

If this happens, fuel can spew out of the tank when the cap is removed, particularly if the cap is unvented or if the vent is plugged. Fuel pouring over a hot engine can be a real fire hazard. The best insurance is to use a well-vented filler cap. Even then, let an ethanol-fired engine cool a few minutes before opening the cap for refueling or to check the fuel level. If any fuel spurts out, immediately retighten the cap, wipe up the spill, and wait until the fuel has evaporated before restarting the engine.

Some characteristics of ethanol, compared with gasoline, must be observed in using the fuel in an internal-combustion engine. The more important of these are vaporization temperature, air-to-fuel ratio, and engine compression.

Gasoline has a mixture of high, medium, and low boiling temperatures. It begins to vaporize at about 85°F, which means easy engine starting, and is completely vaporized at about 400°F, which prevents vapor lock in hot weather. Anhydrous ethanol vaporizes at 173°F.

Gasoline contains little or no oxygen and requires 13 to 15 parts of air (by weight) for each part of gasoline burned. Ethanol contains oxygen and therefore requires less air for complete combustion—about nine parts of air to each part of fuel.

Ethanol has higher octane rating than gasoline, which reduces engine "knock" or preignition. It can be used in engines with higher compression ratios; higher compression results in higher efficiency. The optimum compression ratio for a gasoline engine is about 9:1. With alcohol fuels, the optimum compression ratio is 12:1 to 14:1.

Pure alcohol long has been the fuel for racing engines, because it has higher octane, runs cooler, and delivers more power from a given engine displacement. Also, pure alcohol can be burned with much lower emissions than gasoline and with thermal efficiencies estimated at 25 to 30 percent higher than gasoline. In other words, although alcohol contains fewer Btu's per gallon, each Btu in alcohol performs more efficiently in an engine cylinder than does a Btu in gasoline.

The primary modifications needed to let a standard gasoline engine burn ethanol are to increase the fuel portion of the combustion mixture and to advance the ignition timing by four to six degrees. There are commercial modification kits on the market for both gasoline and diesel engines, but few of them have much engineering data behind them.

Of course, a spark-ignited engine can burn ethanol-gasoline mixtures up to 15 percent 200-proof ethanol, with no engine alteration. With a few minor changes—such as advanced timing, richer fuel mixtures, and possibly some way to preheat the fuel—most gasoline engines operate fairly well on lower-proof ethanol (170 to 190 proof) in lieu of gasoline.

If you're burning 170- to 180-proof ethanol, the main carburetor jet orifices should be about 40 percent oversize. On some carburetors, larger jets can be installed; on others, the existing jets will need to be drilled out. This increases the intake of fuel to maintain a power level somewhat equivalent to gasoline. The carburetor idle jets may need to be opened also, to let the engine idle smoothly when burning alcohol.

Ethanol is a slower-burning fuel than gasoline; combustion in the engine cylinder should start somewhat sooner. Combustion can be started sooner by advancing the ignition timing so that a spark is produced sooner in the cycle. There are no standard timing recommendations for burning ethanol; adjusting the ignition will be a matter of advancing and retarding the distributor until the best position is found. If your engine is adapted for them, you may also want to use "hotter" spark plugs than those recommended for gasoline.

With 200-proof ethanol, there may be some danger of burning exhaust valves in an older engine. However, most engines designed to run on unleaded gasoline have hardened valves and valve seats. Operating an engine on ethanol less than 200 proof may be less hazardous to valves, since the water in the fuel helps "lubricate" the upper cylinder.

Perhaps the handiest way to adapt a road vehicle to ethanol is to install two completely separate fuel systems. This allows the vehicle to be operated on either ethanol or gasoline, whichever is available.

More extensive engine modifications could be made that would increase the performance of ethanol—such as milling the heads to boost compression ratios, altering the valve-cam timing, and installing an electric heating grid between the carburetor and intake manifold to help vaporize the fuel. But these are major—and costly—alterations that not every alcohol burner can justify.

Using alcohol in diesel engines is a somewhat different proposition. Although it can be done, many diesel engine manufacturers are cool to the idea. Basically, there are two approaches: (1) Ethanol and diesel fuel can be blended and used in much the same way Gasohol is used in a spark-ignited engine or (2) ethanol and water can be aspirated into the engine through the air intake, while diesel fuel is injected into the cylinder in the conventional manner.

Blends of 10 percent pure ethanol and 90 percent diesel fuel are being used in diesel engines without altering the engine. Both the diesel fuel and the ethanol must be essentially free of water, however. Any water in the blend will cause the alcohol to separate from the diesel, which could damage the injector pump or the engine itself.

This blend has a viscosity above the minimum for either Number 1 or Number 2 diesel fuel and should provide enough lubrication for the injector pump. However, most injector pumps have composition seals on the low-pressure side and metallic seals on the high-pressure side. The presence of ethanol in the fuel may cause seals to deteriorate more rapidly than normal.

Ethanol has a much lower boiling point than diesel fuel, and this could be a potential problem with ethanol-diesel blends. The ethanol may vaporize (vapor lock) in the fuel-injection system. Pressurizing the fuel tank or cooling the fuel would help prevent this condition.

Injecting ethanol into the airstream of a compression-ignited engine avoids the potential hazards of blending two dissimilar fuels and prevents possible damage to the injector pump or the injectors. At least one company, M&W Gear Company, Gibson City, Illinois, has on the market an attachment to inject ethanol into the air intake.

At this point, it appears that the most efficient engine for burning ethanol would be a sort of hybrid between current diesel and gasoline models. It would incorporate the high compression ratio and perhaps some kind of fuel-injection system akin to diesel engines, but be fired with an ignition system pretty much like that on gasoline engines. Manufacturers of automobiles and farm equipment are studying alternative fuels—including grain alcohol—and how engines perform on them. Some companies, including Ford and GM, are developing ethanol-powered cars for Brazil, where the government is strongly committed to an alcohol fuels program.

But don't expect a totally new tractor engine design next month. Or next year. The engineering problems run somewhat deeper than replacing the injector pump on a compression-fired engine with a distributor, and swapping spark plugs for injector nozzles.

"Even if we already had an engine design tested and ready for production, it would take a minimum of three years just to get the tooling done to shift production lines over," says a spokesman for John Deere, and notes that back when Deere went from three- and four-cylinder diesels to a six-cylinder engine, eight years elapsed from the time the engine design was approved until the first production model was in a dealer's showroom.

No doubt more efficient ethanol-burning engines will be developed in the future. But in the meantime, farmers who are producing their own fuel from grain probably will want to exercise some caution in using the fuel. Check the proof-purity of the ethanol to make sure you know what you're burning.

And, until you have proved to your own satisfaction that your particular fuel can be used successfully in an engine, you may not want to convert every vehicle on the farm to ethanol. It's better to risk a $3000 pickup than a $30,000 tractor. In any case—and with whatever alternative fuel used—check with the manufacturer on equipment still under warranty. Using unapproved fuels in a new tractor or combine can leave you holding the bag for repair bills, whether the fuel caused the breakdown or not.

A WORD ABOUT GASOHOL

The word *Gasohol* has come into such wide usage in the past few years that it is now frequently used to describe fuels that are blends of alcohol and gasoline. Actually, though, Gasohol is a registered trade name for a blend of 90 percent unleaded gasoline and 10 percent ethanol.

To continue borrowing the term, Gasohol is becoming more and more commercially attractive—in large part because of federal and state tax incentives and the steadily climbing prices for petroleum. Currently, the price for unleaded gasoline at the refinery gate in the Midwest averages about 95 cents per gallon. An efficient ethanol plant in the region can produce 200-proof ethanol at a cost of $1.26 per gallon. That puts the wholesale cost of Gasohol at 98 cents per gallon.

However, the federal government exempts 4 cents tax per gallon on Gasohol, which more than makes up the difference in wholesale price between gasoline and Gasohol. To sweeten the pot, several states exempt Gasohol from all or part of the fuel tax per gallon.

So, why aren't more wholesale jobbers handling Gasohol? One big reason is that there isn't enough ethanol to mix with gasoline. In fact, oil companies have been importing ethanol from Brazil to mix with gasoline—at a price close to $2 per gallon.

Highway Oil Company, of Topeka, Kansas, was the first fuel distributor in that state to sell Gasohol. In mid-year, 1980, the firm stopped selling the blend at its 90 service stations because of the poor availability and high cost of ethanol.

With the improved economics of producing Gasohol, it's reasonable to assume that commercial alcohol manufacturers are taking a harder look at the fuel alcohol market. Could farmers get into the act of producing ethanol commercially?

"Drying out" or concentrating the ethanol to 199 or 200 proof on any kind of scale is a higher-capital operation than most individual farmers would be willing to tackle. But farmers are no strangers to the art of cooperation. A central enrichment still could be built —capitalized in much the same way a cooperative grain elevator is financed—and operated to produce 200-proof ethanol from the lower-proof product distilled in the farm alcohol plants.

"Each farmer could sell his alcohol on the basis of proof gallons, and the central distilling plant could contract with wholesale jobbers its production of 200-proof ethanol," says Duane Wessels of Cedar Falls, Iowa. "This kind of arrangement could give grain farmers an optional market for their crops, in the form of lower-proof ethanol, and would keep more of the income produced in the community."

When Wessels talks about *proof gallons,* he refers to the pricing of the product on the basis of the ethanol percentage contained. In other words, the standard market price would be based on 200-proof ethanol, with lower-proof ethanol discounted from this standard. For example, if 200-proof ethanol were priced at $1.25 per gallon, 160-proof ethanol (80 percent ethanol) would sell for $1 per gallon.

Critics of Gasohol point out that even if all fuel for spark-ignited engines were blends of 10 percent ethanol and 90 percent gasoline, we'd only save 1 gal out of every 10 now burned—assuming the consumption stays the same. But that is not necessarily true. Earlier, we talked about ethanol's higher octane rating, and that's a potential petroleum saver that has been largely overlooked.

Gasoline refiners say they use about 10 percent more crude oil to make high-octane than low-octane gasoline. The industry reportedly could produce 75-octane gasoline simply by eliminating the extra processing steps that raise octane rating to 88 or more. Ethanol typically has a research octane number of 107 to 108. If blends were made of 15 percent ethanol and 85 percent low-octane gasoline, the octane rating would be somewhere around 80.

This is not a high enough octane to prevent an engine from "knocking," particularly on acceleration. However, it shouldn't be that difficult to come up with some system— perhaps a fuel-injection method—that would admit ethanol directly into the intake manifold during acceleration. The rest of the time, when the engine is being started, idled, or running under level load, the mixture of ethanol and low-octane gasoline would be satisfactory. This use of ethanol would let renewable fuels replace something like 150 million gal of the gasoline we burn every year.

Still other developments show promise of letting alcohol take over an even bigger share of the energy load now carried by petroleum fuels. There's considerable interest in "ethacoal"—a fuel mixture of ethanol and finely ground bituminous coal. The mixture can be handled much like any liquid fuel and burns with less emissions into the atmosphere than soft coal alone produces. In areas where coal is mined and alcohol crops are grown, this combination may grow into an important fuel—and another potential market for grain alcohol.

Don't forget the value of those distillers' by-products as livestock feed. The 17.5 lb of feed from each bushel of grain from an ethanol plant contains about 28 percent protein, equivalent to the protein in 14 lb of soybean meal. If the distillers' grains replace soybean meal costing $200 per ton, the leftover feed from a bushel of corn used to make alcohol is worth $1.40.

On down the road, there's also a potential market for the other major by-product of ethanol making: carbon dioxide. For each gallon of ethanol produced, about 6.3 lb of CO_2 are generated. At present, there isn't much of a market for small quantities of CO_2. But as more ethanol plants are built, this by-product might become a marketable commodity, with perhaps "milk route"-type pickups.

15

Fuels for Diesels

Until recently, ethanol produced by the fermentation and distillation of farm-grown grains received much of the interest as an alternative fuel for internal-combustion engines. Now, however, more farmers are looking at vegetable oils as a substitute for diesel engine fuel, for a number of reasons.

1. Vegetable oils—from soybeans, sunflowers, peanuts, and other oilseed crops—have an energy content of about 16,700 Btu/lb, which compares favorably with Number 2 diesel fuel at 19,500 Btu/lb. Vegetable oil can be used directly in a conventional compression-ignited (diesel) engine, either as the sole fuel or in blends with diesel fuel, with no modification of the engine and little loss of engine horsepower.

2. Fuel-throttled, compression-fired engines are more efficient under partial load than are air-throttled, spark-ignited engines; most heavy-duty farm equipment today is diesel-powered. U.S. farm machinery manufacturers have gone to diesel engines in most of their tractors, combines, and other self-propelled equipment. In fact, some manufacturers—such as John Deere—no longer build gasoline engines for field tractors. This means that a substitute for diesel fuel potentially has more application on farms than does an alternative for gasoline.

3. Diesel fuel substitutes can be made from a variety of oil-content crops, many of which are especially adapted to double cropping. Sunflowers, for example, are an ideal crop to be grown after winter wheat in many areas of the country.

4. Vegetable oil can be obtained with a simple press-type extractor with a much lower capital investment and smaller energy input than an equivalent production of grain alcohol would require. Also, the operation of oil-pressing equipment does not call for as high a degree of skill in such disciplines as chemistry and microbiology.

5. Vegetable oils first used for cooking can be reclaimed and filtered as diesel engine fuel; for example, oil used for deep-fat frying at restaurants and drive-ins can later be recycled as fuel.

On a weight basis, the calorific (heat) value of vegetable oil is about 14 percent lower than that of diesel fuel. But vegetable oil is heavier than diesel, and on a volume basis, the Btu value is only 10 percent or so below Number 2 diesel: 127,900 Btu/gal for vegetable oil compared with 140,000 Btu/gal for Number 2 diesel fuel. That is much closer than ethanol's calorific value is to gasoline's: 84,000 Btu/gal for ethanol compared with gasoline at 123,000 Btu/gal.

Sunflower oil, the most talked-about diesel substitute, compares favorably with diesel in other ways, and we'll focus most of our attention upon oil from sunflower seed varieties that yield 40 to 42 percent oil by weight; however, the principles apply similarly to soybean oil, peanut oil, castor bean oil, and other products.

As Table 15-1 shows, sunflower oil is within the range of qualities of Number 2 diesel fuel, except for viscosity, or resistance to flow, at both high and low temperatures. This is the source of most of the problems with using vegetable fuel, either alone or in blends of more than about 30 percent in diesel. The more viscous vegetable fuel does not burn completely in an engine at low loads (see Fig. 15-1), which can result in serious lubrication oil contamination and "coking," or the buildup of deposits, on injector nozzles.

At this writing, less research data are available on using vegetable oils in modern diesel engines than on using ethanol in spark-ignited engines. However, these alternative fuels have been used in blends up to 30 percent with diesel fuel, with no apparent harm to the engine. With a 30-70 blend, fuel consumption increases slightly, compared with burning straight diesel fuel (see Fig. 15-2).

J. J. Bruwer, an agricultural engineer with the Republic of South Africa's Department of Agriculture and Fisheries, probably has tested sunflower oils in diesel-powered engines for more hours than any other researchers in the western world.

"Our research and on-farm tests with several makes of diesel engines show that sun-

TABLE 15-1   Comparison of Fuel Properties of Sunflower Oil and Number 2 Diesel Fuel

Fuel property	Sunflower oil	No. 2 diesel
Cetane	37	48
Kinematic viscosity at 20°C	73	6
Kinematic viscosity at 80°C	13	2
Combustion pattern	Large droplets	Small droplets
Engine power	97%	100%
Engine knock	Less	More
Fuel consumption	109%	100%
Lube oil contamination	Serious	Moderate

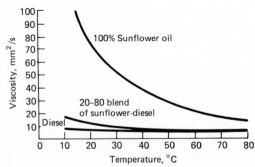

FIG. 15-1 Low viscosity is a diesel fuel property not shared by vegetable oils. Blending vegetable oil in a 20-80 mixture with diesel fuel improves viscosity considerably (lower viscosity numbers indicate better flow).

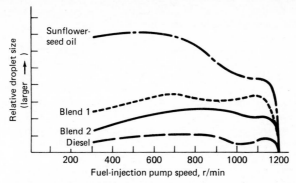

FIG. 15-2 Size of droplets sprayed into the engine cylinder is a reflection of relative viscosity of fuels. Atomization of sunflower seed oil at lower injector pump speeds is poor; in some cases, larger droplet size causes poor combustion and contaminated lubrication oil. Blends 1 and 2 on the graph are blends of sunflower oil with ethyl esters to improve viscosity. *(Source: J. J. Bruwer, Department of Agriculture and Fisheries, Republic of South Africa.)*

flower seed oil, particularly in an ethyl ester mixture, has great potential to extend diesel fuel supplies," says Bruwer. "But there are some practical problems that must be solved before pure vegetable oil can be used successfully."

Bruwer has compared straight sunflower seed oil with straight diesel fuel, as well as blends of sunflower oil with gasoline, diesel, kerosene, and various paraffins. The sunflower oil blends readily with all these materials, at all proportions. Tractor engines started and ran satisfactorily, without any engine modification, and delivered almost full power on all fuels, including 100 percent sunflower oil.

At the start of his trials, Bruwer operated a Ford 5000 tractor for 100 hours, at full power, on 100 percent sunflower oil, with no noticeable adverse results. The same tractor was operated an additional 1004 hours in a farm test, on a blend of 20 percent sunflower oil and 80 percent diesel fuel.

"At the end of this time, the engine delivered eight percent less power than when we started the test," Bruwer reports. "But that is not an abnormal power loss for an engine with that many hours of operation. We replaced the injector nozzles and recalibrated the injector pump, and reduced the power drop to only four percent."

So, what are those "practical problems" Bruwer mentions?

There are two major ones. After testing the tractor for a total of 1104 hours and then replacing injectors, Bruwer coupled the machine to a power-takeoff dynamometer and operated it for 24 hours a day at 70 percent full power on the 20-80 blend of sunflower oil and diesel fuel, for another 278 hours.

"Toward the end of this partial-load trial, we noticed that exhaust smoke increased considerably," he says. "The reason was the injector nozzles had started to carbon, or coke, around the orifices. However, we had similar coking problems with another tractor that was operated on straight diesel fuel at the same time, under the same 70 percent load conditions for the same total hours."

The biggest hazard was found in the crankcase, when the Ford engine was taken apart.

The incomplete combustion of sunflower oil, with the engine run at partial load, had caused severe contamination of the lubricating oil.

"In practical farm operations, it is impossible to avoid part-load conditions on an engine," Bruwer observes. "In extreme cases, the contamination of lubricating oil could result in solidified oil throughout the engine and serious engine damage."

Here is roughly what happens in the engine cylinder: The more viscous (thicker) vegetable oil does not break up into as small droplets as does diesel fuel. Under part-load conditions, unburned oil adheres to the injector nozzles and builds up, or "cokes," as carbon deposits.

As the nozzle becomes clogged, still more incomplete combustion occurs, which allows some of the fuel to enter the lubricating oil. Bruwer recommends using a high-quality lube oil, and probably a shorter oil-change interval, when using sunflower oil or a sunflower-oil–diesel-fuel blend.

Bruwer now is working on ways to modify the chemical properties of sunflower oil so that the fuel will correspond more closely to diesel viscosity. Most promising results to date have been mixing ethyl esters with the oil. At this point, the South African hesitates to recommend pure sunflower oil as a substitute for diesel fuel. Pure vegetable oils are too heavy and viscous for the fuel systems on most diesel engine models.

However, up to 25 percent sunflower oil and 75 percent diesel fuel can be blended to extend short diesel supplies, and this blend causes few problems in most engines. Happily, vegetable oils blend well with diesel fuel at all percentages (see Fig. 15-3).

Even if we solve some of the viscosity problems, a 50-50 blend will probably be about the maximum percentage of vegetable oil in fuels for current engine designs. Diesel engines start and operate on all mixtures, as mentioned earlier—even on 100 percent vegetable oil. But at blends above 50-50 vegetable oil and diesel, horsepower output drops off.

FIG. 15-3 Performance of a diesel engine on diesel fuel, sunflower oil, soybean oil, and blends of the three fuels is demonstrated by agricultural engineers at the University of Missouri. The engine readily starts and runs on all the fuels and blends of all percentages.

EXTRACTING VEGETABLE OILS

In trials at North Dakota State University, Vern Hofman, agricultural engineer, initially used vegetable oil that had been extracted by a commercial refinery. Later, he pressed oil from sunflower seeds in a small screw-press expeller, filtered the oil through a paper filter, and used it in the engine.

"In fact, farm-pressed oil performs better than commercial food-grade oils," Hofman notes. "Most commercial extraction processes use solvents to extract most of the oil from the seed. The screw-press mechanical expeller removes only 75 to 80 percent of the potential oil, so that thicker, gummier oil is left in the seed meal or cake."

While Hofman has worked primarily with sunflower oil, much the same is true of other vegetable oils. For burning in an engine, the oil need not be refined to the same purity as it does for human consumption.

This is encouraging news for farmers who want to try these fuels. Refined, food-grade vegetable oil costs around $2 per gallon, even on a wholesale basis, a price not very competitive with diesel fuel. Farm-produced oil can be made for considerably less than that, particularly if the oilseed crop is grown as a second crop after a main crop of wheat or other small grain.

Farmers with access to "used" vegetable oils, from restaurants or food processors, may be able to line up an economical source of recycled vegetable oil. But many farms do not have this kind of source nearby enough to be practical.

One drawback to making your own diesel fuel substitute from farm-grown oilseed crops is the lack of extracting equipment. There are a few large-scale mechanical extractors made in the United States, but not many small, farm-scale machines are being built here. Smaller, less expensive models are made in Great Britain and Japan and are being imported by farmers and researchers. These machines, powered by an electric motor, literally squeeze the oil out of the seed by mechanical force.

Vern Hofman uses the Japanese-built extractor. It uses about 4000 Btu of energy, in the form of electricity, to produce 1 gal of oil. About 20 percent of the oil is left in the residue, or meal. The extractor presses oil from about 100 lb of seed per hour, which gives it a capacity of about 40 gal of oil for each 10-hour day. The equipment costs about $5000, f.o.b. west coast ports, and is built by Chuo Boeki Goshi Kaisha, Box 8, Ibaraki City, Osaka, Japan.

A similar screw-press expeller is manufactured in Britain, and costs about $7000 f.o.b. east coast ports. This machine is sold by Simon-Rosedowns, Ltd., Cannon Street, Hull, England, HU2 OAD.

Oil produced by both of these machines must be filtered to get rid of bits of hull and other debris before being used as engine fuel. Hofman uses a paper filter that traps particles 4 micrometers and larger in size. Hofman has these comments and recommendations concerning operating a screw-press expeller:

1. Clean the seed first, because any stones or pieces of metal in the seed will damage the equipment. Oil extraction is more efficient if seeds are cracked or broken in a roller mill before being run through the expeller.

2. For higher-protein meal (a livestock feed), remove seed hulls before pressing the oil out. This also reduces the amount of fibrous material that soaks up oil, but dehulling sunflower seed is not always a simple matter.

3. Preheating the seed before pressing lets more of the oil be recovered but increases the energy required to produce the fuel substitute.

4. Even if vegetable oils are to be blended with diesel fuel, some may need to be "de-

gummed" to rid them of paraffins and waxes, to help reduce some of the potential problems associated with fuel filters and injectors. The simplest way to do this is to pass the oil through a water bath to solidify the gum, which can be strained out, then boil off the water by heating the oil.

WHAT WILL IT COST?

Good economic information has not yet been developed on producing alternative diesel fuel from oilseed crops.

As with any capital equipment, there are economies of scale in operating a screw-press extractor. Generally, the more oil you make, the less capital cost per gallon. The cost of the raw material can vary from year to year, and from region to region, of course. Hofman estimates that the cost per acre of growing sunflowers in North Dakota (the major U.S. sunflower-producing state) at $148.40, with $71.18 of that being direct or cash costs.

If the expected seed yield is 1500 lb/ac, a farmer would need a market price of $9.89 per hundredweight (cwt) to cover all costs, as indicated in Table 15-2.

Farther south, where sunflowers can be double-cropped behind wheat, the per-acre charge to sunflowers may be nearer $130 total, when only a share of the capital investment is charged. If the sunflowers yield 1600 lb of seed per acre, and 80 percent of the oil is extracted, that would be the equivalent of about 67 gal of diesel fuel per acre.

That alone is not particularly attractive from an economic standpoint. It puts the price of diesel fuel substitute at nearly $2 per gallon. However, it does not take into account the feeding value of the meal or cake that is left after oil is pressed from sunflower seed.

This residue is a high-protein (22 to 24 percent) feed that can replace soybean oil meal or other protein supplements in many livestock rations. At current feed prices, this meal is worth about $150 per ton in swine rations and $180 per ton in cattle rations.

(*Note:* Sunflower meal is low in some essential amino acids, such as lysine, which must be supplemented in swine rations, thus the lower value when fed to hogs.)

For meal made from dehulled sunflower seed, the protein content may be from 40 to 42 percent. This product could replace soybean meal in cattle rations virtually pound for pound.

At 1600 lb of seed per acre, the feeding value of the leftovers from oil production is worth $90 to $95. Estimating production costs at $135 per acre and a yield of 67 gal of oil, this reduces the cost of the fuel to about 85 cents per gallon, plus the cost of extracting and filtering the oil. The "opportunity cost" of the fuel, naturally, depends also on the market price for the crop (see Table 15-3).

For easier figuring, a screw-press type expeller will remove about 80 percent of the available oil in sunflower seeds. That means each 100 lb of seed should yield 32 lb (4.1 gal) of oil and about 68 lb of sunflower oil meal. By plugging in the current value of the meal as a protein supplement (when compared with, say, soybean meal) and by closely

TABLE 15-2   Effect of Market Value of Raw Sunflower Seed on Cost of Sunflower Oil

	Market value per cwt of sunflower seed (less oil-processing costs)				
	$7.00	$8.00	$9.00	$10.00	$11.00
Value of oil from cwt seed	$3.60	$4.60	$5.60	$6.60	$7.60
Oil cost per gallon	0.88	1.12	1.36	1.61	1.85
Oil cost per pound	0.11	0.14	0.17	0.21	0.24

TABLE 15-3   Effect of Sunflower Oil Price on Value of Sunflower Crop

Sunflower seed yield, lb/ac	800	1200	1600	2000
Oil yield, gal/ac	32.2	49.9	66.5	83.1
When price of oil is:	Combined value of meal and oil in 100 lb sunflower seed is:			
18¢ per lb	$11.52			
22¢ per lb	12.80			
26¢ per lb	14.08			
30¢ per lb	15.36			

estimating the costs of extracting the oil (including the capital costs of the extracting equipment), the value of the oil as a diesel fuel substitute or "extender" can be accurately estimated.

Even if vegetable oil proves out as many researchers and farmers hope, it's doubtful that the fuel will become a major alternative to diesel in industrial engines and over-the-road trucks. Vehicles and equipment in the United States used 52 billion gal of distillate fuels in 1979—some 373 billion lb. Vegetable oil production that year was about 20 billion lb, about 5 percent of the total used, and virtually all of it was in demand for food uses.

Still, for individual farmers who want the security of growing their own diesel fuel, vegetable oils look promising. On a typical crop farm, 10 percent of the acres should provide enough oilseed to supply virtually all the diesel engine fuel needed for the entire farming operation. In the case of sunflowers and soybeans, the land cost of producing the fuel crop by this "tithing" principle could be shared by a preceding crop of winter wheat or other grains.

HOW ABOUT ALCOHOL IN DIESELS?

Mixing ethanol (grain alcohol) with diesel fuel—as can readily be done with vegetable oils —is an idea whose time has not quite come. Not completely. Certainly, the prospect of "diesahol" as a 10 percent alcohol, 90 percent diesel fuel has thornier problems than its Gasohol counterpart used in spark-ignited engines.

For one thing, alcohol is a solvent whereas diesel fuel is a lubricant. Injector pumps and other fuel components on a diesel engine require constant lubrication. Because alcohol is such an effective solvent, some of the lubricating film is washed away, hastening wear.

Another problem—a tough one to solve thus far—is that as little as 0.5 percent of water (in either the ethanol or the diesel fuel) can cause the two dissimilar fuels to separate. Even if pure (200-proof) ethanol is mixed with diesel fuel that is free of water, it is difficult to prevent the mixture from absorbing moisture from the atmosphere.

University of Illinois agricultural engineers have tested a blend of 10 percent ethanol and 90 percent diesel in a three-cylinder tractor engine (see Fig. 15-4). The engineers note that the blend has fuel properties of viscosity and cetane number well within the standards for diesel fuel. The engine started—and ran, for awhile—on the fuel. Then it "vapor-locked" when engine temperature exceeded the boiling point of ethanol, which caused the injector pump to lose suction.

This particular problem may be overcome either by chilling the fuel or by pressurizing it before it enters the injector pump.

But the main bugaboo—water in the mixture—has yet to be solved in practical terms. The Illinois engineers are trying different ways to overcome the bad effects of water. One method under study is to use surfactants to keep the water-alcohol element from separating

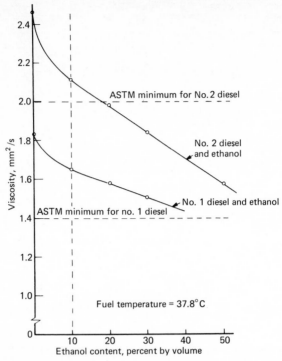

FIG. 15-4 The effect of ethanol concentration on fuel viscosity, with blends of 10 percent ethanol and Number 1 and Number 2 diesel fuel. *(University of Illinois.)*

out of the diesel fuel. However, the cost of these emulsifying agents hampers this approach. Right now, it takes about as much surfactant as ethanol in the fuel blend, which boosts the cost of diesahol by 50 percent or more.

Another approach is to use separate fuel tanks for the alcohol and diesel, then mix the two fuels in-line just before the fuel is burned in the engine. A variation of this idea is to inject the diesel fuel in a normal manner, but to atomize the ethanol into the airstream of the engine. This latter approach is a more immediately available way to go. M&W Gear Company, Gibson City, Illinois, has come up with a system to inject a 50-50 blend of ethanol and water (or 100-proof alcohol) into the intake airstream of a turbocharged engine. M&W calls this system *Aquahol* injection and claims the mixture boosts horsepower, cuts fuel consumption, lowers intake and exhaust temperatures, and promotes cleaner combustion. The company builds kits to convert most popular diesel tractor engines to utilize Aquahol.

With these kits, the ethanol-water mixture is aspirated into the engine only when the tractor is under load. As the mixture enters the airstream, it cools the air by evaporation to let the air carry more oxygen into the engine cylinder. This, according to engineers at M&W Gear Company, produces more power, longer on the piston's power stroke, than does diesel fuel alone. It also lets the engine run cooler. The Aquahol conversion kit can be installed on most tractor models in 5 hours or less and costs about $1000.

However, some engineers are lukewarm to the idea of aspirating alcohol into a turbo-charger, on the grounds that it could damage the charger or interfere with the engine speed governor's performance. Some farm equipment manufacturers are downright cold to the idea of such kits being installed on their tractors, and at least one manufacturer will not warranty a diesel engine into which alcohol is introduced in any manner.

So, as you size up your alternative fuel options—vegetable oils, diesahol or Aquahol —check out the possible hazards as well as the benefits. You may want to go slow about pouring some unknown fuel into the tank of a valuable tractor—at least until more data is in on these promising alternatives to diesel fuel.

16

Methane: Cow Power to Burn

If you're a livestock or poultry producer, you may have walking gas wells on the farm. A growing number of farmers are installing methane digesters to turn manure into burnable gas and stabilized, nitrogen-rich fertilizer.

Methane technology adapts to any size operation, from a digester made from a couple of used oil drums on a very small farm up to a 1.6 million ft³/d plant in the Oklahoma Panhandle. The process itself is simple and requires less equipment than most grain alcohol production plants.

Actually, methane can be produced by anaerobic (without oxygen) fermentation of a great many biomass materials (often called *substrate*)—almost any carbon and nitrogen-containing material that can be broken down by bacterial action to produce *biogas*. Those same kinds of materials, plus a few others such as coal dust and shredded rubber, can be converted to gas under heat in the absence of oxygen, in a process called *pyrolysis*. In practice to date, however, most methane plants use either animal manures or plant materials as raw resources to produce gas.

Despite renewed latter-day interest in methane, the resource is not a new one. In fact, nature has been making methane for eons: Natural gas is virtually pure methane. Organic materials, rotted under conditions out of contact with oxygen, will produce a burnable gas. Marsh gas, produced naturally and ignited by a stray spark or bolt of lightning, is the "fool's fire" or "Will-o'-the-wisp" of legend and literature. As early as 1895, gas from a specially designed septic tank lighted the streets of Exeter, England. India, Korea, and the People's Republic of China have promoted the production and use of methane for years in rural regions, where energy production and distribution are poor.

In this country, huge methane plants have been built adjacent to cattle feedlots in Oklahoma, Colorado, California, and Florida, where the biogas is used as an efficient substitute for natural gas. A plant near Guymon, Oklahoma, operates two 1.2 million ft³/d digesters, pumping the gas into a natural gas pipeline to heat homes and water as far away as Chicago.

In this energy-conscious age, the most immediate and obvious benefit from methane production is the energy value of the gas itself, of course. (See Fig. 16-1.) But for farmers, there is a collateral value in the benefits of waste control, fertilizer, even "recycled" livestock feed, in some cases. We'll talk more about using the by-products of homemade gas later.

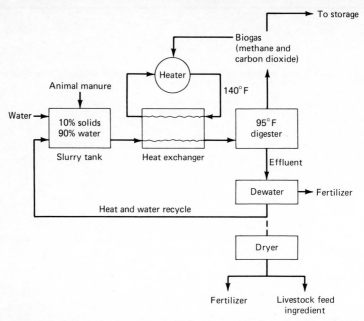

FIG. 16-1   Schematic of an energy-efficient anaerobic digester. Gas produced is used to heat new slurry going into the digester; heat and water are recovered from spent effluent. Optional processing of spent materials can produce fertilizer and cattle feed ingredients.

WHAT IS METHANE?

Pure methane is a colorless, odorless gas that is the principal element in natural gas and has a heat value of about 1000 Btu/ft³. (Natural gas commonly is measured by the *therm,* which is equal to 100,000 ft³.) It heats homes and water, pumps irrigation water, dries crops, performs many industrial processes (among them the manufacture of nitrogen fertilizers), and even powers vehicles for some Americans.

However, the gas produced by anaerobic digestion is only 50 to 70 percent methane (CH_4) and contains 25 to 40 percent carbon dioxide (CO_2), along with small amounts of other gases, including hydrogen sulfide (H_2S). The hydrogen sulfide is the element that imparts a slight "rotten eggs" smell to digester gas, the intensity of the odor a direct effect of the H_2S content of the gas mixture.

Because of the impurities contained in digester gas, many scientists and chemists prefer the term *biogas* to *methane.* As a result of the CO_2 and other elements in digester-produced gas, the material has a heat value of about 600 to 700 Btu/ft³, depending almost directly on the methane fraction of the mixture.

Pure methane has a specific gravity of 0.55 in relation to air, or is about half the weight of air and rises when released to the atmosphere. However, the carbon dioxide weighs twice as much as air, and the percentage of CO_2 in the gas helps govern the weight of the mixture in relation to air.

Generally, digester gas can be used directly in gas-burning appliances for cooking, heating, water heating, etc. Because the product from a digester contains fewer Btu's per volume than natural gas, some adjustments to burners and other equipment often is needed.

PRODUCING GAS BY ANAEROBIC DIGESTION

In the anaerobic digestion process, organic solids are converted to methane and carbon dioxide by specific fermentation bacteria. There are some parallels in the process to the fermentation of grain to make alcohol. Methane bacteria, while differing from the enzymes and yeasts that digest sugars to form alcohol and CO_2, perform much the same functions. The basic gas-forming reaction in a digester converts carbon plus water ($2C + 2H_2O$) to methane plus carbon dioxide ($CH_4 + CO_2$).

The digestion process occurs in two stages. Both of these may take place in the same container or in separate compartments, depending on the design of the methane plant. For examples in this chapter, we'll consider methane made from animal manures, which contain a portion of *volatile* solids—fats, carbohydrates, proteins, and other nutrients— that are food sources for the specialized bacteria that produce methane.

In the process, a given amount of gas is produced by each pound of volatile solids broken down or degraded by the bacteria. This is the gas *yield* of a material, and average values are given in Table 16-1.

Assuming an efficient digester, Table 16-1 shows approximate cubic feet of methane produced per animal per day, and the energy production per animal per *hour*—in terms of Btu's of available energy. The digestible portion of total volatile solids is 75 to 85 percent. This is the figure to be used when estimating the gas yield from a given number of animals of a certain weight. For example, a 150-lb pig will excrete about 7 lb of manure each day. The manure is 80 percent water. On a dry-weight basis, the manure contains about 1.4 lb of total solids. However, only 85 percent of that is volatile or potentially digestible solids.

To complicate things more, an efficient digester will convert only half or so of the volatile solids to methane. From the 0.6 lb of *digested* volatile solids, the methane bacteria can produce about 8 ft³ of gas per day. (Generally, 1 lb of volatile solids will produce about 12 ft³ of biogas.)

In most methane digesters, raw manure is adjusted to about 10 percent total solids by adding or draining off water, before this slurry is admitted to the digester. Obviously, the production of methane by anaerobic digestion depends greatly on the kind and amount of raw material used. The more biodegradable (digestible) the organic matter, the more methane that can be produced from it. Table 16-2 gives the estimated daily production of manure, total solids, volatile solids, and potential biogas from various kinds and weights of animals.

Initially, volatile solids in manure are broken down by bacteria to a series of fatty acids, most notably acetic acid. This first stage of the digestion process can be likened to the liquefaction and saccharification processes performed by enzymes in alcohol production.

TABLE 16-1 Average Methane Yield and Energy Production of Various Animals

Animal	Methane yield, ft³/d	Energy production, Btu/h
Cow, dairy (1200 lb)	23	568
Steer, beef (1000 lb)	31	775
Swine (150 lb)	4	103
Poultry (4 lb)	0.21	5.25

TABLE 16-2   Estimated Daily Manure and Biogas Production of Various Animals

Animal	Manure, lb	Total solids, lb	Volatile solids, lb	Biogas, ft^3
Cattle:				
Bull (1600–1800 lb)	88	17.6	14	168
Cow (1200–1300 lb)	67	13.5	11	132
Steer (900–1000 lb)	52	10.0	8	96
Calf (500–600 lb)	27	6.0	5	60
Swine:				
Boar, Sow (300 lb)	16	3.2	2.7	32
Pig (160–200 lb)	9	1.8	1.5	18
Pig (40–50 lb)	2	0.4	0.4	5
Horses:				
Heavy (1100–1200 lb)	42	8.4	6.7	80
Medium (800–900 lb)	32	6.4	5.1	61
Light (pony)	22	4.4	3.5	38
Sheep:				
Ewe, ram (70–80 lb)	4	0.8	0.6	7
Lamb (30–40 lb)	2	0.4	0.3	3
Poultry:				
Turkey (15–17 lb)	0.5	0.1	0.08	0.9
Goose (14–16 lb)	0.5	0.1	0.07	0.8
Broiler, layer (4 lb)	0.1	0.08	0.06	0.6

These first-stage organisms prepare the organic material for the next set of bacteria, which do the actual converting to methane.

In the second stage, other specialized bacteria—the *methane formers*—digest the fatty acids to produce methane, carbon dioxide, and other minor gases. Methane-forming bacteria are more sensitive to temperature, relative acidity, and minerals in the organic matter to be digested than are acid-forming bacteria. These organisms can tolerate no oxygen and function best at about 95°F, in a narrow pH range of 7.0 to 7.5, slightly on the alkaline side of neutral. They are also sensitive to materials in some livestock rations that may show up in the manure: feed ingredients such as salts, heavy metals, and antibiotics.

As with alcohol fermentation, anaerobic digestion is a *biological* process. Living microorganisms consume digestible raw material and give off waste products in the form of methane and carbon dioxide. Management of an anaerobic digester involves maintaining a critical balance between the two types of bacteria—the acid formers and the methane formers—in the digestion vessel. Of the two, methane formers are slower growing, and much more fastidious about their environment.

When organic materials are digested, only part of the material is converted to methane and other gases. The rest is indigestible by the bacteria and either collects in the digester or is piped out with the spent sludge. For example, 100 lb of fresh poultry manure contains 75 to 80 lb of water and 20 to 25 lb of total solids. Of these, only 15 to 22 lb are available for digestion. These volatile solids are the potential fraction of that 100 lb of manur~ that could be converted to biogas.

Even in a digester that is 100 percent efficient, not all the volatile solids are digested. The material that is left in the digester either forms a scum on top of the liquid or settles to the bottom in a layer of muddy sludge—depending on whether the undigested material is lighter or heavier than water.

As the bacteria go about their business, they consume both carbon (from the carbohydrates) and nitrogen (from the protein) in the raw material. However, they consume carbon much faster than nitrogen, so the ratio of carbon to nitrogen bears heavily on efficient digestion. If there is too much carbon in relation to nitrogen, the nitrogen will be consumed first, leaving some of the carbon (methane potential) to be wasted. If there is relatively more nitrogen, the carbon is used up first and digestion stops, as the bacteria in effect starve. A ratio of about 30 to 40 parts of digestible carbon to each part of digestible nitrogen appears to be the optimum balance. Happily, most animal manures fall into the acceptable carbon-nitrogen ratio for efficient methane production.

In a well-designed, well-operated digester, the conversion of biomass materials to methane should be efficient. On the average, about 40 ft^3 of methane, with a heating value of about 25,000 Btu, can be produced daily from each 1000 lb of animal. Some of the energy produced in most farm-scale digesters typically is used to heat the plant itself. But even if 40 percent of the gas is burned to provide the heat needed in a digester, a biogas plant is more energy-efficient than some farm-fuel options.

DIGESTER DESIGN

Designing a biogas digester for a specific livestock operation is beyond the scope of this book. Essentially, a methane digester is a big fermentation vat with inlets for new raw materials and outlets for the biogas and digested slurry (see Fig. 16-2). Efficiency, convenience, and the uses to which the gas is to be applied usually dictate other components and equipment, of course.

There are several ways to approach digester size and design. How much gas is needed, and for what purposes? How much capital can be invested in equipment? However, as a methane digester is a clean, efficient method of handling a potential problem (livestock or poultry wastes), most farmers likely will approach design from the standpoint of how much manure is to be gotten rid of, from what kinds and sizes of animals.

Two chief considerations govern the designed size or volume of an anaerobic digester: How much total solids must be treated daily, and how long must the material remain in the digester? Other design factors will influence the choice of digester type and the amount of supporting equipment that is needed.

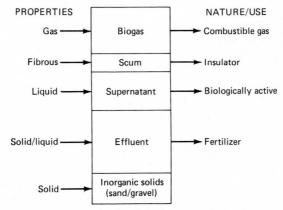

FIG. 16-2 Materials in a biogas digester tend to stratify into layers, unless the contents are agitated regularly.

1. How handily can the raw materials be gathered and processed?
2. What is the average particle size of the raw material?
3. What are the heating requirements and heat source?
4. What kind of mixing equipment will be needed?
5. What construction materials are available at least cost?

All these questions influence the amount of gas produced and the methane content of the gas, as well. Materials that degrade easily will be digested faster and will stabilize more quickly, once in the digester. Since the size of a methane digester is determined by the volume of raw materials to be loaded daily, multiplied by the detention time, any feature that speeds up the digestion process has the effect of reducing digester size. Particle size influences digestion rate; production of the same amount of gas can take place faster and in a smaller digester when easily digested materials are fed as small particles. Also, smaller particles flow more smoothly into and out of the digester, with less sediment and scum produced, and fewer clogged inlet and outlet pipes.

For efficient gas production in most regions, some supplemental heat is needed to maintain a constant workable temperature of about 95°F. Maintaining temperature within a narrow range in the digester is easier if the plant incorporates some method to heat the raw materials before they are admitted.

The process also can be speeded up by some method to mix or stir the contents of the digester, so that new surfaces are exposed to bacteria. In larger, continuous-feed digesters, some way to agitate the slurry is almost essential.

As to building materials, a methane digester can be built of almost any material that can be made watertight and airtight: concrete, plastic, steel, or a combination of materials. The hydrogen sulfide in biogas is highly corrosive to metals, a factor which should be taken into account when comparing one building material with another.

The simplest digester is a batch-type container that can be sealed off from the atmosphere, with openings to fill and empty the digester and some method of collecting the gas produced. In this type of plant, the digester commonly is filled with a mixture of water and manure, then sealed and maintained at 95°F until gas production is completed. When the contents have stabilized (when digestion stops), the digester is emptied and refilled with a new batch of materials.

In a batch-type digester, a start-up period of about two weeks is required to allow aerobic bacteria to use up oxygen in the material. After this, gas production slowly begins and continues for 75 to 80 days.

Because of the time required for each load of materials to digest, batch-type digesters are not very popular. More common are continuous-feed digesters supplied with a small quantity of manure each day. The manure typically is mixed with water to form a pumpable slurry of 8 to 10 percent total solids. The consistency of the proper mix is about that of light cream. The rate of gas production is more or less continuous, with a given loading of manure requiring 12 to 20 days to be digested, then removed from the vessel.

Most continuous-flow digesters are cylindrical in shape, of the *plug-flow* type, and have some method of mixing or stirring the material. A plug-flow digester is designed so that, with each loading of new material, a like volume of spent slurry, or *effluent,* overflows from the digester. Some plants employ a single digestion chamber, others a two-compartment chamber, and still others have two separate containers connected with an overflow pipe.

Many digesters are upright; that is, the cylindrical digester is built in a vertical position as shown in Fig. 16-3. In some respects, however, a horizontal displacement-type digester is superior to the vertical design. For one thing, as the organic material is digested, the slurry

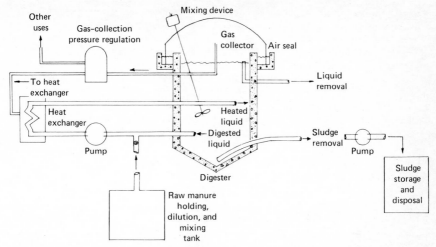

FIG. 16-3 Schematic of a plug-flow-type digester. Note the mechanical mixing device. *(Agricultural Engineering Department, University of Missouri.)*

gradually is displaced from the loading end to the discharge end of the vessel. This allows the rate of flow to be controlled for maximum or most efficient digestion and gas formation. Also, since the horizontal design has more surface area, the formation of scum on top of the liquid is easier to prevent.

Digester tanks should have some provision for draining the entire contents. Any continuous-feed digester eventually will accumulate scum and sediment that need to be cleaned out. At times, toxic materials or acidity can build up in a digester to the point that the contents must be emptied and the digester cleaned so that the process can be started over.

With a well-designed digester, about a quarter-pound of volatile solids can be fed daily for each cubic foot of digester volume. Suppose you need to dispose of the manure from 120 dairy cows each day, fed in dry lots where virtually all the manure will be collected. Those cows will produce about 1200 lb of total manure solids each day, on a dry-matter basis. That's a daily loading of some 960 lb of volatile solids. Using the "quarter-pound" rule of thumb, 960 lb of volatile solids added daily would require a digester volume of some 3840 ft^3.

Here's why: If the detention time is 12 days, the digester would need to be large enough to hold 11,520 lb of volatile solids (960 lb × 12 days). If the manure solids make up 8 percent of the total slurry, the digester would need to contain about 172,800 lb of material, or about 21,600 gal. Just to contain the liquid would occupy 2800 ft^3 of space (each cubic foot contains 7.5 gal). Adding another 30 percent to the volume for "head space" brings the designed size of the digester to about 3800 ft^3.

How much will it cost to build a gas plant? That is a hard figure to pin down and depends on the materials used, to a great extent. Generally, the initial cost of a methane digester will run $10 to $15 per cubic foot of capacity. Ted Landers, who heads up the Rural Gasification Project, has come up with economical digester designs for small farms, using heavy-weight plastic bags and plastic piping. Several of these digesters have been built on swine and poultry farms.

Each digester built by Landers and his colleagues is slightly different from others in design and operation. But all are built with easily obtainable materials and are designed for nearly automatic operation. One such digester, designed to handle the manure from 14,000 lb of hogs daily, produces 240,000 Btu of heat energy in biogas daily. Landers computes the payback on this model at about $1 per cubic foot of capacity per year—at current costs of conventional fuel.

A farmer who can use the slurry to fertilize crops or pastures probably can realize even more value from that resource than from the gas as fuel.

As noted, Landers uses PVC plastic pipe throughout the plant, even for gas service piping (see Fig. 16-4). Plastic is noncorrosive, lightweight, and relatively low in cost. However, Landers points out that plastic pipe may be prohibited for gas service by some building codes.

OPERATING AND MAINTAINING A DIGESTER

The main functions of a methane digester—converting manure to biogas and high-quality fertilizer—require primarily that the methane-producing bacteria be treated with tender, loving care. This means controlling the acidity, temperature, and toxicity of the slurry, all of which require no great deal of attention in a healthy, well-designed digester. Temperature control, mixing or agitation, and even piping in a new load of raw materials and removing a like volume of spent effluent are jobs that lend themselves to a fair degree of automation.

Handling and using the gas and fertilizer likewise are chores that permit some automation, in most cases. Pressure switches can handle the equipment required to pressurize the

FIG. 16-4 PVC plastic pipe is noncorrosive, lightweight, and relatively low in cost. Sight-glass tubing helps keep liquid levels of a digester in view.

gas for distribution. Thermostats and other controllers can regulate the burning of the gas in most appliances. The fertilizer slurry can be spread on fields by "honey-wagon" liquid spreaders or through irrigation piping.

Since a methane digester involves a microbiological process by a specific group of bacteria, it is common practice to "seed" the start-up plant with a culture of both acid-forming and methane-forming organisms. Sludge from a municipal anaerobic treatment plant or slurry from a livestock manure pit or other seed sources generally can be used.

In a continuous-feed digester, the seeding material should be about twice the volume of the fresh manure slurry during the start-up period. Successful anaerobic digestion depends upon a rather critical balance between the acid- and methane-forming organisms. The acid formers are hardier and usually more abundant in seeding materials. If too much fresh raw material is loaded into the digester before the two types of bacteria "balance out," the slurry may become too acid for the methane-producing bacteria. This problem usually can be corrected by adding lime or some other buffering agent. The best practice, in most cases, is to give the digester plenty of time to get the "bug" population in balance.

The slurry in a continuous- or daily-feed digester usually needs to be stirred or agitated daily. There are several ways to do this. A mechanical stirring device can be installed and operated by hand or by a small electric motor. The inlet piping for the new slurry can be positioned to create some agitation or turbulence of the contents each time a new load of material is admitted.

Or, the biogas itself can be used to agitate the slurry. One digester design uses a small rotary-blower-type gas pump to move gas from above the digester liquid to a point near the bottom of the digester vessel (see Fig. 16-5). The gas is released through a diffuser, and the action of the rising gas bubbles provides the needed mixing. The pump is controlled

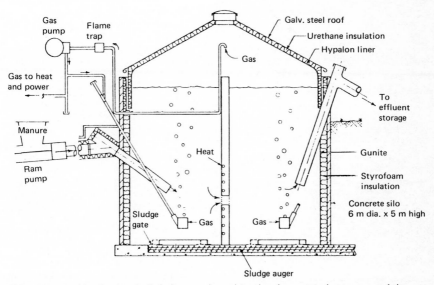

FIG. 16-5 In this digester, built below grade and insulated to retain heat, some of the gas produced is pumped to the bottom of the digester vessel and released. As the gas bubbles up through the slurry, it effectively stirs and mixes the contents. *(Pennsylvania State University.)*

by a timer to run 45 min every 8 hours. This type of agitation also breaks up scum layers that tend to form on top of the slurry.

As mentioned earlier, the quantity of gas produced in a digester each day depends on the amount of waste material fed per unit of digester capacity. The concentration of solids in the influent slurry (the new material entering the digester) should be about 10 percent of the total mix.

An early indication that the methane-forming bacteria may be getting into trouble is a decrease in the pH level. In an operating digester, a lower pH usually means an increase in volatile acid concentration, either because of "slug" loading the digester with raw material or because the acid-forming bacteria and the more temperamental methane-forming bacteria are out of balance. Testing the effluent with litmus paper can help spot acidity before it becomes so severe as to result in a "stuck" digester. The paper is available at most drugstores; it turns red in acid solutions, blue in alkaline solutions.

As also mentioned, some elements in biogas—particularly hydrogen sulfide—are corrosive to metal. For that matter, undigested manure is fairly corrosive too. PVC plastics are resistant to corrosion and are cheaper to use for low-pressure piping than stainless steel or copper.

Finally, as a safety note, always remember that methane will burn. It is no more hazardous than natural gas or LP gas, but any gas leaks in the system are potentially dangerous. To be on the safe side, don't install open-flame burners in the same compartment with the digester, and use mercury-type electrical switches that reduce the likelihood of an arcing spark.

USING BIOGAS

Biogas is most easily used in burning appliances that can use the gas directly from low-pressure sources. Table 16-3 shows quantities of gas required for various equipment. As mentioned earlier, because of the "less-than-pure" nature of biogas, burner orifices may need to be oversized to allow a greater volume of gas.

Digester gas also can be used to fuel internal-combusion engines with a compression ratio of 8:1 or higher. Methane has an octane rating of 120. Because the gas from a digester contains some oxygen in the CO_2, biogas requires about 9 to 10 parts of air (volume basis) for each part of gas burned. The gas burns cleanly and emits fewer pollutants than does gasoline.

Several companies make conversion kits to adapt standard gasoline engines to operate on either gasoline or methane (see App. D). Diesel engines also can be modified to run on part or all biogas. However, it takes a good deal of biogas to operate an engine—about 18 ft³/(h)(hp). Operating a 40-hp engine for 10 hours each day would require all the gas digested from the manure of 140 to 150 mature cows.

While the gas works very well in a stationary engine, the suitability of methane as a

TABLE 16-3   Quantities of Gas Required for Various Gas-Burning Equipment

Use	Quantity of gas required, ft³/h
Cooking, heating	16.5 per 4-in burner
	22.5 per 6-in burner
Lighting	2.5 per mantle
Gasoline engine	18.0 per horsepower
Diesel engine	16.5 per horsepower

motor vehicle fuel is limited. The performance of biogas as an engine fuel depends on the methane content. Or, to put it another way, the amount of CO_2 and other materials in biogas limit its usefulness.

Methane is one of the lighter gases and can be liquefied only under pressures of nearly 5000 psi. (By comparison, LP gas liquefies at about 250 psi.) That limits the traveling range with a methane-powered vehicle—either because not enough energy can be carried in low-pressure gas or because the weight of a tank or cylinder needed to withstand 5000 psi would seriously decrease fuel mileage. Also, the safety implications of riding in traffic astride a tank pressurized to 5000 psi would not be comforting to most motorists.

There are other problems. Hydrogen sulfide is corrosive, particularly to bearings and other engine working parts. The hydrogen sulfide content of biogas can be reduced by passing the gas through a selective filtering agent—such as steel wool or iron filings—but this boosts the cost of the fuel.

The CO_2 content of the gas presents another drawback to its use as motor fuel. While CO_2 does not necessarily harm an engine, it does dilute the gas to reduce the heat content and thus lowers the operating efficiency. Also, if the gas is to be compressed, the CO_2 must be "scrubbed" out, because carbon dioxide liquefies at lower pressure than methane and can wreak havoc with compressor parts.

Still, biogas as a motor fuel has its champions who point out that 40 percent or more of all automobile travel is in short trips, as drivers commute to work, go shopping, etc., within a few miles of home.

Obviously, such short-trip driving would allow fuel tanks to be filled more frequently, as would be necessary if low-pressure biogas were to be used as motor fuel. Many farmers make farm-to-market hauls or commute to an in-town job, of course. But the potential for refilling the gas tank more often would be greater for a farmer with his own gas plant.

The fuel equivalency of methane is about 132 ft³ to 1 gal of gasoline. Biogas could be partially compressed and stored in a cylinder in the trunk of an automobile or the bed of a pickup. A gas cylinder 9 in in diameter and 50 in long, designed to withstand 3500 psi, would weigh about 190 lb. At this pressure, about 500 ft³ of methane could be stored in the cylinder—the equivalent of 3.7 gal of gasoline. An automobile with a fuel mileage of 20 mpg would have a range of about 75 mi per filling. Accepting these assumptions, a methane-fueled auto could perform most of the short-trip driving that is done. Longer traveling ranges could be achieved by using a vehicle with a dual-fuel modification, to allow the engine to burn both methane and gasoline.

PRODUCING GAS BY PYROLYSIS

A heat-and-pressure system, similar to that used in charcoal manufacture, can produce a biogas that is burnable, although not of as high heat value as methane from anaerobic digestion.

In this process, raw organic material—anything from cornstalks to wood chips to shredded auto tires—is ground up and fed into a kilnlike gasifier, where the material is heated rapidly in an oxygen-excluding chamber. Part of the carbon in the raw material combines with hydrogen to form methane, which is drawn off for various uses.

The idea is not new. World War II veterans may recall seeing trucks in Japan and Europe carrying their own gasifiers along with them for making fuel as they drove—usually from wood chips, charcoal, or dried kelp.

Scientists at Kansas State University are bringing back this wartime technology, with several refinements. They are producing a low-energy gas from corn stover, as an alternative fuel for spark-ignited engines. The gas is produced in a fluidized-bed gasifier, and the engineers hope to come up with a substitute fuel for natural gas used in irrigation pumpers.

The fluidized bed (as diagramed in Fig. 16-6) is a layer of sand in the gasifier chamber, heated to about 1300°F and slightly pressurized by steam. When raw materials are fed into this chamber, three types of products are made almost instantly: (1) gas, consisting of various combinations of methane (CH_4), carbon monoxide (CO), carbon dioxide (CO_2), hydrogen (H_2), and some other components; (2) tars and liquids; and (3) char, or the solid residue, of which carbon and ash are the principal components.

The proportion of the three types of products depends primarily on the temperature and heating rate. With low temperatures and slower heating, char is the main product, with relatively less gas and liquid produced. With high temperatures and rapid heating of the raw material, more gas is produced in relation to char and liquids.

The gasifier being tested in Kansas (shown in Fig. 16-7) produces gas in three processing steps:

1. Pretreatment of the residues (corn stover, in this case) involves grinding the material to a fine particle size, to speed the heating rate and the rate of gas production.

2. The gasification step involves heating the material rapidly, in the absence of oxygen, in the fluidized-bed reactor.

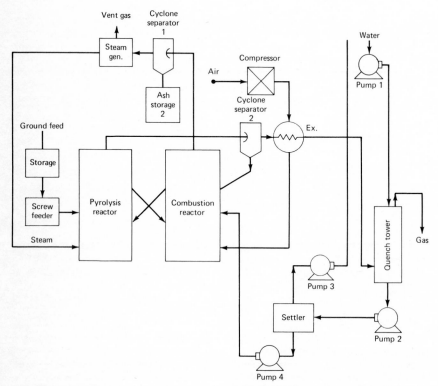

FIG. 16-6 Flow diagram of a two-reactor gasifier, similar to that in use at Kansas State University.

FIG. 16-7 Fluidized-bed gasifier at Kansas State University. Ground crop stover enters from the hopper tank at top and is converted to gas and other products at about 1300°F in the insulated chamber just below.

3. Gas cleanup consists of scrubbing and centrifuging the gas to remove particles and condensable materials from the stream of gas produced. Part of this step also involves recovering heat from the gas, which helps heat the fluidized bed. Part of the char material also is recovered to be used as fuel.

The gas produced has a heating value of about 400 Btu/ft³, only 40 percent of the Btu's in natural gas and about two-thirds of the heat value of biogas from an anaerobic digester. Kansas State researchers have run a spark-ignited engine on the gas and found that a blower needs to be used to *supercharge* the fuel slightly, to get more fuel into the cylinder. Also, a conventional carburetor (air-throttled) often cannot be adjusted to provide a rich enough fuel-air mixture.

In tests, the engine operated very well on the reactor gas, but produced only about 60 percent of the power expected with gasoline. Also, deposits on intake valves and in the engine cylinders have been a problem. With better methods to screen impurities out of the gas, a fluidized-bed gasifier could be more flexible in operation than an anaerobic digester, however. The plant at Kansas State University goes from a cold start to full operation in about 4 hours, whereas an anaerobic digester takes several weeks to reach full operation.

The plant can be made portable, also. The gasifier can be mounted on a trailer or skids and moved to where the gas is needed. As the raw material is fed into the gasifier dry, there's not as much problem with storage and handling as with liquid slurries.

The technology is here now to convert animal manure, crop residue, and other organic material into energy-loaded biogas—and then into heat, mechanical power, or electricity. Any farm with an abundance of raw materials and a steady demand for gas might well benefit from biogas.

The price ticket is fairly high for either an anaerobic digester or a fluidized-bed gasifier, built on any kind of scale. The cost for a gas-producing plant of the Kansas State design that could fuel a 25-hp engine might cost from $125,000 on up, in terms of 1980 dollars. If 50 or more of these units could be produced, by a farm building firm or equipment manufacturer, the cost per unit no doubt could be reduced considerably.

An anaerobic digester and related equipment to produce an equivalent volume of gas might cost half that, depending on materials used and the amount of construction work done by farm labor. Naturally, energy costs and capital investment are big factors in computing whether a gas plant will be a paying proposition.

17

Hydrogen: Fuel of the Future?

The energy problems facing the world are not likely to be solved by any one big break-through: not solar, or grain alcohol, or any other single solution now in view.

Among the more promising sources of fuel at first glance is hydrogen, a clean-burning, high-energy material potentially as abundant as water. On paper at least, hydrogen appears to be one of the better alternatives to petroleum fuels. It is available worldwide; it can be used to fuel airplanes, vehicles, and farm equipment; it can heat homes, factories, and water; it can even produce electricity.

Furthermore, most of the technologies needed to produce, transport, store, and use hydrogen are available now. Plants have been constructed to produce hydrogen from coal. Every high school physics student knows how to separate the hydrogen and oxygen in water with electrolysis.

Washington is anxious to promote fuels that might help solve the energy crisis by the mid-1980s. Still, little public research and development money has been allocated to study hydrogen, compared with the funds made available to test other alternatives.

OBJECTIONS TO HYDROGEN

To date, the biggest objection to hydrogen is cost—both the cost of equipment to produce and store it and the cost in energy to manufacture it. The primary cost involved in a hydrogen energy system is the initial investment required for equipment and facilities, that is, if the fuel is to be produced in any volume. Also, while the heating value of hydrogen is high, compared with petroleum fuels, the energy needed to release hydrogen from water is even higher. Considerable heat is generated during electrolysis—much of which is dissipated and wasted—so the energy efficiency of the molecular electrolysis process is 70 percent or less.

In other words, to get 70 Btu of energy in the form of hydrogen, it is necessary to invest 100 Btu in the form of electrical energy.

The production of 1 lb of hydrogen (with an energy value equal to about a half-gallon of regular gasoline) requires 35 kWh of electricity. At current costs of commercial power, that is not a particularly cheap source of fuel. Hydrogen looks even less promising when you consider that the equipment to make 5 lb of hydrogen per day may run $15,000 or more, plus another $3500 or so for a storage system.

Negative safety aspects plague hydrogen, also. The fuel has a reputation as an explosive, unstable, and unsafe gas—based on such disasters as the Hindenburg dirigible fire in the

1930s. And it can be hazardous, unless proper materials are used for storage equipment and transmission lines.

ENERGY POTENTIAL OF HYDROGEN

With the gloomier features of hydrogen out of the way, let's look at the amazing potential of the fuel, in terms of energy. On the basis of Btu's per pound, liquid hydrogen tops the fuels list, with 61,000 Btu/lb—well ahead of pure methane at 23,900 Btu/lb. (There's another wrench in the works where *liquid* hydrogen is concerned, however, and we'll discuss this more thoroughly shortly.)

Some other properties of hydrogen are:

Specific gravity	0.0696 (air equals 1.0)
Lb/ft^3	0.00561
Boiling point	$-253°C$ $(-423.4°F)$

There are several processes for manufacturing hydrogen, but the most promising, from a farm energy standpoint, is separating hydrogen from oxygen in water, among the more common materials on earth. Here, briefly, is how hydrogen is produced from plain old H_2O.

The process of conducting an electrical current through a liquid and the resulting chemical change is called *electrolysis.* This is what makes an electrical storage battery work, and it's the process most often used in the plating of metals. When two electrodes from a direct-current electrical source, such as a battery or generator, are inserted into a tank of water, electrical current flows between them.

The electrode connected to the positive terminal is called the *anode;* the other electrode, from the negative terminal, is called the *cathode* (see Fig. 17-1). Electron movement (current flow) through the water is from the cathode toward the anode. Water molecules ionize to form positively charged hydrogen ions (H+) and negatively charged hydroxide ions (OH−).

However, pure water is not a very good electrical conductor. In actual hydrolysis, some material—such as caustic soda or potash—is added to the water to improve its conductivity. Depending on the material added, more complex gases than hydrogen and hydroxides may be produced, but the principle is the same. During electrolysis, oxygen is liberated at the anode, while hydrogen is released at the cathode. These two gases bubble to the surface and are collected. Some material, often a diaphragm of asbestos or plastic, is placed between the electrodes to keep the gases separated.

As mentioned earlier, the big drawback to producing hydrogen by electrolysis is the cost of electricity. At least 1.3 V is required to break down water into oxygen and hydrogen, even when some supplemental heat is added to the liquid. Most industrial electrolyzers operate at about 2 V per cell.

Obviously, buying electricity "off the pole" to manufacture hydrogen would make this a most expensive fuel. However, hydrogen-generating equipment can be powered by wind, water, or other alternative energy sources; it can also provide a way to use off-peak electricity for both private and commercial power generators.

The use of wind energy to produce hydrogen by electrolysis is a proven concept, and a few U.S. farmers now have plants in operation. John Lorenzen, an Iowa farmer and inventor who has produced his own electricity for many decades, has developed a hydrogen "battery" that uses electricity to separate hydrogen and oxygen from water (see Fig.

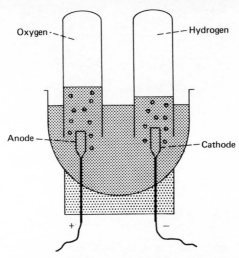

FIG. 17-1 During electrolysis of water, oxygen is released at the anode (positive electrical lead) while hydrogen is released at the cathode (negative electrical lead).

17-2). He then burns the 99 percent pure hydrogen in an ordinary gasoline engine. The hydrogen, burned in mixture with gasoline in the engine cylinder, gives up to 35 percent better performance than gasoline alone (see Fig. 17-3).

Lorenzen feels it would be possible to develop a completely hydrogen-powered engine, and he is working toward that end. But, in the meantime, he believes there is a great

FIG. 17-2 Hydrogen "battery" developed by Iowan John Lorenzen uses stainless steel plates as anodes and cathodes. *(Photo: Ralph Watkins.)*

FIG. 17-3 Hydrogen gas is metered into the intake manifold of this gasoline engine, to supplement gasoline fuel. *(Photo: Ralph Watkins.)*

potential for hydrogen as an extender for petroleum fuels, and he sees his system of manufacturing hydrogen "on the go" as being safer than storing the fuel in quantity in a vehicle.

Lorenzen's hydrogen generator is a series of stainless steel plates and plastic dividers enclosed in a compartment slightly larger than an ordinary car battery. The sheet stainless steel plates—50 of them—are about 2.5 by 5 in, and are spaced $\frac{3}{32}$ in apart. Each pair of plates, one serving as cathode, the other as anode, is divided by plastic strips. The case or tank of the generator is filled with water. When the electrodes are energized, hydrogen gas is driven off the negative plates; oxygen is released at the positive plates. The plastic dividers are shaped to channel the gases off separately.

Lorenzen pumps the hydrogen through copper tubing to an LP-gas storage tank just outside his farm shop. Check valves and relief valves are used throughout the system for safety. In his machine shop, Lorenzen pipes the hydrogen to the intake manifold of a small internal-combustion engine. He can operate the engine on either a mixture of hydrogen and gasoline or on hydrogen alone.

It would take considerably more electricity than Lorenzen uses to make hydrogen in quantity. The Iowan uses direct current from a wind-powered generator or from storage batteries to power his hydrogen-making system. But he believes that an automobile equipped with a large generator could manufacture enough hydrogen to make a net gain in energy, when the hydrogen is used to supplement gasoline to power the car's engine.

Perhaps 10 percent of the engine's horsepower would be required to manufacture the hydrogen, but Lorenzen believes the increased mileage when the hydrogen was mixed with gasoline to fuel the engine would be a net gain. Some engineers are skeptical about such hydrogen-producing systems, claiming that there is no way to produce net energy from water and a vehicle's electrical generation system. John Lorenzen does not claim to have invented perpetual motion with his on-the-go hydrogen battery, and he notes that not all of a car's generator output is required to keep an electrical storage battery fully charged; the surplus electricity could be used to manufacture hydrogen, which would increase the mileage from each gallon of gasoline.

While pointing out that safety precautions must be taken, some engineers endorse the idea of using farm-generated electrical energy to produce hydrogen by electrolysis. Such a system is potentially practical, say agricultural engineers at Virginia Polytechnic Institute. In fact, the interaction of a wind generator and the electrolyzing cells could improve the overall efficiency of the system, they say.

Here's how: Wind machines frequently do not operate fast enough to generate usable power below a wind velocity of 5 to 7 mph (depending on the machine's design), because the output voltage is too low. However, any voltage in excess of about 2 V would be usable for a hydrogen-generation system (see Fig. 17-4). So, even when winds are intermittent or light, some hydrogen could be generated.

During periods when the wind is higher, additional cells could be automatically cycled into the system to utilize a higher rate of energy output, or batteries could store surplus electricity for later use to manufacture hydrogen.

The idea is intriguing. A rated 10-kW wind plant operating for 10 hours at only 60 percent efficiency (6 kW output) could produce about 424 ft^3 of hydrogen, with a heating value of 1.4 million Btu.

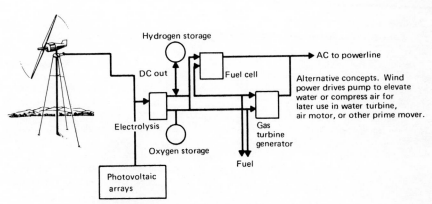

FIG. 17-4 Typical schematic of a wind plant for dual uses. In this case, direct current is provided to the electrolysis cells by either the wind generator or photovoltaic solar cells. *(Drawing: National Science Foundation.)*

STORING THE FUEL

The fuel aspects of hydrogen are somewhat complicated by the storage requirements. Storing hydrogen can be a hazardous, expensive proposition. Basically, hydrogen is stored in one of three ways:

1. As a gas
2. As a liquid
3. In a metal hydride

For most farm-scale applications, liquid hydrogen can be written off as too impractical. For one thing, the hydrogen must be held at $-235°C$. Also, liquefying the gas would consume at least a third of the potential energy in hydrogen.

Gaseous hydrogen can be stored in tanks, at pressures up to about 2000 psi. However, not just any old tank will serve the purpose. Hydrogen is a light, slippery gas that is hard to contain. It can be pressurized into a storage tank by a pressure-type electrolyzer cell or by a mechanical compressor. Generally, pressure-type electrolyzers are too expensive to be considered for on-farm applications. A tank 6 ft in diameter and 13 ft long can store 50,000 ft³ of hydrogen at about 2000 psi. This volume of hydrogen has a heating value of some 166 million Btu.

However, ordinary LP-gas tanks are not strong enough to contain hydrogen at these pressures. Without some treatment of the tank interior, LP-gas tanks may not be suitable for storing hydrogen at *any* pressure. The elusive gas can penetrate the metal and will leak through most conventional valves and fittings.

Of course, storing hydrogen as a gas assumes that the fuel can be used in gaseous form.

The most promising way to contain hydrogen at usable pressures and volumes is in metal hydrides, principally iron-titanium alloys. Hydrides have been used as fuel carriers for experimental vehicles. They are a safe, convenient way to store and use hydrogen. The metal hydrides can be charged and discharged through hundreds of cycles without breaking down.

Metal hydrides release heat when hydrogen is absorbed and absorb heat when hydrogen is driven off. Typically, the hydrogen is stored in a tank with iron-titanium hydride, which ties up the hydrogen. Then, waste heat from the engine is used to drive off the hydrogen for fuel.

The metal hydrides absorb or release hydrogen at various rates, depending on temperature. A tank of iron-titanium hydride maintained at 10°C will give off a continuous flow of hydrogen until all the gas has been extracted. Then, the hydride is recharged by lowering the temperature and feeding in new hydrogen.

An added benefit of using metal hydrides is that more hydrogen can be contained than would be contained in the same volume of liquid hydrogen.

One company reportedly has developed a liquid hydride solution, composed of ammonia and other elements.

There's little doubt that hydrogen has a great deal of energy potential. However, it is likely that problems remain to be solved with this controversial fuel—particularly in the areas of safe storage and use.

Whether there is a "hydrogen revolution" on the energy horizon is beyond the vision of this author. But, with so much known promise from hydrogen as a fuel, and with such an available source (water) from which to make the fuel, energy planners might be spending more time and money trying to learn whether hydrogen is the key to energy salvation, or a rather dangerous brand of snake oil better left alone.

18

Producing Electricity on the Farm

Electricity supplies a vital energy ingredient for virtually every American. It is an essential support of our standard of living, as well as much of our business and industry.

Despite higher costs, more electricity is being used on farms and in homes today than ever before. Electricity is among the easiest forms of energy to control. But it is becoming more expensive to buy, and a real possibility exists that the supply of power to some areas may be limited before long. There is growing interest among farmers in generating at least part of their electrical energy needs from close-at-hand renewable resources: wind, water, direct solar, and engines driven by farm-made fuels. There are probably as many different power-generating ideas as people interested in applying them.

ECONOMICS OF HOMEMADE ELECTRICITY

There are no *cheap* ways to produce electricity. Even with off-the-pole power costing 10 cents per kilowatthour (kWh) or more, most private generating systems are hard to justify strictly on the basis of economics. At best, you're looking at $3000 per 1000 W of generating capacity for a dependable system.

Of course, business and energy tax credits are allowed on the purchase or construction of most power-generating equipment. We'll go into that economic aspect more thoroughly in Chap. 23. But for example's sake, let's say a Kansas farmer plans an electricity-generating plant (probably wind-powered, in Kansas) that is rated at 8000 W peak output. Depending on the engineering costs involved, how tall a tower is needed, and other variables, a plant with that capacity will cost in the neighborhood of $25,000.

Amortizing the equipment investment over 20 years, the capital outlay breaks down to $1250 per year. If the wind plant delivers 1000 kWh per month, the power costs just over 9 cents per kilowatthour. Most rural Kansans can buy utility power for less than that. Also, if the farmer borrows the $25,000—or part of it—there are interest payments to make, and farmers are all too aware of today's interest rates. Adding another 2 cents per kilowatthour to cover interest and operating and maintenance costs brings the rate of homemade power to about 11 cents per kilowatthour—twice the cost of "boughten" electricity.

But, the picture is not necessarily that gloomy. For one thing, the federal government allows up to a $2200 income tax credit on the purchase of wind-powered generating equipment. (Actually, wind generators installed for household electricity generation qualify for a 30 percent tax credit; whether that expanded credit also applies to farm businesses is sort of up in the air at this writing, but the portion that applies to the family home's share of power used probably qualifies.)

Further, the wind plant would qualify for a 10 percent investment tax credit. And the state of Kansas allows an income tax credit of 25 percent, up to a maximum of $1000, on wind energy equipment. When you add up all the income tax credits (and these credits are dollar-for-dollar reductions in the amount of tax owed), they make a dent in the initial cost of the wind plant:

Federal energy tax credit	$2200
Investment tax credit	2500
State income tax credit	1000
Total tax credits	$5700

That brings the cost of the wind plant down to $19,300 and reduces the cost of each kilowatthour of electricity generated to about 8 cents, plus the allowance for maintenance and repair. It's still a lot of money, but there may be other ways to make a private generating plant more financially attractive.

For example, the National Public Utilities Regulatory Act of 1978 requires that public utilities, in many cases, buy "excess" power generated by private plants. This may help offset the cost of generating your own electricity; however, the power fed into the utility grid may be purchased at a fairly low wholesale unit rate. (Before you attach any local power-generating equipment to a utility's meter or service line, get the company's advice. The utility will require that positive automatic disconnects are installed to protect crews working on company lines or equipment.)

MAKE A LOAD ANALYSIS

Planning a private generating plant begins with a *load analysis* of all the electricity-consuming devices you use, how much you use them, and when you use them. Electrical power is measured in watts (W); kilowatts (kW), or thousands of watts; and megawatts (MW), or millions of watts. You'll want to determine your average consumption of watts over a specific period of time—usually one hour or one day. The terms *watthour* (Wh) and *kilowatthour* describe the accumulated use of energy for that period of time. For example, 1000 W used for 1 hour equals 1 kWh.

The following pages set forth estimated amounts of electricity used by selected farm and home equipment, based on data obtained in Iowa and Missouri tests. However, the best way—the only *sure* way—to pinpoint your electrical requirements is to multiply watts used times period of use for each light, motor, and appliance. Every light bulb, electrical appliance, and motor has a power rating—usually indicated in the watts required to power it. A 100-W bulb, not surprisingly, requires 100 W. If you burn the bulb for 3 hours each day (or night), you will require 300 Wh per day or 9 kWh (9000 Wh) per month.

Some electric motors may show the rating in amps (A), rather than in watts. Don't let this throw you. Watts equal amps times volts (V). A 115-V appliance with a 10-A rating requires 115 times 10, or 1150 W. This same formula,

$$\text{Watts} = \text{amps} \times \text{volts}$$

can be used to find any one of the missing values. For example, a 75-W bulb requires 0.65 A of 115-V power, but requires 6.25 A in a 12-V system.

You'll also need to compute the *peak* demand for power. The system must be adequate to meet the *starting* demand of AC motors, as well as the continuous running load. Start-up requirements for motors (called the *locked-rotor amperage*) may be three to six times greater than their normal running requirements. Motors on refrigeration units, ventilation and heating systems, pumps, and other automatically controlled equipment can "surge" the power requirement temporarily well above the average operating level.

For heavier-duty electrical equipment, you can figure the power consumption at 1000 W per horsepower. This rule of thumb doesn't always apply exactly, but it's a fair approximation.

Table 18-1 shows average electrical loads required by farm and household equipment. This is the place to start when you are considering a power-generating system, of whatever kind. How much electricity do you use? When do you use it? What equipment or appliances does it power? How much does it cost? Will you need three-phase power? Will the system provide a substantial part of your load requirement?

By going through this exercise and examining old utility bills, you should be able to estimate the kilowatt hours of electricity used on a month-by-month basis. Once you've analyzed the electrical load on your farm, you'll be in a better position to decide whether to install a generating plant to produce all or part of the power needed—or whether to install a power station at all.

And, while you're doing the homework, look into power *co-generation* with your local rural electric cooperative or utility company. More and more utilities are making agreements to buy power from private generators during peak periods at favorable rates and then sell power back during nonpeak periods at reduced rates.

In fact, whether you're installing a generating plant or not, you may want to check on the possibility of buying cheaper power during off-peak times. The rate break could be substantial enough to justify shifting some high-power-use chores to a different time of the day.

In the next three chapters, we'll discuss the on-farm generation of power from wind, water, and other sources.

TABLE 18-1   Average Electrical Requirements of Farm and Household Equipment

Equipment	Common size power load	Average kWh used
General:		
Concrete mixer	$\frac{1}{3}$–10 hp	0.5 per yd^3
Electric fencer	7–10 W	7 per month
Electric welder	7500–8500 W	45–60 per year
Ensilage blower	5–10 hp	1 per ton
Ensilage cutter	5–10 hp	7 per ton
Fanning mill	$\frac{1}{4}$–$\frac{1}{2}$ hp	1.5 per 100 bu
Motors	Up to 7$\frac{1}{2}$ hp	1 per hp-h
Silo unloader	3–5 hp	1 to 2.5 per ton
Stock tank heater (insulated)	550 W	300 per season
Tank heater (uninsulated)	1200 W	1100 per season
Pump, deep well	550–1100 W	2.5 to 2.8 per 1000 gal
Crop-drying equipment:		
Drying baled hay (not including fuel)	5 hp	20 per ton
Drying grain in bin:		
With electric heater	21–27 kW	1400 per 1000 bu
With fuel heaters	3–5 hp	550 per 1000 bu
No-heat drying (aeration)	5–7 hp	1 per hp-h

TABLE 18-1   (con't)

Equipment	Common size power load	Average kWh used
Feed-processing and handling equipment:		
Automatic blender-grinder	2–5 hp	3.5–4 per ton
Crusher, ear corn	1–5 hp	5 per ton
Feed mixer	1–7.5 hp	1 per ton
Roller mill, automatic	2–5 hp	2.5–3.5 per ton
Corn sheller	1–1.5 hp	5 per 100 bu
Auger, 9-in horizontal, 25 ft	5 hp	2.5 per 10 ton
Auger, 9-in horizontal, 50 ft	5 hp	2.85 per 10 ton
Auger, 9-in horizontal, 90 ft	5 hp	3.5 per 10 ton
Auger, 9-in horizontal, 120 ft	5 hp	15 per 100 steers per month
Dairy equipment:		
Barn cleaner	1–3 hp	0.7–1 per cow per month
Livestock waterer	600–700 W	340 per season
Milk cooler, bulk	500-gal tank	11 per 100 gal
Milking machine (no pipeline)	350–500 W	1.5–2 per cow per month
Milking machine (pipeline)	750 W	3.3 per cow per month
Swine-production equipment:		
Floor cable, farrowing bldg.		
Spring-fall seasons	30 W/ft^2	50 per litter
Winter season	30 W/ft^2	110 per litter
Heat lamp, farrowing	250 W/lamp	6 per day
Waterer, indoors	750–1250 W	200 per season
Waterer, outdoors	750–1250 W	400 per season
Ventilation, confinement bldg.	Varies	9.25 per cwt pork
Poultry equipment:		
Automatic water heater	100–600 W	40–70 per season
Brooder, hover type	Varies	20–70 per 100 chicks
Egg grader-sizer	¼ hp	1 per 8000 eggs
Egg-gathering belt	⅓ hp	1 per 4 h
Egg washer	⅙ to ½ hp	1 per 2000 eggs
Egg cooler	0.5 hp	2 per case
Feeder, automatic	½ hp	20 per month
Lighting (layers—14-h day)	60 W/200 ft^2	5 per 100 hens per month
Ventilation fans		
Broilers (up to 8 weeks)	⅓–½ hp	30–35 per 100 birds
Layers	⅓–½ hp	180 per 100 hens per season
Household equipment:		
Air conditioner, room	1300–2760 W	1.5–3 per hour
Air conditioner, central	5000–8000 W	5–8 per hour
Blanket, electric	100–200 W	10–20 per month
Blender	230–450 W	5 per year
Can opener	100–200 W	1 per year
Clothes drier (electric)	4300–6000 W	55 per month
Coffee maker	600–1000 W	4 per month
Dishwasher	500–1600 W	2–4 per month

TABLE 18-1 (con't)

Equipment	Common size power load	Average kWh used
Household equipment *(cont.):*		
Electric clock	1–10 W	2–4 per month
Fan, attic (summer)	⅓ hp	45 per month
Fan, kitchen vent.	100–150 W	0.1 per hour
Fan, portable	50–200 W	0.5–2 per 10 h
Frying pan	1100 W	15 per month
Food waste disposer	¼–½ hp	2–3 per month
Freezer, home	300–800 W	5 per ft³/month
Furnace blower	⅙–½ hp	150 per season
Heating pad	50–100 W	0.1 per hour
Hot plate	600–1650 W	7–30 per month
Iron, hand	600–1500 W	11 per month
Iron, automatic	1000–3000 W	20–40 per month
Lighting, home	Varies	20–40 per month
Mixer, food	50–200 W	1 per month
Portable heater	1500 W	1.5 per month
Radio, console	150–300 W	0.2 per hour
Radio, table	40–150 W	0.1 per hour
Range, electric	6500–11,500 W	80–160 per month
Record player	100 W	0.1 per hour
Refrigerator	200–600 W	4 per ft³/month
Sump pump	250–500 W	0.25–0.5 per hour
Television	200–350 W	0.2–0.35 per hour
Toaster	400–1200 W	2–4 per month
Vacuum cleaner	165–750 W	1–2 per month
Washing machine	250–750 W	3–6 per month
Water heater	1500–4500 W	250–300 per month

19

Power Blowin' in the Wind

Wind is a stepchild of solar energy. Air is set in motion as the sun does a rather uneven job of heating the earth's surface, and air on the move carries a great deal of energy.

Forty years ago, wind plants were common fixtures on U.S. farms, harvesting "free" power by generating electricity or pumping water for rural families (note the contrast in Fig. 19-1). Then, the Rural Electrification Administration brought relatively cheap power to the countryside, and most of the old Wind Kings, Windchargers, RuralLites, and Jacobs plants were no longer needed. The manufacturers of these durable, low-voltage plants disappeared along with their product, for the most part. Even makers of water-pumper windmills fell on hard times, as farmers found commercial electrical power a more convenient force for bigger, faster pumps.

In the past few years, the winds of change in the world's energy situation have blown new interest into wind-generated electricity. Plant designs and manufacturers are mushrooming across the country, with more efficient equipment than our grandfathers could buy at any price.

There's a great deal of potential energy even in slow-moving winds. Efficient electricity generation depends not so much on *high* winds as on *steady* winds. In fact, the operating range for most small wind-energy conversion systems (SWECS) is from about 8 to 30 mph —well below gale force.

Obviously, the way to start planning a SWECS is by measuring the resource: the wind itself. (Actually, the *first* step is to analyze the farm's electrical load, but we'll assume that's already been done.) That means monitoring the wind speed at the best potential site or sites for a wind plant.

It's a good idea to understand something about the wind and the power it carries beforehand. Wind (air in motion) flows from high- to low-pressure areas, as indicated in Fig. 19-2. Over wide areas, the airstreams follow general patterns, such as the easterly and westerly prevailing winds, or "trade winds" of old sailing ship days. On a more regional scale, wind speed and direction are influenced by geographical and landscape features.

Theoretically, it's possible to extract only about 60 percent of the potential energy in the wind, and perhaps half of this is lost to slippage, gearing, drive mechanisms, and design faults in the equipment. Still, that leaves a lot of energy from a resource that is free and ever renewable. A great deal of potential electricity rides breezes that blow only 10 to 12 mph.

Wind propellers of whatever type function by geometric law. If a 20-ft-diameter propeller (also called *rotor* or *turbine*) produces 2 kW in a 10-mph wind, a 40-ft propeller of the same design in the same wind should produce 8 kW—a fourfold increase in power,

FIG. 19-1 Old and new versions of wind-harvesting equipment stand side by side near Clayton, New Mexico. The familiar water-pumper windmill in the foreground is contrasted with a 200-kW high-speed electricity-generating turbine. *(Photo: U.S. Department of Energy.)*

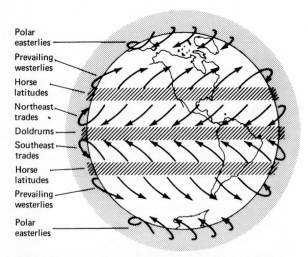

Polar easterlies

Prevailing westerlies

Horse latitudes

Northeast trades

Doldrums

Southeast trades

Horse latitudes

Prevailing westerlies

Polar easterlies

FIG. 19-2 The earth's rotation affects wind direction at different latitudes north and south of the equator.

rather than merely twice as much. The speed of the wind has an even greater effect on the potential energy that can be harvested: Power available in the wind is the *cube* of the speed. In other words, a propeller that delivers 1000 W in a 10-mph wind will produce about 8000 W in a 20-mph wind. Doubling the size of the propeller can result in the production of nearly four times as much power. But doubling the speed of the wind results in an *eightfold* increase in power.

Since the performance of a wind generator is so affected by wind speed (up to the designed maximum of the particular plant used), it's important to locate the SWECS where it will take best advantage of the prevailing winds. In most areas of the country, the wind blows from the same—or nearly the same—direction 65 to 70 percent of the time. These are prevailing winds. For a site to be suitable for a SWECS, the prevailing winds should average at least 10 mph.

However, the other 30 to 35 percent of the time, storm winds gust from other than the prevailing direction. These winds, although they blow only two days out of seven, on the average, contain about two-thirds of the potential power in wind. Rotor diameter ordinarily is sized to produce the maximum output in the more powerful, though less frequent winds.

EVALUATING THE SITE

Do you have a good site, with enough wind—enough *steady* wind—to operate a SWECS? As noted earlier, an area with an average wind speed of 10 mph is a good candidate. But even that statement assumes that the wind will blow more or less steadily most of the time —say from 8 to 15 mph. An area that is calm for six days, then has a 70-mph gale on the seventh day is not ideal, although the wind speed may average 10 mph.

An early step in selecting a good wind power site is to eliminate the clearly unsuitable ones. At any given location, the wind speed and direction are influenced by anything that breaks the path of the air flow (see Fig. 19-3). Trees, buildings, and other obstacles cause wind turbulence; so do land contours, such as cliffs and steep, rough hills. More promising sites are near flat, level ground or open water, where the wind is unimpeded by physical obstacles.

However, some topographical features can *increase* wind speed, and you should keep them in mind as you search for the best site for your SWECS. Mountaintops and hilltops

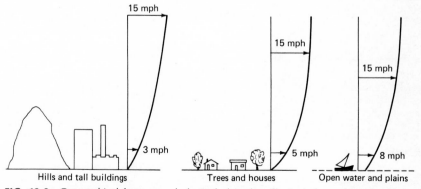

FIG. 19-3 Geographical features and physical obstacles affect wind speed locally; friction with the earth's surface slows wind near ground level.

are windy because the airstream picks up speed as it passes over them. A valley that stretches in the direction of prevailing winds between two hills can "funnel" the wind and increase its speed in a sort of venturi effect.

The wind also picks up speed with altitude. Because of friction with the earth, wind speed is reduced near the ground surface, and increases with height, almost in a linear curve.

For example, if the wind at eye level of a person standing on the ground is 10 mph, the speed at 40 ft. above the ground will be about 13½ mph, and at 90 ft, about 15 mph. This is important to keep in mind. Remember, the same wind plant will produce almost four times as much power in a 15-mph wind as in a 10-mph wind.

The surest way to evaluate a site is to measure the wind potential accurately, over a period of time, at or very close to the rotor height. In other words, if you plan to install a wind plant on the hill above the barn, on a 40-ft tower, measure the wind there, rather than on or near the ground at the planned tower site. A handy way to measure and compare wind speeds is to use an anemometer and a recording odometer positioned at the site. These devices cost about $150 (a list of distributors is included in App. D), but some wind plant manufacturers rent them to potential customers. The odometer records one count for each mile of wind that runs through the anemometer's cups. Simply divide the number of counts per day by 24 to find the average wind speed in miles per hour.

Naturally, the longer period of time you measure wind speed at the site, the more accurate planning information you have. It would not be very realistic to measure the rainfall in April, then multiply by 12 to find the yearly precipitation. By the same token, it's not accurate to measure wind speed for only a week or a month, then multiply by 52 or 12 to find out how much power-producing wind you can expect in a year.

The next best route is to monitor the wind speed at your site at specified times each day—say once in the morning, at noon, and once in the evening—at the same times each day, then compare your findings with the nearest weather station, or stations. The Weather Service of the U.S. Department of Commerce hourly records wind speed and direction at several stations around the country. Local airports, weather bureaus, and military bases also can be helpful in providing general wind information.

You'll need to monitor the wind speed and direction at your site long enough (at least a month) to establish a correlation between your winds and those recorded by the weather station for the same times; then you can interpolate for the rest of the year. This is not as accurate as a full year's readings at the site, but it takes much less time and work than measuring the wind for a full year. And, in most cases, this will be a close enough estimate for planning purposes—particularly if the recording station is fairly nearby, is at about the same elevation above sea level, and has similar topographical features.

The local weather station may also have information on maximum winds and gusts to be expected—useful as you plan the kind and size of wind plant to install.

Here's how to take the historical data provided by a weather station and adjust them for the difference in wind speed you have observed at your site and tower height. By adjusting your observed wind speeds to those from the weather station, you can make a wind distribution forecast for each month of the year.

Suppose you record wind speeds at a 40-ft height every day for a month and find that winds at the site average 20 percent more than those recorded by the weather station's anemometer at a 30-ft height. Table 19-1 shows how the wind distribution forecast for the site might appear for a 30-day month.

With a SWECS that starts generating electricity at 8.5 mph or higher, there would be 286 hours during this month that no power could be generated. This kind of information is important to help determine the battery storage capacity needed. If the 286 hours are distributed more or less evenly through the month, you can get along with much less battery capacity than if the light winds come all in one period.

TABLE 19-1   Sample Wind Distribution Forecast

Weather station wind speed*, mph	No. hours per month	Wind speed at your site†, mph
0–7	286	0–8.4
8	49	9.6
9	42	10.8
10	55	12.0
11	21	13.2
12	65	14.4
13	24	15.6
14	34	16.8
15	41	18.0
16	14	19.2
17	12	20.4
18	10	21.6
19	8	22.8
20	7	24.0
21	6	25.2
22	6	26.4
23	7	27.6
24 and above	33	29.0 and above

*Anemometer at 30 ft.
†Figures increased by 20 percent.

WHICH PLANT DESIGN?

Once you have analyzed electrical power needs, calculated annual average wind speed and interpolated a year's worth of data from the nearby weather station, you are ready to start comparing wind plants. Each SWECS manufacturer computes an *output* curve for his machine. With the above figures on wind speed and the output curve for various wind plants, you can forecast the kilowatthours each plant would produce per month.

Table 19-2 shows the power that can be expected from wind plants with an overall system efficiency of 30 percent, which is probably on the high side of average for most SWECS designs. With most equipment and most wind distribution patterns, the biggest portion of power is produced by medium-strength winds of 15 to 20 mph.

The heart of a SWECS is the turbine-generator combination. Devices on the market today to convert the mechanical power in the wind to electrical energy are varied in shape, design, and appearance, but they all perform basically the same functions in much the same

TABLE 19-2   Expected Power from Plant with 30 Percent Efficiency

Rotor diameter, ft	Watts of power generated at average wind speeds of:				
	5 mph	10 mph	15 mph	20 mph	25 mph
4	2	19	64	152	296
6	5	43	144	341	665
8	10	76	256	605	1183
10	15	119	399	947	1848
12	21	170	575	1363	2661
14	30	232	782	1855	3623
16	38	307	1022	2423	4731
18	48	383	1293	3066	5988
20	59	473	1597	3785	7393

way. Propellers are classified into two general categories: vertical axis and horizontal axis. Rotors with their shafts parallel to the wind stream are horizontal-axis machines and typically employ a tail vane or rudder to keep the rotor facing into the wind. These plants have from one to five or more blades. The spinning propeller acts somewhat like a gyroscope and does not track changing winds quickly.

Vertical-axis rotors have their shafts perpendicular to the wind flow, at right angles to the earth's surface. These rotors need no tail vanes, since they can accept the wind from any direction. This lets them make more use of the extra power in gusty, shifting winds.

All SWECS machines have some sort of overspeed protection. Without some way to govern the rotor speed, strong winds would spin the rotor past its designed speed and possibly damage the equipment. Older wind machines—and some newer ones—have a simple, spring-loaded governor that is actuated by wind speed. When the designed rotor speed is reached, centrifugal force causes either air scoops to extend or brake shoes to engage a drum on the shaft, slowing the machine. Many newer rotors use a "feathering" system to twist the rotor blades so that they intercept less wind when the upper wind speed limit is reached.

Multiblade turbines are those most often associated with the familiar water-pumping windmill. The horizontal-axis rotor has 14 to 40 blades and is designed to develop high starting torque but low maximum speed and low end horsepower. The high starting torque makes this machine well suited for pumping water, as the turbine must begin pumping on the first revolution, but it's not particularly efficient for generating electricity.

High-speed horizontal-axis machines are made with two or three (sometimes more) airfoil blades of varying diameters. The shaft power developed by wind moving the blades is transferred directly or through a gearbox to a generator. The propeller blades resemble those of an airplane, but are aerodynamically designed to grab energy from the wind rather than propel an aircraft through it. Three-bladed propellers are easier to balance and are most stable in shifting winds. This type of machine has low starting torque, but high horsepower at high speeds. Both the high-speed propeller and the multiblade turbine require some means—a tail vane or rudder—to keep the rotor turned into the wind.

Super-speed turbines resemble multiblade turbines somewhat. However, these machines have airfoil blades designed to grab and dump air quickly, so that "used" wind from one blade does not interfere with the blade next behind. Some of these machines drive the generator with a V belt around the perimeter of the turbine.

Vertical-axis machines are of two general types, represented by the Savonius and Darrieus designs. Savonius rotors have high starting torque but fairly low top end speeds. Many Savonius rotors are homemade, from oil drums split to form the two halves of the rotor.

Darrieus rotors, also vertical-axis, have two or three blades that somewhat resemble an egg beater turned upside-down. This machine has high efficiency but low starting torque. In fact, Darrieus rotors usually require incorporation of a small Savonius rotor as a starter, or they are started electrically by motorizing the rotor. Vertical-axis rotors are always positioned to receive the wind, thus need no vanes or rudders to align them. These rotors can be stacked on the same shaft for extra generating power.

THE GENERATOR

On the electrical end of a SWECS, there are basically two ways to go: battery storage or tie-in with a utility grid through a synchronous inverter or induction-type generator. There are advantages to both systems, depending on the situation, but for a farm already served by an electric utility, the latter is often the better bet. We'll look briefly at both systems.

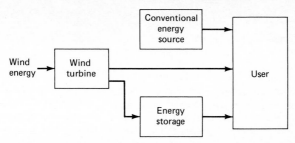

FIG. 19-4 Flow of electricity to load with a typical battery storage system.

A battery storage system is basically a battery-charging plant. The generator charges a bank of batteries with DC power, through a voltage regulator, much as the generator on an automobile charges the battery (see Fig. 19-5). The DC power then is changed to 115- (or 220-) V, 60-Hz alternating current through an inverter, as the electrical load demands power from the batteries.

Batteries are expensive, however. So are inverters, the equipment needed to impose a sine wave on DC power to convert it to 60-Hz alternating current. For example, a 110-V, 360-ampere-hour (Ah) battery costs in the neighborhood of $3000. Inverters cost about $2 per watt capacity, and require nearly a third of their capacity just to run themselves.

A storage battery is required for generating plants that utilize a power source that is intermittent or insufficient to meet the load on demand. For example, lights and equipment often are needed when the wind is not blowing.

Generally, lead-acid batteries are most economical, but automobile batteries aren't well suited to the charge/discharge patterns of a wind plant. A battery with a low self-discharge, such as a pure lead cell or lead-calcium grid, is better.

Batteries are built with a value of 2.4 V per cell. Thus, a 12-V battery would require six cells; a 24-V battery would have 12 cells; and so on. Sizing batteries at a rounded-off 2 V per cell allows for a charge/discharge efficiency of 90 percent. Battery capacity is rated in ampere-hours, whatever the combined voltage of the cells. For example, a 50-Ah battery can discharge 1 A per hour for 50 hours or—theoretically, at least—50 A for 1 hour. While the voltage of a battery is determined by the number of cells, the ampere-hour rating depends on the number of plates per cell and the size of the plates.

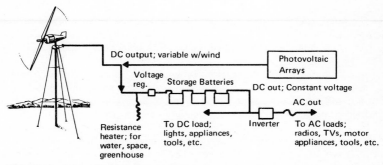

FIG. 19-5 Schematic of battery storage system.

From a safety standpoint, batteries should be stored where there are no sparks or open flames in the same compartment. Batteries undergoing charge contain an electrolytic process that produces hydrogen gas, and we've already talked about the flammability of hydrogen. Also, batteries should be stored where the temperature can be kept well above freezing. There are two reasons for this: (1) Batteries deliver more of their stored power quicker when the electrolyte fluid in them is warm and (2) if batteries discharge below about 50 percent, there's always a danger that the fluid will freeze and destroy the battery.

It's essential that a voltage regulator be used to prevent batteries from being overcharged, too. Overcharging can literally destroy a battery. The regulator limits the charge voltage and reduces the current through the battery cells as the peak charge is reached. Also, a voltage regulator allows a generator to maintain steady voltage as the turbine speeds up or slows down in fluctuating winds.

The DC electricity stored in batteries must be converted to 60-Hz AC power for most appliances and electrical equipment. The device that does this is called a *DC–AC inverter.* (*Rectifiers* operate in just the opposite way, to flatten out the sine wave of AC power and produce steady, DC-like power.) Three types of inverters are used with various SWECS to convert DC power to the voltage and waveform frequency needed.

Rotary inverters are battery-powered alternators that produce 115-V AC power. A DC motor turns a shaft that drives an alternator. Rotary inverters are not especially efficient —perhaps 60 percent, on the average—but they do impose good voltage and frequency control. They are available to about 2000 W output, but cannot handle "surge" loads such as fan motors, pumps, refrigeration compressors, etc.

Vibrator inverters are more efficient than mechanical rotary devices. In this type, DC power drives a small vibrator assembly that imposes the sine wave to convert the power to 60 Hz AC. Vibrator inverters are probably the most economical to use in low-power systems; however, they are limited to less than 1000 W continuous load.

Electronic inverters use solid-state, semiconductor technology to invert DC to AC in high-power systems, with good efficiency. These controllers are fairly expensive, compared with other types—costing up to $2 per watt capacity.

There's more equipment, more cost, and potentially more trouble with a battery storage system; that's why most modern SWECS users with utility power available elect to go with a system that can be interconnected with the utility grid. With these systems, power from the utility makes up any shortfalls from the wind plant. This eliminates the need for storage batteries, DC-to-AC inverters, and a backup engine-powered generator to be used when the wind fails to blow.

In the past several years, equipment has been developed that makes wind-generated electricity compatible with commercial alternating current. Called *synchronous inverters,* these solid-state electronic devices use the incoming commercial voltage and frequency as a reference to convert power from the wind plant to a current that is synchronized to the utility's power (see Fig. 19-6). During slack wind periods, the interconnection allows the utility to deliver power to the load in a normal fashion. When the SWECS is putting out more power than the electrical load requires, the excess flows into the utility power grid (this is often called *co-generation*).

Compared with a battery storage system, the wind-plant–synchronous-inverter setup operates at higher overall efficiency, provides uninterrupted power (from either the wind or the utility, or both), and can be installed for substantially less capital outlay.

Unhappily, since the synchronous inverter is tied directly to the power grid, the output of the generator cannot be used when the utility is out of operation—downed power lines also put the wind plant out of commission. This is an automatic built-in safety feature of this kind of system, for both the generator and any employees who may be working on utility lines. The wind plant cannot generate power without the reference voltage and

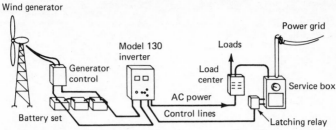

FIG. 19-6   In this system, the wind plant is tied to the utility grid in normal operation. When grid power is lost, the inverter automatically switches to become a battery storage system. *(Drawing: Real Gas and Electric Company.)*

frequency from the utility. The power plant could be designed with battery storage and inverters for use in emergencies, and at least one company makes a synchronous inverter with automatic switching relays to shift to battery power when the utility goes dead. But this boosts the cost of the total system considerably.

A system similar to the synchronous inverter is the *induction generator* powered by a wind turbine. Such a system must be tied in with utility grid power, as the induction generator takes its excitation from the utility. The induction generator idea is not a new one. This type of generator operates on the principle that an electric motor, when driven faster than its normal operating speed, starts to generate current. For years, equipment such as elevators and oil-well pumpers have used electric motors that regenerate as the equipment overdrives the motor on the "coasting" stroke.

However, modern induction generators are more efficient than electric motors that are mechanically driven past their motorized rpm's. Units such as the Entertech 1500 wind plant can be plugged directly into a 20-A circuit, where the power generated flows directly to lights and appliances. When the wind doesn't blow, utility power automatically picks up the load.

The SWECS you might choose will depend on, among other things, the amount and kind of uses you have for electrical power, the speed and patterns of winds in your area, and —most likely—your budget. What can you expect to pay? A 115- to 220-V SWECS with a peak output of 8000 W would have an 18- to 22-ft-diameter rotor and should supply 600 to 1100 kWh of power per month, with average wind speeds of 12 to 15 mph. This SWECS would cost $16,000 to $20,000, depending on the amount of engineering work involved, tower height, etc. At 10 cents per kilowatthour, this system would amortize over a 12- to 14-year period.

An "independent" SWECS, including batteries and inverter, would cost between $2.50 and $4.50 per watt of peak generating capacity. If the system were amortized at 14 cents per kilowatthour, it would require about 15 years to pay back the initial capital investment.

OTHER USES OF WIND POWER

While direct electricity is the most common end product of a wind plant, there are other ways to store and use the power. We mentioned the possibility of making hydrogen fuel with a wind generator in Chap. 17. See Fig. 19-7 for other ideas.

Another option is *heat*. Heating of farm structures and water consumes about 10

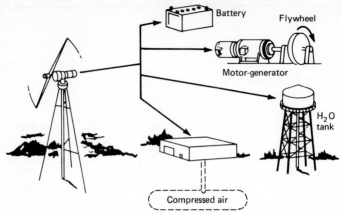

FIG. 19-7   Alternative ways of using and storing wind energy: in storage batteries, as mechanical energy in a spinning flywheel, as water pressure, and as compressed air. *(National Science Foundation.)*

percent of the electrical energy used in agricultural production. Wind energy could provide a substantial amount needed for space and water heating. This could also help reduce the peak electrical load of rural electric distribution systems.

L. H. Soderholm, USDA researcher at Ames, Iowa, notes that wind energy in the form of heat can be used in several ways:

1. The SWECS could be sized so that its output only met the heating requirements on the farm. This way the heat supplied from the wind would be only a fraction of the heat load carried by the entire electrical system and so no additional storage would be needed.

2. The SWECS could be sized to carry most of the heating load, with surplus heat stored in, say, water, to be used during periods that wind energy is insufficient to carry the heating load—much as is done with some solar systems.

3. The SWECS can be designed to carry the entire heating load, with any excess power stored or converted for other uses.

When wind energy is converted to electric power, the use of resistance heating elements—either for space or water heating—makes a simple, low-cost system. By using resistance heating, the generator can be allowed to run "wild cycle," with variable voltages and frequencies of AC power. The only controls needed are those required for overspeed protection of the rotor and to keep the generator within maximum ratings.

For space heating, water heated by wind energy can be circulated through a water-to-air heat exchanger in a forced-draft air-heating system. Another method of using wind energy to reduce the utility load is to operate a heat pump in conjunction with the heated-water storage. For water heating, wind-generated electricity can be interfaced directly with the water system by using immersion heaters, or heat exchangers, as shown in Fig. 19-8.

"In the temperature range of 140 to 200 degrees F, for space heating, water probably is the least expensive method of heat storage currently available," says Soderholm. "Four cubic meters of water, with a temperature differential of 60°C, will store 279 kWh of

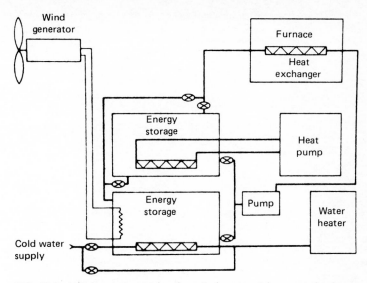

FIG. 19-8   Wind energy stored as heat (in hot water) lets a wind generator operate at peak efficiency and requires less control equipment than with an electricity-generating system. *(Courtesy: L. H. Soderholm, USDA-SEA Agricultural Engineer, Ames, Iowa.)*

energy. The size of the energy-storage system would be determined by the wind pattern, the capacity of the wind generator, the heat load and the carryover period."

Could you justify the cost of installing a SWECS to heat space and water? It would be more likely if you have a year-round demand for hot water—for dairy washdown, or underfloor heat for baby pigs, for example. The only sure way to know whether a wind-powered heating system would pay is to total up what you now spend to heat structures and water, compute what percentage of this load could be handled by a wind plant, and estimate how well the unit would pay for itself—and over how long a time.

Another possible method of storing wind energy is as compressed air. The SWECS operates an air compressor to fill banks of air cylinders. The compressed air then can be used to operate air-motor-powered equipment, to pump water, or to generate electricity.

While an air-compressor system might have potential for some applications, it seems a trifle unhandy for most farm operations. Still, it's another option to think about.

Finally, but not least, wind power can still be used to pump water—either directly, with a compressed-air pumping system, or in tandem with other power sources, as we described the wind-assisted irrigation pumpers in Chap. 12. Since windmill pumpers first came on the scene in the 1860s, farmers and ranchers have put more than 5 million of them over wells—and probably a quarter-million of them are still pumping water today.

There's a new design of water-pumping windmill on the market now. It drives an air compressor to pressurize water in the well and force it into a storage tank aboveground. Unlike other windmills, the air-compressor model does not have to stand over the well or water source. It can be positioned anywhere the wind is best and steadiest, up to a quarter-mile from the wellhead (see Fig. 19-9).

Before you invest in any wind energy conversion system, ask questions—a lot of questions. We've raised several in this chapter; you'll no doubt think of more. Getting the facts is the first major step. Then, go through the evaluation suggested in the foregoing pages: Analyze your load, monitor your wind resource, and compute the expected payback on various systems and applications.

There's a lot of power blowing in the breeze, but it takes a major investment to harvest it.

FIG. 19-9 This water-pumping windmill drives a small air compressor; the compressed air pumps water. With this system, the wind plant can be located up to a quarter-mile away from the well. *(Photo: Bowjon Corp.)*

20

Hydroelectric Power

Like wind energy, the power in moving water is an indirect product of solar energy. The heat of the sun evaporates water from oceans, seas, and lakes; wind currents carry the moisture over upland regions, where it falls to earth as precipitation. The water then makes the trip back downhill to oceans, seas, and lakes, and the hydrologic cycle begins all over again.

Water power was a vital energy source for early American settlers, who tapped the mechanical power in streams to mill grain, saw lumber, spin and weave cloth, and perform other essential functions.

Today, water power generates about 11 percent of the electricity Americans use and could potentially produce a lot more. In a study done during President Carter's administration, the U.S. Army Corps of Engineers estimated that there is a potential power supply of 54.8 *billion* W at existing dams in the country. Most of these are small "low-head" (less than 65 ft high) dams that have been abandoned as power sites or were never developed in the first place. Of about 50,000 existing dams with a power potential of 25 kW or less, only 2000 are now generating electricity.

For farmers who have it, hydroelectric potential is a happy accident of geography. Some power can be generated from almost any stream with a year-round flow, but not every rural acreage has a stream that can be harnessed and put to work *profitably*.

Volume, gradient, and steadiness of a stream's flow are important considerations for evaluating any site as an electricity-producing resource. Individual hydroelectric plants have been built with output capacities as low as 500 W. However, the cost of engineering, damming, siting, and locating equipment usually dictates building a plant that will produce considerably more electricity than that. Economies of scale bear more on hydroelectric power than on some alternative energy methods. Even for a relatively large plant—say over 2000 kW capacity—the cost of construction can run $1000 to $1500 per kilowatt capacity. Smaller plants typically cost more per unit of power.

Still, in some instances, where one property owner has a suitable site for hydroelectric power to serve several nearby farms, there are cooperative or joint ownership possibilities. A joint power-producing venture is not always convenient to administer—to everyone's satisfaction. Establishing a local ownership utility might be a more practical and workable way to go, in most cases.

That is being done by several irrigation districts in western states, where water flowing through irrigation canals is used to generate power during the irrigation season. Through purchase agreements with electrical utilities, the irrigation district's plant is tied in with the regional grid.

Still, if you're one of the fortunate few with hydropower sites that could be practically developed, there's a lot of potential electricity flowing down the creek. Once the site preparation is completed and the equipment is installed, the main expense would be interest on the capital investment. Well-designed, carefully installed hydroelectric plants are reliable, low-maintenance sources of electricity—some in use today have been generating power for 50 years or more, with little down time for repairs and maintenance.

EVALUATING THE SITE

The power theoretically available from a water source depends on the weight of water flowing in a given time, multiplied by the drop in elevation (the *head*) of the water. In terms of horsepower, the theoretically available power can be expressed by this formula:

$$hp = \frac{V \times H \times 62.4}{33,000}$$

where V = volume of water, cfm
   H = head or vertical distance between dam or water source and power plant
   62.4 = weight of 1 ft³ of water, lb

The product of multiplying volume times head times 62.4 is then divided by 33,000—the foot-pounds per minute in 1 hp of work.

The end result of this formula, as stated, is the theoretical power available. Actual power will be less by some factor, depending on friction and slippage in the entire hydroelectric system. Reducing the original power computation by about 20 percent may be close to the actual power produced.

Several facts about the water resource need to be gathered before the decision to build a hydropower plant is made. With some of these, qualified engineering help can make better assessments than can most lay persons.

1. Maximum and minimum flows of the stream

2. Head or fall of water

3. Length of pipe (penstock) needed to get the needed head

4. Water condition (clear, muddy, acid, alkaline)

5. Soil condition

6. Minimum tailwater elevation (below the plant)

7. Area and depth of storage pond behind the dam, if any

8. Horizontal distance from water source to power site

9. Distance from power plant to point of use of the electricity

Measuring the water flow is a necessary early step in evaluating a hydroelectric site. The rate of flow of a stream, ditch, or canal is measured either in cubic feet per minute, in gallons per minute, or both. Cubic feet per minute is the more common measurement for determining power potential; gallons per minute is more commonly used when measuring a water resource for irrigation. One cfm is a flow of water equal to a stream one foot wide and one foot deep, flowing at a velocity of one foot per minute. One cubic foot per minute equals 7.5 gallons per minute.

MEASURING STREAM FLOW

There are several ways to measure flow. Here are those most commonly used.

1. For a small stream, you can use a temporary dam and a container to estimate flow. Divert the stream channel with a dam so that the entire quantity of water can be caught and measured. You may want to install a pipe in the temporary dam to draw off the water. Use a container of known volume, and time the period it takes for the water to fill it.

Repeat this process several times, and then average the results. Divide the quantity of water (in gallons) by the number of seconds it takes to fill the container, and then multiply that result by 60 to get gallons per minute of flow. Divide this figure by 7.5 to find the flow in cubic feet per minute. Obviously, this method of measuring flow can be used only on rather small, slow-flowing streams.

2. For a larger stream with a known type of bottom, the flow rate can be easily estimated by the *float* method. Use this procedure when high accuracy is not needed and when more expensive measuring techniques are not justified. Select a section of stream 100 ft or more in length that has fairly uniform depth and width. Measure the surface width of the water, and estimate the average depth by making depth measurements at several places across the stream (see Fig. 20-1). Stretch a line tightly across the channel at the beginning and ending points and measure the distance between the two lines.

Choose a windless day and place a small float in the stream a few feet upstream from the beginning line. Time the period it takes for the float to pass between the two lines. Use a float that rides high in the water—you may want to attach a flag to make the float easier to keep in sight—and time the float for several trips to get an average time of travel. Correct the average reading you get by a factor of 0.8 for a stream with smooth bed and banks, and by 0.6 for a rock-strewn, hilly stream.

To compute the flow, divide the length of the trial section in feet by the time in seconds required for the float to travel that distance. This gives velocity in feet per second. Multiply the velocity by the estimated stream cross-sectional area in square feet to determine the flow rate in cubic feet per second. Then, multiply that result by 60 to get the reading in cubic feet per minute.

Measure the stream several times through the season, to establish a seasonal pattern of flow. If this is not convenient, measure the stream in the season that has the lowest natural flow rate—usually late summer.

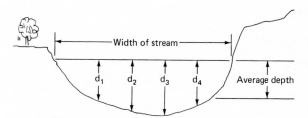

FIG. 20-1 Stream cross-section. Make depth measurements at d_1 through d_4 and average results; then multiply by stream width.

A tracer method, using colored dyes or salts, can be used in much the same way as the float measurement process. However, because the dye is often diffused throughout the stream, determine the velocity of the first and last portions of the dye and then take an average.

3. Using a *velocity head rod* is another fast, inexpensive way to measure stream flow. The rod, about 5 ft long, can be made from wood. If it will be used often, a 26-gauge copper sheet can be fastened to the cutting edge to protect it from damage. Mark a scale in half-inch increments on the rod, starting with zero at the bottom of the rod and stopping at 18 in.

To find flow rate with a velocity head rod, place the rod in the water with the cutting edge upstream and mark the stream depth on the scale. Then, place the rod sideways in the water, broadside to the current. This causes turbulence and the water will rise, or "jump," above its normal depth. Stream velocity is proportional to the jump. Measure the depth of the turbulent water on the upstream side of the rod, as shown in Fig. 20-2.

Subtract the normal stream depth from the turbulent depth to find the "jump height," or velocity head, in inches. Then, find the stream velocity in feet per second from Table 20-1. Determine the stream velocity at intervals across the current and average them to get an average stream velocity. Multiply this average velocity by the

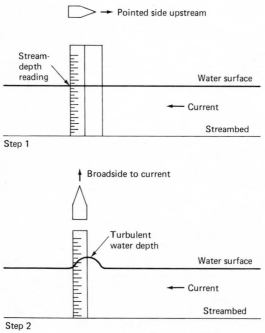

FIG. 20-2 How to use a velocity head rod: Step 1, measure stream depth with pointed side of the rod upstream; step 2, turn the rod perpendicular to current flow and read turbulent water at highest point on upstream side.

TABLE 20-1   Stream Velocities from Head Rod

Head, in	0	½	1	2	3	4	5	6	7	8	9	10	11	12	15	18
Velocity, ft/s	0	1.6	2.3	3.3	4.0	4.6	5.2	5.7	6.1	6.5	6.9	7.3	7.7	8.0	9.0	9.8

stream cross-sectional area, as mentioned earlier, to find the flow rate in cubic feet per second. Multiply by 60 to get flow in cubic feet per minute. This method is practical for estimating flows in streams with an average rate of flow. It is not accurate for velocities less than 1 ft/s or for very fast currents. Obviously, it cannot be used in channels more than about 12 to 14 in in depth.

4. A *current meter* can be used to determine the velocity of a stream of almost any size and flow rate. This device is the standard one used by the U.S. Geological Survey and by many engineering firms to gauge stream flow. This commercially made instrument has a revolving wheel or vane that is turned by water movement. To use the meter, mount it on a rod or lower it into the water with a cable. Rod mountings can be used in shallow streams, but cables may be needed in deeper water.

    Take meter readings at several locations across a stream's channel, then determine the average velocity from the instrument's calibration curve. Multiply the measured velocity by the cross-sectional area in square feet to obtain the cubic feet per minute of flow of each section measured, then sum the flows from each section to obtain the total flow rate of the stream.

5. An accurate flow-measuring device is the *weir,* a notch of a specific size and shape through which water flows. Using a weir also involves the most work, unless your stream already has a dam constructed. Weirs require enough slope in a ditch or stream to allow the water to be partially held back and spill over the weir. An air space is necessary under the falling sheet of water for accurate flow measurement.

    A V-notched, or triangular weir is useful only on fairly small streams, with flows of about 1 cfs. The bottom of the notch over which water flows is the *crest* of a weir. To use a weir, follow these steps:

    a. Build the weir with plywood, sheet metal, or 1-in and 2- by 4-in lumber. Make the crest and sides of the weir notch no more than ⅛ in thick. You can fasten metal strips to plywood or planks to form a sturdy crest. The crest should be fairly sharp on the upstream side.

    b. Install the weir structure in a uniform channel so that it will provide a long, deep pool on the upstream side, wide enough to permit a uniform current with a slow velocity. The height of the weir crest above the channel bottom should be at least twice the estimated head (select the probable head from Tables 20-2 to 20-4), which means you'll need some idea of the maximum flow rate of the stream to construct a weir and crest length with sufficient capacity to handle the estimated flow. In some cases, the banks of the stream above the weir may need to be raised to hold more water. Set the weir at right angles to the direction of stream flow, with the crest straight and level.

    c. Measure the depth of water flowing over the weir far enough away from the notch to be unaffected by the sharp downward curve of the water as it approaches the crest (see Fig. 20-3). Drive a stake about 3 ft upstream from the weir, with the top of the stake exactly level with the weir crest. Measure the head of water over the weir by placing a ruler on the stake, as shown in Fig. 20-3. A second method is to measure the depth of water right next to the weir dam, but far enough to one side of the crest to be in still water. Drive a nail on the upstream side of the weir, so that the top of

TABLE 20-2   Approximate Flow over 90° Triangular Weirs

Head, in	Flow, gpm	Flow, ac-in/h
3	36	0.08
4	74	0.16
5	126	0.28
6	200	0.44
7	294	0.65
8	405	0.89
9	548	1.21
10	714	1.58
11	895	1.98
12	1118	2.48
13	1365	3.05
13.5	1495	3.34
14	1630	3.63

TABLE 20-3   Approximate Flow over Rectangular Weirs

	Crest length							
	1 ft		2 ft		3 ft		4 ft	
Head, in	Flow, gpm	Flow, ac-in/h	Flow, gpm	Flow, ac-in/h	Flow gpm	Flow, ac-in/h	Flow, gpm	Flow, ac-in/h
2	98	0.22	198	0.44	298	0.66	398	0.88
3	181	0.40	366	0.81	552	1.22	738	1.63
4	278	0.62	560	1.24	852	1.88	1140	2.52
5			772	1.70	1164	2.58	1560	3.45
6			1010	2.22	1535	3.40	2055	4.54
7			1270	2.80	1980	4.27	2590	5.75
8			1540	3.40	2330	5.18	3120	6.90

TABLE 20-4   Approximate Flow over Trapezoidal Weirs

	Crest Length							
	1 ft		2 ft		3 ft		4 ft	
Head, in	Flow, gpm	Flow, ac-in/h	Flow, gpm	Flow, ac-in/h	Flow, gpm	Flow, ac-in/h	Flow, gpm	Flow, ac-in/h
2	101	0.22	202	0.45	302	0.67	404	0.89
3	190	0.42	376	0.83	560	1.24	750	1.66
4	296	0.65	580	1.28	864	1.91	1160	2.56
5	–	–	802	1.77	1196	2.66	1500	3.52
6	–	–	1062	2.34	1580	3.50	2100	4.64
7	–	–	1350	2.98	2000	4.42	2660	5.88
8	–	–	1638	3.62	2430	5.38	3220	7.14

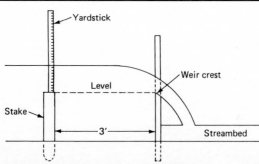

FIG. 20-3   Measuring head on the weir.

the nail is level with the weir crest, and then measure the head over the weir with a ruler. Refer to Tables 20-2 to 20-4 to find stream flow rates for head in inches.

See App. B for methods of measuring water flow in pipes.

MEASURING HEAD

Head—the distance water falls from a dam or catch basin to the power turbine—has a direct affect on the amount of power produced. As you'll note from the formula to compute horsepower earlier in the chapter, potential power depends not only on flow rate—the amount of water available—but also on head—the vertical elevation from which the water falls. A high head produces the most power at least cost, and at least consumption of water. In fact, head (which translates into water pressure, as 1 ft of water exerts a pressure of 0.43 psi) in a hydroelectric plant has much the same kind of linear effect on power output that wind speed has on a wind generating plant.

High-speed, high-head turbines utilize the velocity of water directed through specially designed jets or nozzles to get an amazing amount of work from relatively small turbines. For example, the importance of head in a hydroelectric plant is illustrated in Table 20-5, which gives performance figures for a 6-in-diameter Peltech wheel developed by Small Hydroelectric Systems and Equipment, of Arlington, Washington. You'll notice that doubling the head, from 50 to 100 ft, results in a threefold increase in the electrical power generated.

Head can be measured with a surveyor's transit and pole or with a hand level. Two people generally are required: one to hold the pole, the other to sight through the transit or read the level. Sight to a point on the pole, measure the height—including the height from the ground to the instrument in the initial measurement—and then move to the next lower elevation.

Repeat the process in stair-step fashion, as illustrated in Fig. 20-4. Total all measurements taken to establish the total, or gross, head. Measurements need not be taken along the stream's course, so long as starting and ending points correspond to the stream level at the desired points. For that matter, the water conducted to a power site often does not follow the stream, but is led through buried or aboveground penstock by the most convenient route.

Depending on the region of the country and the water-use laws that apply, dams can be built to get a higher head than the stream itself offers and to level out the supply of water to the turbine. When you build a dam in a stream, a storage reservoir is created that can be used to good advantage to conserve water during times when the turbine is consuming more water than is flowing in the stream or to supply more water than the stream flows during peak demand periods. The load on any plant is seldom (if ever) fixed; it varies with the needs of the power consumer. Dams also provide a settling basin, where trash and

TABLE 20-5   Performance Figures for 6-in-Diameter Peltech Wheel

Head, ft	Water flow, cfs	Horsepower	Turbine Speed, rpm	Watts generated
20	0.196	0.3	643	187.5
40	0.276	0.8	910	500
50	0.309	1.2	1018	750
80	0.392	2.5	1286	1,000
100	0.438	3.4	1440	2,125
200	0.618	11.0	2040	6,875
300	0.757	20.1	2490	12,562

Total vertical distances from A to B

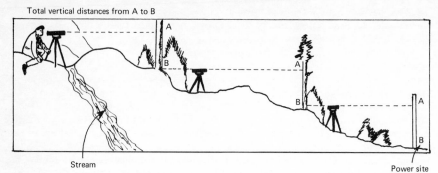

Stream

Power site

FIG. 20-4 To find total head, add all vertical distances from A to B. Elevation measurements need not be made along the stream, but should begin at water source and end at the power site.

water-borne material can settle out of water that is run through the turbine. Dams can be built of earth, stone, concrete, timber, or combinations of materials, depending on the nature of the site and which materials are handiest and lowest in cost. Small streams can be dammed fairly easily, but larger streams and those in areas subject to heavy flooding need professional engineering to prevent costly mistakes.

In virtually all states, streams that flow year-round are considered public waterways. Permission is customarily required of one or more government agencies before a stream is dammed. In more humid regions, where riparian water rights prevail, obtaining permission to build a dam may be a fairly simple bureaucratic exercise. In many western states, impoundments on flowing streams by private individuals are generally prohibited. However, part of the stream usually can be diverted into a catch basin or power channel, with some type of low diversion structure.

Pump-back storage is a growing idea with small hydroelectric systems where water is impounded. Pumped storage systems use the same principles as conventional plants, but have a tailwater pool below the power plant to catch water that has done its work through the turbine. Then, during periods of low load, the water is pumped back into the upper reservoir to be used again. It takes more equipment, but it's a way to get more use from a limited flow of water.

WHAT KIND OF TURBINE?

The hydraulic "engines"—the turbines—that convert the kinetic energy in falling water into mechanical-shaft horsepower are designed with blades that allow most of the potential energy to be transmitted to the rotor. They generally are classified by types—either reaction or impulse—or are named for their developers:

1. Francis reaction turbine

2. Pelton impulse turbine

3. Reaction propeller turbine

Most Francis turbines are descendants of simple reaction turbines used in Egypt about 100 B.C. They have been improved in the past 2000 years, of course. They are often called *full-admission* turbines, as the water enters the rotor around its complete circumference.

The direction of water flow changes through a Francis turbine; it enters tangentially at the circumference of the rotor and discharges parallel to the axis of rotor rotation. Francis-type turbines can be mounted either vertically or horizontally.

The Pelton wheel is an impulse-type turbine (see Fig. 20-5). Whereas the complex direction changes of the water through a Francis turbine radically lower the water pressure, there is little pressure change across the Pelton rotor. The potential energy of the water is converted to kinetic (mechanical) energy in a specially designed nozzle that directs a high-velocity stream of water onto hemispherical buckets or blades. The increase in velocity, and corresponding rapid pressure drop, occurs in the nozzle, rather than in the turbine itself. Improved Pelton designs use more than one nozzle to partially overcome the partial-admission limitation (water impacts on only part of the wheel at any one time).

Reaction propeller turbines are of two general types: fixed-blade and variable-pitch. These turbines resemble a propeller-type pump or a ship's propeller. The water is admitted around the complete circumference of the rotor (full-admission principle) and flows parallel to the axis of rotation. The Kaplan-type propeller has blades with variable pitch to keep the speed of the turbine constant despite variable loads. This lets the rotor operate at greater efficiency.

Head plays an important role in virtually all turbine designs; however, reaction-type turbines can generally operate at lower head than can impulse turbines. Some manufacturers, such as James Leffel and Company, Springfield, Ohio, market self-contained turbine-generator units that perform well in low-head situations (see Fig. 20-6).

As with wind-powered generators, hydroelectric plants can generate either AC or DC power. Generally, a site that will develop 1500 W or more is equipped to deliver 60-Hz AC power. AC generators (alternators) must be run at a steady speed, to produce the 60-Hz power needed. For example, a four-pole alternator driven by a water turbine must run at exactly 1800 rpm to deliver power at standard voltage and frequency. (The number of poles determines the alternator's speed of rotation: a two-pole unit would need to run at 3600 rpm, whereas a six-pole alternator would run at 1200 rpm.)

Both mechanical and electronic speed-governing devices are built to be used with hydroelectric plants. Woodward-type governors that work by centrifugal force are reliable, but costly. New solid-state electronic load-diversion systems work very well with AC

FIG. 20-5   Improved versions of the Pelton wheel are driving members in an integrated turbine-generator-controller unit. *(Photo: Small Hydroelectric Systems and Equipment.)*

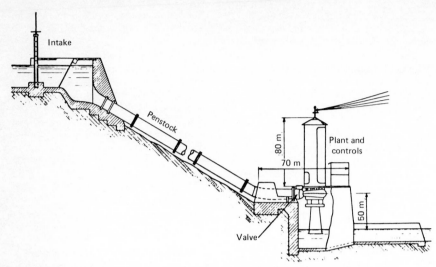

FIG. 20-6   Typical layout of low-head, self-contained turbine-generator unit. *(National Academy of Sciences.)*

plants. Overspeeding is no hazard with these systems when they employ an *over/under* cycle switch and an automatic *blow-by* or by-pass valve. The plant shuts down automatically if the AC cycles move out of a preset range. The blow-by valve is held closed by a solenoid powered by the alternator, and opens instantly if power is interrupted.

Hydroelectric systems also can be hooked in with utility grids through synchronous inverters or AC induction generating systems. The principle is the same as described in Chap. 15: Once grid power is lost, the generator shuts down automatically.

Not every farm owner, as we mentioned, has a flowing water source that will generate electrical power. And for those who do, small-scale hydropower plants built with durability and efficiency do not come cheap. The cost of building a hydroelectric plant will depend almost directly on the physical features of your site. If the site has a fairly high head, so less expensive turbines can be used and large dams are not required, the water-power option may be worth considering.

But get the advice and counsel of a qualified engineer before you start building dams and buying equipment.

Sun-Made Electricity and Standby Power

Direct solar-generated electricity, while still out of practical reach for most farm operations, is steadily coming into the range of an economically feasible source of power. You may want to watch new developments (they're coming fast) that could put solar-generated power in the ball park for some applications.

It's a great idea: producing electricity directly from sunlight. It's inexhaustible, "free," intermittently available everywhere, and can be harvested with equipment that has no moving parts to wear out. Solar electricity also has some rather glaring drawbacks, chief among them the fact that the sun doesn't shine 24 hours a day and the high cost per watt of electricity generated.

PHOTOVOLTAIC CELLS

As with most other so-called "new" alternative energy sources, the simple photovoltaic cell has been around for a century or more. Early models used selenium to turn sunlight into electrical current, through an electrochemical reaction of certain materials to light.

In the 1950s, with gasoline selling for 30 cents a gallon or less, the U.S. space industry began fairly intensive research into photovoltaic cells. These early cells that powered instruments on spacecraft were expensive—in the neighborhood of $2000 per peak watt of output (the maximum power generated in full sunlight). Then, with the decline of interest and funding in the space program, solar energy research fell on hard times, until the energy squeeze of the mid- to late-1970s.

Today, gasoline and other petroleum-based fuels have climbed to four times 1950s prices, and mass-produced silicon photovoltaic cells cost $8 to $10 for each watt they deliver at high noon on a clear day. That's still out of line with the cost for more conventional sources of electricity, but it's well below the investment in solar-generated power just a few years ago and represents dramatic strides in making an energy source less expensive.

The promise of direct solar-generated electricity also is evidenced by the fact that many small, pioneering solar firms have been bought or eclipsed by big oil companies—Mobil, Shell—or big electronics firms—RCA, Honeywell, Texas Instruments. These companies have the bucks and technological expertise to move the technology forward rapidly, and that is happening. Recent breakthroughs in materials and manufacturing processes promise to trim the cost to $1 or less per watt of electricity generated.

The most commonly used material in solar cells is silicon, a chief component of common sand and therefore relatively abundant. However, ordinary silicon crystals won't work. The silicon must be refined to virtually 100 percent purity in an electric furnace, then "grown" into crystalline rods with a coherent molecular structure. A few atoms of other materials—often boron and phosphorus—are added by chemical bonding to form positive and negative (P-type and N-type) cells. Then, the silicon rods are sawed into waferlike sections only a few hundredths of an inch thick. In the process, more than half of the tediously made material is lost as sawdust.

A typical solar cell contains several specially coated layers of silicon or other light-sensitive semiconductor material, with external wires attached. When light strikes the cell, electrons are released and an electric current results from the flow of these dislodged electrons. Because of the molecular makeup of the semiconductor material, each photon of light energy that strikes the cell liberates one electron. The more light, the more electrical activity in the cell (see Fig. 21-1). A portion of that electricity can be drawn off and used.

Individual solar cells are connected electrically to form solar modules or building blocks —called *arrays*. For example, 40 solar cells could be connected together in an array that would provide enough electricity to charge a 12-V automobile battery. An array about 200 ft² could generate 3500 W of electricity in bright sunlight. See Fig. 21-2.

New Developments in Photovoltaic Cells

Photovoltaic cells built to this date are not especially efficient. On the average, only 8 to 10 percent of the sun's energy is harnessed and converted to electricity. The efficiency of converting light to electrical power in a silicon cell rarely exceeds 12 to 15 percent. Because of high manufacturing costs and relatively low efficiency of silicon cells, most research emphasis today is on two fronts: bringing down the cost of making silicon cells and finding other, more efficient semiconductor materials.

Cells made from gallium-arsenide–arsenide, copper-sulfide–cadmium, and other combinations, while no less expensive to produce than silicon cells (in fact, some are more costly) are more efficient at converting sunshine to electrical power. Gallium-arsenide cells can transform 22 to 25 percent of sunlight into power.

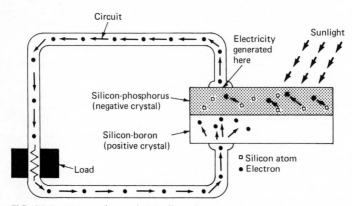

FIG. 21-1   How a photovoltaic cell works.

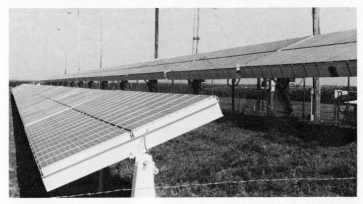

FIG. 21-2   This Nebraska solar cell array takes up a third of an acre; it generates power to pump irrigation water on 80 acres of crop land.

Some companies have developed an automated, continuous-flow process of building gallium-aluminum-arsenide–gallium-arsenide cells. The completed cells are placed in a series of strips and sealed in freon-bearing tubes, which are placed at the focal point of a concentrating solar collector. The sun reflects from the concentrating collector onto the cells, to generate electricity at about 25 percent efficiency. Meanwhile, another 25 percent of the sun's potential energy is converted to freon steam to propel a turbine generator. According to the company, this dual system captures 50 percent of the sun's energy and generates power at about $690 per peak kilowatt.

Other researchers are putting emphasis on new manufacturing processes that bring the cost of silicon cells down. Rather than grow the silicon in long, pure crystalline rods, Honeywell, Inc., has developed a process to coat long strips of ceramics with silicon. Radio Corporation of America (RCA) has come up with a similar "sandwich" approach to making solar cells.

While some of these processes produce photovoltaic cells that are somewhat less efficient at capturing solar energy, the greatly reduced cost of manufacture is usually more than a fair trade-off. These techniques generally involve dipping, coating, or spraying silicon on strips, sheets, or ribbons of ceramics, graphite, or other material. At this writing, few —if any—of these new-process cells are on the market.

But the photovoltaic power industry is still an infant, for the most part, and will probably suffer some growing pains before it comes of age. The potential exists for direct solar-generated electricity to become an important source of power. The big question right now is: *when?*

Solar cells produce DC power, which must be inverted to AC power before it can be used to operate most conventional electrical equipment. The problems of storing, inverting, and using solar-generated electricity are very similar to those described for wind-generated battery-charging systems.

You'll recall that we described the University of Nebraska's experimental solar-powered irrigation system in Chap. 12. That system works, technically—although it is far from practical for a commercial farm operation at this point. The Nebraska researchers are using "surplus" solar power to make fertilizer and operate grain dryers, lights, and fans. As costs, efficiency, and reliability of photovoltaics get more in line, the first farm operations to justify the capital investments will no doubt be those that can utilize the energy output completely, to minimize unit energy costs and make solar power as economically viable as possible.

For example, if direct solar-generated power is used primarily to pump irrigation water in summer, the solar plant could be utilized more fully throughout the year by crop drying or livestock operations that have relatively low energy requirements during the irrigation season.

The idea, of course, is to keep the expensive solar power plant fully employed in order to spread the initial cost of equipment over as many kilowatthours of usable electricity as possible. But then, that's the major justification for any on-farm electricity-generating system.

STANDBY ELECTRIC GENERATORS

Whether you produce your own electricity or buy it off the pole, a standby electric generator can eliminate most of the discomfort and inconvenience during a power interruption on the farm. If you have a dairy operation, a climate-controlled greenhouse, or a fully confined, environmentally controlled livestock unit, standby power can let you sidestep substantial financial losses during power outages.

A standby generator should be capable of providing adequate power at the correct voltages for essential equipment. Most farms have single-phase power with 120/240 dual voltage. If three-phase power is needed, this may call for special planning and equipment.

It's a good idea, usually, to consult the local power supplier before buying auxiliary generating equipment. You can get advice about correct, safe electrical connections and possibly about generating equipment that is compatible with your electrical load.

Generators for farm use are of two general types: direct-connected engine-driven units and tractor-driven generators.

Self-contained, direct-connected engine generators may be either manual start or automatic start. In situations where immediate start-up is required after a power failure, an automatic start unit should be chosen. This type of generator is connected through a control panel that automatically starts the generator when the main power source is interrupted.

An auto-start unit should have the capacity to start and run all the equipment connected to it. Manual-start generators are essentially the same as the auto-start models, but must be started by an operator when a power interruption occurs. These are less expensive to buy and less complicated to install, of course.

Direct-connected engine-driven generators should be run about once each month to ensure their dependability. Some auto-start units have controls that automatically start the engine for "exercise" runs at predetermined intervals.

Tractor-driven generators usually consist of an alternator mounted on a trailer or three-point hitch, with power supplied through the power takeoff (PTO) shaft from the tractor. They are available in sizes to meet most farm needs.

Generally, the tractor should have a horsepower rating of at least twice the kilowatt capacity of the generator. For example, a 15-kW generator should require at least a 30-hp tractor to drive it at full load. A PTO-driven generator mounted on a trailer or three-point hitch makes a handy portable power plant for operating a welder or other electric tools in the field.

To determine a standby generator size, compute the load to be served, as outlined earlier. If the generator is too small for the job, electric motors can't be started or will burn out on lower voltages. Electronic equipment won't function properly if AC cycles or voltage is below normal.

When sizing a standby generator, remember that electric motors draw three to six times more power at starting than when running, even under full load. So, even though a 5000-W generator might run a 5-hp motor, it cannot produce the 15,000 to 17,500 W needed to start it. Table 21-1 shows the power required at typical electric loads for most 240-V motors.

Standby generators are rated in watts or kilowatts of output. The ampere rating of equipment can be converted to watts by multiplying the voltage times amperage shown on the equipment's nameplate, as we outlined earlier.

Standby generators normally are connected to the main service panel (or panels) through a double-throw switch located just below the electric meter. When properly installed, this positive lock-out type of switch prevents interconnecting the main power source and the standby generator. The proper wiring of this type of switch is shown in Fig. 21-3.

(*Note:* Three-phase power would have three poles on the switch, rather than two, as shown.)

The transfer switch must have enough capacity to carry the total load of the farm or building it feeds, even though the standby generator may have less capacity. For instance, a farm with a 200-A main service should have a 200-A transfer switch, even if the standby generator is capable of producing only 50 A to serve a "stripped-down" load.

Emergency standby generators should be grounded, to prevent electrical shock to people and livestock that may come in contact with the equipment. It's also necessary that engine-driven generators be ventilated with enough air for engine combustion and cooling. Duct work can be used to supply a standby generator housed in a building. When an engine-driven generator is installed inside a building or other enclosure, some means should be provided for exhausting the gases from the engine exhaust system.

Can your farm weather a power outage? If you need reliable electricity to make milking machines milk, incubating eggs hatch, and pumps, fans, and other equipment stay on the job, take a long look at standby electric generators. Buy as much generating capacity as your farm requires for safety and comfort; however, you probably won't need an auxiliary unit that can power every electric motor and light bulb on the place. It is, after all, an *emergency* piece of equipment, meant to power necessary electrical devices until normal power is restored.

TABLE 21-1   Power Required at Typical Electric Loads

	Motors		
Horsepower	Amperage (240 V)	Running watts	Starting watts
7.5	40	7500	26,250
5.0	28	5000	17,500
3.0	17	3000	11,500
2.0	12	2000	7,000
1.5	10	1500	5,250
1.0	8	1000	3,500
¾	6.9	800	2,800
½	4.9	600	2,100
¼	2.9	400	1,400

Appliances		
		Watts
Coffee maker		800
Iron		1000
Portable heater		1350
Range		8000
Toaster		1000
Water heater		4500

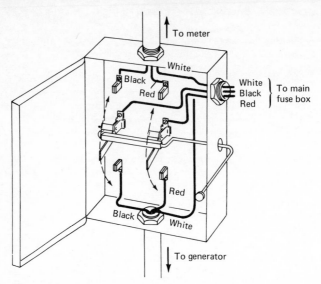

FIG. 21-3 Typical wiring of a positive-disconnect transfer switch. *(University of Missouri.)*

22

Keeping out of Trouble

Rising energy costs are causing almost everyone—farmers and nonfarmers alike—to consider investments that will reduce energy expenditures.

These investments are expected to result in fewer cash expenses for fuel and power, but often result in increased capital, labor, management, and maintenance costs. There's usually a trade-off: spending dollars now to save dollars in the future; using limited capital (credit) and labor to cut back on current dollar outlays.

A farm operator who invests in energy-producing or -conserving equipment and practices needs to know not only that the move is *profitable,* but that it is *technically feasible* and *legal.* No one man can be expert in all subjects, and much of the alternative energy technology being adapted gets into brand-new fields. The two big tasks involved in the successful completion and operation of any new facility are choosing the right system in the first place and hiring the right help to build or install it—neither of which is a simple matter where alternative energy systems are concerned.

Most commercial farmers have experience at working with farm building contractors: Well over half of the farm buildings constructed are of the "packaged" or "preengineered" variety. Even when farm labor is used for most of the construction work, a good deal of the design chores have been done by professional engineers who standardize the basic building package. With alternative energy systems, new dimensions of design and construction are encountered. Some knowledge of plumbing, steam engineering, microbiology, and fuel technology are required to build, as well as to operate, a fuel alcohol plant. Similarly, special skills and talents are called for in designing and building biogas plants, wind-energy systems, hydroelectric plants, and solar heating systems. The need for these new technical skills not only limits a farmer's ability to design and build his own energy systems, but narrows the circle of contractors qualified to perform the tasks required.

IS IT LEGAL?

For some alternative energy systems—notably, grain alcohol plants—there are legal requirements to meet, operating permits to obtain, federal records to keep, etc. Depending on the laws of a state or locality, statutes and building codes may specify material standards for methane digesters, solar collectors, wood-burning apparatus, and other natural energy systems. Therefore, it's vital that engineers and contractors be aware of the rules and regulations that apply, as well as the technical requirements.

While we're on the subject of keeping alternative energy legal, the Bureau of Alcohol,

Tobacco, and Firearms (BATF) of the U.S. Treasury Department has eased up on the rules for fuel alcohol plants. A permit is still required, but the bond formerly required for small-scale plants has been eliminated, and the amount of the bond for larger commercial plants has been reduced. New rules that went into effect July 1, 1980, require that anyone who intends to produce ethyl alcohol first obtain a permit from BATF, but many of the earlier legal obstacles have been removed. Briefly, here's what's involved:

A detailed letter describing the distillery equipment and purpose should be sent to the nearest BATF office. The letter should give the exact location of the still and describe what equipment is to be used, as well as the feedstocks and process planned. The still must be located where it can be made secure from theft or unauthorized use. The product must be stored in a locked container. If the alcohol is to be sold or used off the plant premises as fuel, it must be *denatured* by one of several formulas provided by the BATF. Denaturing often is done by adding 5 gal of kerosene to each 100 gal of ethanol. Records are required to show the volume of production and distribution of the alcohol.

For small-scale plants (not more than 10,000 proof-gallons per year) no bond is required. After the initial letter is received, BATF will forward special forms for permits and annual reports. Generally, BATF wants to know how the by-products of alcohol production will be disposed of, too.

Medium-sized plants, producing from 10,000 to 500,000 proof-gallons per year, are required to file a bond in an amount of about 20 to 25 percent of that normally required for a beverage alcohol producer of this volume. The bond runs about 0.1 to 0.2 cent per proof-gallon of alcohol produced. Some additional information is required on intended markets and by-product disposal for these intermediate-sized plants, and reports to BATF are required semiannually.

Large alcohol plants—over 500,000 proof-gallons capacity—also must post a bond, and are required to file quarterly reports. More information and forms required can be obtained from the BATF office serving your state:

1. Bureau of Alcohol, Tobacco, and Firearms
   230 S. Dearborn St.
   Chicago, IL 60604
   (312) 353-3883

   Serves *Illinois, Iowa, Kansas, Minnesota, Missouri, Nebraska, North Dakota, South Dakota, and Wisconsin.*

2. Bureau of Alcohol, Tobacco, and Firearms
   500 Main St.
   Cincinnati, OH 45202
   (513) 684-3337

   Serves *Indiana, Kentucky, Michigan, Ohio, and West Virginia.*

3. Bureau of Alcohol, Tobacco, and Firearms
   Two Penn Center Plaza
   Philadelphia, PA 19102

   Serves *Delaware, Maryland, New Jersey, Pennsylvania, and Virginia.*

4. Bureau of Alcohol, Tobacco, and Firearms
   Six World Trade Center
   P.O. Box 15, Church Street Station
   New York, NY 10008

   Serves *Connecticut, Maine, Massachusetts, New Hampshire, New York, Rhode Island, Vermont, Puerto Rico, and Virgin Islands.*

5. Bureau of Alcohol, Tobacco, and Firearms
   3835 Northeast Expressway
   Atlanta, GA 30301
   (404) 455-2631

   Serves *Alabama, Florida, Georgia, Mississippi, North Carolina, South Carolina, and Tennessee.*

6. Bureau of Alcohol, Tobacco, and Firearms
   Main Tower, Room 345
   1200 Main Street
   Dallas, TX 75202

   Serves *Arkansas, Colorado, Louisiana, New Mexico, Oklahoma, Texas, and Wyoming.*

7. Bureau of Alcohol, Tobacco, and Firearms
   525 Market Street
   San Francisco, CA 94105

   Serves *Alaska, Arizona, California, Hawaii, Idaho, Montana, Nevada, Oregon, Utah, and Washington.*

The BATF retains the right to enter and inspect any alcohol plant at any time. In addition to federal requirements, most states require permits for the production of alcohol; some require fees and bonding, as well.

MAKING DECISIONS

The right energy system (or systems) for your farm is the one that provides for your needs at the least cost. With most alternative energy systems, you're dealing with rather large capital outlays. That makes early planning vital, if you're to select a system or combination of systems that will perform satisfactorily. Good decisions are based on good information. Throughout this book, we have tried to provide information of a kind to let you evaluate various alternative energy options as they apply to your operation, presenting both the strengths and drawbacks of different alternatives.

The fact remains, however, that no outsider can decide which system best fits an individual operation. That is reserved for the person who does the managing—with perhaps some influential counseling from the lender. The purchase of an energy system can be compared with that of a tractor, new livestock building, or any other major farm investment. There are wide differences in prices. There's also a certain amount of faddishness connected with much energy hardware. Optional accessories may or may not add to the successful operation of the equipment in a given installation, but they will certainly add to the cost.

Once you have made the basic decisions regarding the kind, size, function, and application of an alternative energy system, you're ready to start shopping for the hardware and a contractor to install it. This is not quite as simple as finding someone who can build fences or put up a machine shed with plumb walls.

As noted earlier, special design and construction skills are needed if the entry into alternative energy is to be successful and durable. The basic process of finding a qualified contractor for an alternative energy system is much the same as finding one for a farm building project. Several turnkey alternative energy systems are available. Appendix D lists several firms providing engineering, design, and construction services, as well as equipment in nearly all areas of alternative energy. Check local advertising media: telephone yellow pages, newspapers, and farm magazines. Try to find other farmers who have installed the same or similar energy equipment; their recommendations and cautions can help keep you out of trouble.

Don't sign with the first contractor you discover; don't buy the first energy system you hear about. Take a long look at a contractor's track record. Any good contractor should be happy to provide a list of references, including past customers. Investigate the contractor's workmanship, timeliness, and reliability in living up to agreements. Given a certain level of equipment reliability, the choice of a contractor may be more important than the choice of a particular piece of equipment.

Neil F. Meador, agricultural engineer, University of Missouri, recommends that each of the following items be thoroughly discussed with prospective contractors and that all be specified in writing before the contract is signed.

Bid Alternatives

In some cases, it may be desirable to have separate bids on parts of a complete project, as well as bids on the entire job. When money is limited, a farmer may wish to use his own labor to do site preparation or some equipment installation to keep costs down. Alternative bids allow selecting those jobs that can save the most money.

Duties of the Contractor

What is the contractor expected to do? On most projects, the contractor will supply most or all of the labor, equipment, and materials to complete the project.

Duties of the Owner

Any work or materials to be supplied by the farmer should be spelled out in the contract. Usually, the farmer provides electrical power, telephone service, and water required during construction. The farmer or his representative also should be available at specified times for consultation, to interpret plans and specifications, etc.

Drawings and Specifications

No building or system should be started without a complete set of plans and written specifications. These may be supplied by the owner, the contractor, or the equipment supplier and should be included as part of the written contract.

Shop Drawings for Fabricated Equipment

Many energy systems contain equipment that is designed and built for a specific installation. To facilitate service after construction, the farmer should have a complete set of plans for any nonstandard items of this type.

Laws, Permits, and Regulations

The design and construction of any system should conform to applicable laws and regulations. Usually, it is the contractor's responsibility to adhere to these, although the required permits may be obtained by either the farmer or the contractor. Make sure the contract spells out who is responsible.

Changes and Modifications

Few building projects go exactly according to original plans and specifications. Both the farmer and the contractor need to agree on procedures to be followed in accomplishing changes. The contract should include details on the initiation of changes, revisions in plans and specifications, and contract price revisions needed to make the changes.

Substitutions

Events beyond the control of either the owner or contractor—delivery schedules, equipment model changes, price changes—often require substitutions during construction. Substitutions should be subject to the approval of the owner before being incorporated into the project.

Insurance

The contract should specify whether the owner or contractor is responsible for securing adequate risk protection. Four general types of insurance coverage are generally required for protection during a major construction project.

1. **Workmen's compensation** covers injury to employees working at the construction site. It usually is carried by the contractor.
2. **Public liability and property damage insurance** protects the contractor and subcontractors from claims for personal injury, death, and property damage. This coverage normally is provided by the contractor.
3. **Protective liability** protects the owner in the event of claims arising from the construction project. It may be provided by either the farmer or the contractor.
4. **Builder's risk insurance** protects on-site materials in the event of loss or damage by fire or other casualties. This usually is provided by the contractor, but may be a responsibility of the owner in cost-plus contracts.

Payment

The contract should spell out the method, terms, and time of payment for the project. It's common for large projects to call for payment of portions of the contracted price at specific points during construction, with final payment due on completion and acceptance. Be sure the contract specifies who is responsible for paying subcontractors.

Storage of Materials

The responsibility for providing weatherproof, on-site storage of construction materials should be indicated in the contract.

Cleanup

Upon completion of the construction, the contractor should be required to clear the site of construction debris and to clean up building surfaces. If the owner assumes this responsibility, the contract should so specify.

Utility Connections

Responsibility for connection to electric, water, sewer, and gas lines as required should be specified in the contract. Extension of the utilities to the site normally is included under the "duties of the owner" section of the contract.

Warranties

Terms of any contractor-guaranteed work or equipment should be specified in the contract. Provisions also should be made to transfer to the owner any warranties provided by manufacturers or suppliers of equipment

Service Manuals and Operating Instructions

The contractor should be responsible for providing the owner with complete operational and service manuals for component equipment. The contractor should also provide instruction in proper operation of any equipment not familiar to the owner.

Time Schedule for Completion

For some construction projects, particularly those involving animal housing, it is essential that a completion date be known well in advance. If time is a critical factor, make sure both the owner and contractor understand—in writing—when a project is to be ready for owner acceptance.

The contract may indicate test performance standards, with payment terms tied to a satisfactory shakedown period.

If the farmer chooses a reputable, competent contractor, the odds are excellent that the dealings will result in the satisfactory completion of the project, Meador notes. When troubles do happen, they usually are related to poor communication between the owner and builder. That's where a specific, clear written contract can benefit both parties. Discuss each of the above contract points with prospective bidders, plus any additional points that bear on the specific job to be done. If the potential bidder hedges on a point or two during the preliminary discussion, do some more looking.

With most alternative energy systems, you're generally talking about a big-ticket kind of capital investment. Some, such as solar collectors incorporated as part of the structure on a new livestock building, will not add a great deal of cost to the total project. But you're spending a lot of money for the total building—which makes lining up the right contractor a critical ingredient to success.

23

There's Money Available, But . . .

To date, there has been more talk than hard cash in the federal energy program, particularly where grants and loans to individual farmers are concerned.

While smaller, farm-sized plants (alcohol stills, wind-energy systems, etc.) require less financial backing, generally, than larger, commercial plants, Congress and the U.S. Department of Energy (DOE) have focused mainly on large-scale operations that might make bigger dents in our imported petroleum fuel. The picture is changing somewhat, where grant and loan money is concerned. Both DOE and the U.S. Department of Agriculture (USDA) are putting some money into small-scale energy financing programs, as are other federal agencies, such as the Department of Commerce and the Small Business Administration (SBA).

Sorting through the paperwork to find out what's available, who's eligible for it, and how to get it is a formidable—and unending—task. Describing government funding programs available at this writing and guessing which will be in effect or fully funded by the time you read this is about as hazardous as trying to forecast the weather that far in advance. Some federal assistance programs have a notoriously short shelf life—you'll need to check on current provisions.

However, changes (although less money) at USDA should make it easier for individual farmers to learn about and apply for loans from that agency. In short, money is available to help finance farm-scale alternative energy systems; getting it is often another story. Here are federal incentives, loans, and grants open to farm operators and their cooperatives, "at this point in time," as they say in Washington.

INCOME TAX CREDITS

Tax incentives are probably the best and most generally available form of assistance to farmers who build alternative energy systems. It's an after-the-fact method of funding, true, but the Windfall Profits Tax Act of 1980 boosted energy tax credits and also added special credits for alternative fuels produced and used on the farm.

Since the energy tax credit (and the investment tax credit which usually applies to energy equipment) comes off the top of your federal income tax bill, it's worth some extra effort to make sure your energy system qualifies:

- The energy tax credit was raised from 10 to 15 percent for solar, wind, and most geothermal equipment that provides heat, hot water, cooling, or electricity, including storage and distribution systems. For example, if you install solar collectors to replace some form of conventional energy in a farm operation, and the equipment has a useful life of seven years or more, you can take a straight 25 percent of the purchase price off your federal income taxes (15 percent energy credit, plus 10 percent investment tax credit).

- Small-scale hydroelectric plants (with a capacity of 25 MW or less) may qualify for an energy credit of 11 percent.

- For farmers who produce *and use* alcohol fuels (either ethanol or methanol) in their farming operations between October 1, 1980, and December 31, 1992, a new income tax credit is allowed. The credit amounts to 40 cents per gallon on alcohol of at least 190 proof, and 30 cents per gallon on alcohol of at least 150 proof but less than 190 proof. The alcohol must be made from renewable feedstocks, in processes that use energy sources other than oil or natural gas.

- Equipment that converts biomass to energy (methane digesters, alcohol distillation plants, and storage and handling equipment) is eligible for a 10 percent energy tax credit through 1985.

- Co-generation systems are eligible for a 10 percent energy tax credit for three years (the credit can be carried back three years or forward seven years). This credit applies to equipment that produces more than one form of energy, such as a boiler that produces both electricity and space heat.

- Farm residences qualify for a 40 percent tax credit on up to $10,000 in expenditures for certain solar and wind-energy equipment. The law is not especially clear on how this credit would apply to energy systems that serve both the farm business and the family residence, such as a wind generator that provides electrical power for both. You may want to get a current copy of IRS Publication 572 from your nearest Internal Revenue Service office, and go over the tax credit rules with your tax adviser.

Energy tax credits are nonrefundable; that is, if the tax credit totals more than your total tax liability, you cannot receive a federal tax refund for the difference. Tax credits are also taxable, which means that the amount of the credit claimed must be added to gross income; then the taxes owed are calculated on that total income and the tax credit is subtracted from taxes owed.

Here's how these tax credits might apply to a rather idealized farm energy setup. Let's say a farmer spends $7500 to build a methane digester to handle manure from a swine finishing house. Biogas from the digester will be used to fire an ethanol still, provide heat for the digester itself, and heat an in-floor hot-water system for the hog building or for a farrowing-nursery building. The alcohol plant costs $12,500 to build and will produce 5000 gal of 170-proof alcohol per year—all of which will be used to power machinery on the farm. In addition, a solar space- and water-heating system is installed on the family home, at a cost of $8000.

The combined energy equipment would probably qualify for tax credits of the magnitude shown in Fig. 23-1.

At this point, many farmers might be wondering who earns enough net taxable income to owe $8700 in taxes to which these credits could be applied—or wondering where all that energy construction money is coming from in the first place. (We said this is an *idealized* situation, remember?)

Methane digester (10% biomass credit)	$ 750
Alcohol plant (10% biomass credit)	1250
Investment credit on above (10%)	2000
Farm fuel credit (30¢/gal)	1500
Solar residence credit (40%)	3200
Total energy tax credits	$8700

FIG. 23-1   Typical tax credits for integrated energy system.

How these credits would apply to taxes due in any particular tax year is beyond the scope of this book, of course. But the farm fuel credit of 30 cents per gallon (40 cents per gallon for alcohol of more than 190 proof) could be taken on taxes due each year through 1992. The investment tax credit assumes that the energy equipment has an expected life of at least seven years. On most systems, farmers could carry forward the excess value of tax credits to taxes due through 1994.

In addition to the above credits, the IRS exempts Gasohol (90 percent gasoline, 10 percent ethanol) from the federal excise tax of 4 cents per gallon.

LOANS

Your best first choice for an alternative energy equipment loan is Farmers Home Administration (FmHA), the rural loan agency of USDA. Most loan funds available to farmers are administered by USDA, although some additional money originates in DOE and other government agencies.

The FmHA both lends money directly and guarantees loans made through commercial lending institutions. For the most part, FmHA puts emphasis on alcohol fuel plants, but methane gas and wood-energy projects also are eligible, under Title II of the Energy Security Act of 1980.

Regulations provide that FmHA make insured direct loans of $1 million or less per project, and agency policy puts first priority on projects using nonpetroleum fuels for primary heat in the production process. The USDA credit agency makes and guarantees loans only for construction of new facilities or for conversion and expansion of existing plants, not for operating expenses. FmHA will make or guarantee loans for up to 30 years or the useful life of the equipment—whichever is less—at interest rates set by the cost of money to the government.

Farmers can get more details on energy loans at county FmHA offices.

Rural electric cooperatives now have funding authority to make low-interest loans to members to weatherize dwellings. Loans are limited to $3000 per single-family dwelling to pay for labor and materials for caulking, weatherstripping, insulation, and other energy-saving work. Contact your local rural electric cooperative.

Cooperatives, rural communities, and other not-for-profit groups may be eligible to borrow money from FmHA for hydroelectric plants. Essentially, these low-interest loans go to restore deactivated dams and hydroelectric generators, to enlarge or improve existing plants, or to construct new hydroelectric facilities. Under the provisions of this program, FmHA could also finance connecting lines to the nearest utility grid.

The Commodity Credit Corporation of USDA makes loans to finance solar grain-drying equipment and high-moisture grain-storage structures, through Agricultural Stabilization and Conservation Service (ASCS) offices. Generally, loans are made to a limit of 85 percent of the total cost, up to a maximum of $50,000. Interest rates are pegged to the government's borrowing cost. County ASCS offices have details.

Outside USDA, a few other agencies provide capital for farmers wishing to construct alcohol plants and certain other energy-producing systems. The Small Business Administration has three types of financial assistance available.

1. Direct loans have a maximum limit of $300,000 per borrower and have a "floating" interest rate pegged quarterly to the commercial cost of money.

2. Immediate participation loans also have a $300,000 limit and are a joint-funding venture between SBA and local lending agencies. Interest rates are weighted on the basis of the percentage of the loan from each lender.

3. Guaranteed loans by SBA are similar to loans guaranteed by FmHA. The agency guarantees 90 percent of the total loan (up to a $50,000 maximum) made by a local commercial lender.

The Office of Resource Management of the DOE occasionally makes loan guarantees to "developers of alternative fuels," which presumably would include farmers. More information on this program can be had from U.S. Department of Energy, 20 Massachusetts Avenue, N.W., Washington, DC 20565.

GRANTS-IN-AID

Outright grants are available for energy research, demonstration, and development. However, as most farmers are interested first in energy systems that fit their operations and reduce dollar outlays for conventional energy, with research and demonstration being at best a secondary motive, not all farm-built energy systems qualify.

It's worth a try, though. At the worst, you'll be out some time and paperwork. At the best, Uncle Sam will pay the cost of building your energy-producing system, perhaps with the provision that you conduct regular tours of the setup.

The DOE programs on small technology and resources management are perhaps the best prospects for direct grants. These are made to individuals, as well as to institutions and businesses. Contact DOE at the Washington address given earlier, or the regional DOE office that serves your state.

The Community Service Administration (CSA) has some technical assistance grant money available for rural low-income areas, as well as some pass-through funds from DOE. Contact the Energy Program Office, Community Service Administration, 1200 19th Street, N.W., Washington, DC 20506.

If you come up with a truly unique, innovative energy-related idea, you may be awarded a direct grant or contract for development from the Office of Energy-Related Innovations, National Bureau of Standards, Washington, DC 20234.

Some private organizations and foundations also make grants and loans to individuals.

STATE PROGRAMS

Several states have enacted energy-related legislation that provides tax incentives, loans, and grants for residences and businesses. Again, new statutes come on the scene almost weekly; it's likely that any listing will be out of date before long. Nevertheless, here's a summary of state energy legislation as of this writing. Contact your state's appropriate agencies to bring yourself up to date on tax credits, grants, loans, and other programs:

Alaska   Residential conservation credit of 10 percent up to $200 for expenses on insulation, storm windows, and solar, wind, tidal, and geothermal energy sources.
State Department of Revenue, State Office Bldg., Juneau, AK 99811

Arizona   Tax credits for solar energy to 30 percent of the cost, to a maximum of $1000. Solar energy devices also exempt from property, transaction, and use taxes. Solar Energy Commission of state government funds some ethanol projects.
Department of Revenue, Box 29002, Phoenix, AZ 85038

Arkansas   Individuals can deduct the cost of solar equipment, biogas systems, wind energy, and some wood-burning equipment from taxable income. Also, motor fuel containing at least 10 percent alcohol produced in Arkansas qualifies for a tax exemption of 9½ cents per gallon.
Department of Revenue, 7th and Wolfe, Little Rock, AR 72201

California   The Solar Energy Demonstration Loan program provides $2000 in interest-free loans for solar space and water heating. The Department of General Services has $10 million for alternative transportation fuels. Income tax credits of 55 percent of the cost of solar equipment, up to $3000 limit.
Franchise Tax Board, Sacramento, CA 95807

Colorado   Individuals can deduct the cost of solar, wind, and geothermal equipment from taxable income. Alcohol and Gasohol fuels qualify for a tax exemption of 5 cents per gallon.
Department of Revenue, 1375 Sherman St., Denver, CO 80261

Connecticut   Solar collectors were exempt from sales tax through October 1, 1982. Gasohol earns a tax exemption of 1 cent per gallon.
State Tax Department, Farmington Ave., Hartford, CT 06115

Delaware   Income tax credits to $200 for solar hot-water systems.
Division of Revenue, 820 French St., Wilmington, DE 19801

Florida   Solar energy systems are exempt from sales tax through June 30, 1984.
Department of Revenue, Carlton Bldg., Tallahassee, FL 32304

Idaho   Geothermal, solar, wind, and wood systems qualify for tax deductions up to $5000 in any one year, at the rate of 40 percent of cost the first year and 20 percent of cost in each of the next three years.
State Tax Commission, 5257 Fairview, Boise, ID 83772

Indiana   Fuels with at least 10 percent alcohol qualify for a tax exemption of 2 to 3 cents per gallon.
Indiana Department of Commerce, 440 N. Meridian, Indianapolis, IN 46204

Iowa   Gasohol is tax-exempt through July 1984. Iowa Development Commission funds programs to promote Gasohol and related by-products.
Iowa Development Commission, 250 Jewett Bldg., Des Moines, IA 50300

Kansas   Income tax credit of 25 percent of cost of solar and wind energy systems, to a maximum of $1000. Rebate of 4 cents per gallon on gasoline used to make Gasohol.
Department of Revenue, P.O. Box 692, Topeka, KS 66601

Louisiana   Tax exemption of 8 cents per gallon for Gasohol.
Department of Natural Resources, P.O. Box 44156, Baton Rouge, LA 70804

Maine   Solar space- and water-heating systems exempt from property tax for five years after installation.

Maryland   State tax exemption of 1 cent per gallon on Gasohol.

Michigan   Income tax credits for solar, wind, and water energy equipment, at 25 percent of first $2000, plus 15 percent of next $8000 of cost.
State Tax Commission, State Capitol Bldg., Lansing, MI 48922

Minnesota   Income tax credits of 20 percent of first $10,000 spent on renewable energy equipment installed before January 1, 1984.
Department of Revenue, 658 Cedar St., St. Paul, MN 55145

Montana   Income tax credits of 10 percent for first $1000 and 5 percent of next $3000 spent on solar, wind, solid waste, biogas,wood, and hydroelectric equipment. Gasohol produced in state earns tax exemption of 7 cents per gallon until April 1985.
Department of Revenue, Mitchell Bldg., Helena, MT 59601

Nebraska   Gasohol tax exemption of 5 cents per gallon.

New Hampshire   Gasohol made in state exempt from 5 cents per gallon of tax.

New Mexico   Income tax credit of 25 percent of cost of solar and solar irrigation systems, up to $1000.
Department of Taxation and Revenue, P.O. Box 630, Santa Fe, NM 87503

North Carolina   Income tax credit of 25 percent of the cost of solar systems, up to $1000.
Department of Revenue, P.O. Box 25000, Raleigh, NC 27640

North Dakota   Income tax credits of 5 percent per year for two years on solar and wind energy equipment; 10 percent blends of ethanol and gasoline receive tax exemption of 6.5 cents per gallon.
State Tax Commission, Capitol Bldg., Bismarck, ND 58505

Ohio   Income credit of 10 percent of cost of solar, wind, and hydrothermal systems, up to $1000 maximum.
Ohio Tax Commission, 1030 Freeway Drive, Columbus, OH 43229

Oklahoma   Income tax credits for solar devices of 25 percent of cost, to a maximum of $2000; total credit can be applied on taxes for up to three years. Tax exemption of 6.5 cents per gallon for Gasohol.
State Tax Commission, 2501 Lincoln Blvd., Oklahoma City, OK 73194

Oregon   Income tax credits of 25 percent of cost of solar, wind, and geothermal systems, to a maximum credit of $1000.
Department of Revenue, State Office Bldg., Salem, OR 97310

South Carolina   Fuel tax exemption of 4 cents per gallon for Gasohol, through July 1985.
Office of Energy Resources, 1122 Lady St., Columbia, SC 29201

South Dakota   Tax credit for Gasohol of 4 cents per gallon.

Tennessee   State program makes loans to low- and medium-income persons for energy-saving improvements in residences.
Housing Development Authority, Hamilton Bank Bldg., Nashville, TN 37219

Texas   Solar equipment for heating, cooling, and electric power is exempt from sales tax.

Vermont   Income tax credit of 25 percent of cost of solar, wind, and wood-burning devices, to a maximum of $1000.
State Tax Department, Montpelier, VT 05602

Wisconsin   Direct-subsidy program refunds part of cost of alternative energy systems— 12 percent in 1983 and 1984—on up to $10,000 of cost.
Department of Industry, Labor and Human Relations, 201 E. Washington, Madison, WI 53702

Wyoming   Gasohol tax exemption of 4 cents per gallon until June 30, 1984.

Integrating Energy Systems

There's not much doubt that America *can* solve its energy problems, given a few years' time and a conscientious effort by everyone involved—which is everyone, period.

For farmers, energy self-sufficiency is a possibility right now. Whether a petroleum-independent farmstead is practical may be open to question, however. No single alternative energy system will solve all problems; it will take integration of several systems into an overall energy conversion plan.

Basically, three factors come into play in designing an integrated energy system. You must know your energy requirements, peak-demand periods, and totals for each form of energy; we've discussed that earlier. You must know what energy resources are available on your farm and which ones can be most profitably developed. And you must know what you can afford.

With capital tight and interest rates high, it's impractical for most farmers to set up an energy-independent system from scratch. However, as elements of an overall energy system are planned and built, it's usually possible to allow for expansion and additions later on. It's the same idea of planning a grain-handling and -storage setup: You build those bins, augers, and dryers you need now, but plan for additional structures later on.

Good planning is the key to successful integration of different energy sources. In the ideal energy setup, no energy would be wasted: Heat, electricity, fuel, mechanical energy —all would be fully utilized, as would the by-products of energy production. In Chap. 25, we go into utilizing by-products (perhaps a better term is *coproducts*) of energy-producing plants.

The mix of energy resources varies from farm to farm, and from region to region. Perhaps the idea of energy integration can best be illustrated by examples.

One example of energy recycling exists on the farm of Harry and Mary Lou Nienaber, near Lindsay, Nebraska. The Nienabers use biogas fermented from livestock manure to power their ethanol distillery. The 180-proof alcohol from the still is burned to power irrigation equipment to help grow their corn crop, which is fermented to make the alcohol. The Nienabers feed wet distillers' grains to hogs and dairy cattle. Dried residue from the methane digester is used for livestock bedding and for fertilizer to raise part of the 240 acres of corn they grow.

Another development that bears watching is a petroleum-independent research project getting underway at the University of Nebraska. The ultimate objective of this four-year study will be to eliminate any need for petroleum fuels, including lubricants, on the farm. The planned farm will be a totally integrated crop and livestock system. Energy resources used include grain alcohol, solar, wind, and methane.

THE MISSOURI ENERGY COMPLEX

Something near a perpetual energy plan is in the works at the University of Missouri swine research farm. Scientists are building facilities to distill alcohol from grain, feed the stillage to livestock, process the livestock manure into methane, use the biogas to power an internal-combustion engine to drive an electric generator, and use the generator to run the alcohol plant and provide the electricity to power the farm.

The integrated energy research facility is now in place and operating. A confinement farrow-to-finish pork production unit provides manure to operate the digester. Gas is collected and stored in a huge Army-surplus rubberized fuel storage bag. An engine-driven electrical generator is installed, with heat exchangers to carry exhaust heat and cooling water to the methane digester and to farrowing and nursery buildings. Figure 24-1 shows the layout of the system.

"The real key to methane efficiency is to use it to operate an internal-combustion engine to drive a generator," says Neil Meador, agricultural engineer, University of Missouri, noting that the energy cost to pressurize and store methane uses much of the energy in the gas. "This way, the methane can be used as it is produced. As a bonus, the manure processed through the digester has very little odor, but retains most of the fertilizer nutrients."

Meador and his colleagues—Charles Fulhage, James Fischer, and F. D. Harris—have designed a model that shows that a 3200-hog confinement operation could be virtually independent of outside sources of energy for much of the year. In the Missouri plan, a Waukesha model VRG 155U spark-ignited engine drives a 20-kW induction generator. The engine operates at 1200 rpm, and heat from the exhaust and water cooling jacket is recovered to help heat farrowing and nursery buildings, as well as to preheat manure slurry going into the digester.

Heat from the engine alone supplies enough thermal energy for both the swine operation and the digester until outside temperatures drop to about 28°F (see Fig. 24-2). Below that temperature, space heaters powered by the generator are used to keep farrowing and nursery quarters warm. With careful operation, the generator can produce all the electricity needed by the hog operation.

Manure to be turned into methane doesn't "keep" well. Bacteria begin to break down the volatile solids immediately after the manure is voided. The Missouri scientists load manure into the digester daily. Two scrapers in each building remove the manure to a collection-settling basin. The basin stores manure from all buildings until it is pumped into the digester.

The animals produce about 775 lb of volatile solids per day. To handle this volume of manure, a cylindrical digester (6 ft 10 in in diameter and 21 ft 9 in high) is bolted and sealed to a flat concrete slab. The digester container is made of fiberglass coated with urethane insulation. The efficiency of the digester is increased by "back-flowing" spent effluent from the digester into the manure in the collection basin. The slurry comes out of the digester at about 95°F and adds heat (through heat exchangers) to the new batch of manure waiting to be pumped into the digester. This cuts in half the thermal energy needed to heat the fresh manure.

The 20-kW generator produces enough power to operate the hog complex, if the electricity and its use are managed carefully and distributed throughout the day. For example, manure scrapers are set to operate automatically at night, when little other electrical energy is being used.

Similarly, digester loading and operating functions can be sequenced manually or with timers to maintain a level demand on the generator. Coordination and timing of energy-related operations is crucial to the success of an integrated system. In the Missouri design,

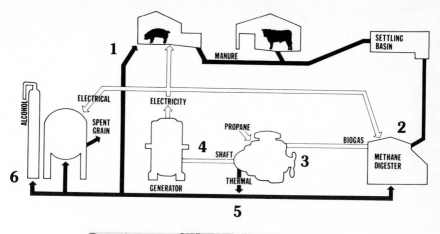

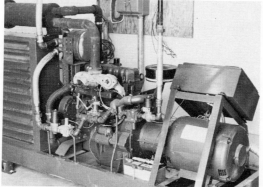

FIG. 24-1 *(a)* Layout for energy self-sufficient swine and cattle farm under study at University of Missouri: (1) livestock eat spent grains from ethanol distillery and produce manure; (2) methane digester produces biogas from the manure; (3) biogas fuels internal combustion engine, which powers electrical induction generator (4); at step (5), the engine's thermal energy helps run the ethanol plant and heats methane digester; (6) distillery produces fuel for farm vehicles. Spent sludge from the methane digester is used to fertilizer crop fields. *(b)* Modified internal combustion engine drives an induction generator. The generator is tied to the utility grid and takes its reference current (excitation) from the utility power line. When generator output exceeds farm power demand, the excess automatically flows into the utility grid. When the farm load exceeds generator output, the balance or "makeup" electricity automatically flows from the grid.

the electrical load can be leveled throughout the day, and throughout the year, fairly easily, by scheduling power-using operations.

But how about the thermal load? If heat from the methane-fired engine-generator is to be fully utilized (see Fig. 24-3), the alcohol-making operation might need to be scheduled for warmer weather, when farrowing and nursery buildings require less heat. Also, the temperature difference between the 95°F required in the digester and the outside air would be less during summer, which would make smaller demands on the engine's by-product heat.

The key to making most integrated energy setups work is to make energy production

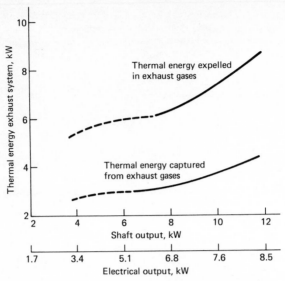

FIG. 24-2 Thermal energy produced by generator engine exhaust, and amount of that heat captured by heat exchangers for useful work.

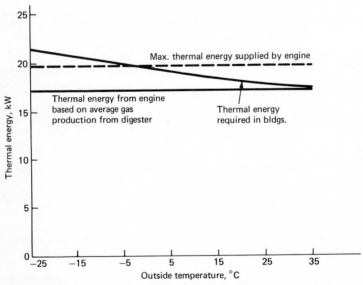

FIG. 24-3 Thermal energy required by the Missouri experimental farm can be supplied by generator engine exhaust until outside temperatures drop below freezing.

match demand peaks. In farming, there are seasonal as well as daily peaks. This makes planning vital if the energy produced is to meet the load, without a great deal of energy being wasted.

Truly integrated alternative energy systems are, at this point, mostly hypothetical. But more and more farmers—and researchers—are getting there, utilizing the primary sources of energy available in their region and on their acres.

Building an integrated energy system takes planning, and running one would require management changes. Managing an integrated, multisource energy program that is super-imposed on the basic farming operation would make big demands on management, talent, labor, and time.

25

Using the By-Products

Some of the information being distributed on alternative energy systems appears to be based on rather faulty assumptions. Admittedly, well-proven figures are in short supply where many systems are concerned. But some of the economic analyses of various energy projects—particularly fuel-making plants—stretch credibility.

For example, do any of these statements sound familiar?

"With our system, you can make alcohol fuel for 25 (or 35 or 45) cents per gallon, because you can use damaged, moldy grain."

The error here is twofold: First, not many farmers *plan* on producing damaged, moldy grain. Secondly, grain that is in such poor condition as to be totally useless for anything else is also too far gone to be a very good feedstock for an alcohol plant.

"With this type of methane digester, you can not only replace all the natural gas you now use but also provide fertilizer worth X dollars per ton."

The assumption here seems to be that the livestock producer is not now using manure for fertilizer and that this feedstock for a methane plant has no value whatsoever. The second error in this type of statement is the assumption that all farmers have natural gas available. Some do, but many do not. The realistic approach to evaluating an alternative fuel's value is in terms of the fuel it replaces in your operation.

"When gasoline gets to $1.50 (or $1.75, or $2) per gallon, you'll be sitting in the catbird seat with your own alcohol still."

Admittedly, energy costs are among the faster-rising costs today, but inflation is pushing *all* costs upward, and the idea that you can justify a fuel alcohol plant by applying increased costs only to fuel is in error. Land, labor, interest rates, machinery, seed, fertilizer, farm chemicals—all the resources used to grow a crop are increasing in dollar terms, which means the cost of the raw feedstock will increase by some measure, right along with the cost of conventional fuels.

"You can produce enough alcohol to replace all the gasoline you now use, just by operating the still in your spare time."

Time (or labor) and capital are among the more limited resources on most farms. The idea that a farmer can produce his fuel needs in his "spare time" may be valid, but the assumption needs to be challenged.

Every time OPEC raises the price of oil another dollar per barrel, the incentives to develop alternative energy sources gain momentum, but no equipment thus far on the market will let you make a "silk purse out of a sow's ear" where energy is concerned. The material, labor, capital, and energy invested in alternative energy production have some opportunity cost—they would be worth something in other uses.

A farmer shouldn't let himself be stampeded into making a rash investment in high-capital energy equipment. The decision to purchase or build alternative energy equipment should be made only after a careful analysis of how the coproducts of energy production can be used and what they will be worth in a particular operation. A pork producer who already knifes liquid hog manure into crop soil could not realistically credit a methane digester with many dollars for the fertilizer produced.

Like the alternative energy that is produced, the coproducts should be evaluated on the basis of the purchased inputs they will replace. The problem is that even less research has been conducted on the use of alternative fuel by-products than on the uses of the fuels themselves.

A major by-product of all three principal biomass fuels (alcohol, vegetable oil, and biogas) is a concentrated high-protein livestock feed ingredient. On a dry-matter basis, the leftover mash from making ethanol has 28 to 35 percent protein, depending on the type of grains used. The effluent from a methane digester contains about 30 percent usable protein, again on a dry-matter basis. And the meal or cake from farm-pressed sunflower oil contains about 28 percent protein. Potentially, these protein-rich feedstuffs are valuable ingredients in livestock rations, their value measured by the equivalent cost of the protein supplement replaced.

Based on a spot check of feed ingredient prices, Table 25-1 shows equivalent protein percentages and costs for more conventional feeds used in cattle rations.

To be fair about it, distillers' wet grains or sunflower seed meal that replaces, say, soybean oil meal in a ration should be credited with what the soybean meal would have cost—no more, no less—regardless of what fraction of the cost of energy production falls to the by-product. For example, a pound of actual protein from an energy plant by-product is worth 37 cents if it replaces soybean meal, and worth only 29 cents if it replaces the protein in alfalfa hay—strictly on the basis of the protein value.

There are special problems with each of these farm fuel by-products, however; we'll go into some of them now.

WHOLE STILLAGE

For each bushel of grain used for making alcohol, there are from 16 to 30 gal of liquid by-product that must be disposed of or used in some way. The economics of producing ethanol dictate that this material be used, and the most convenient use is in livestock feeds. See Table 25-2.

There is a great deal of research on feeding *dry* distillers' grains, but little on using the wet product. And wet is the form this potential feed ingredient will be in most farm-scale operations—from 90 to 95 percent water. That's wet!

"Centrifuges, evaporators and dryers are not likely to be incorporated into smaller on-farm stills that are being built or contemplated, because of their cost," says Robert McEllhiney, grain scientist at Kansas State University. "So, we must be concerned about

TABLE 25-1   Equivalent Protein Percentages and Costs

Feed	Protein, %	Cost per ton	Pounds needed for 1 lb of protein	Protein cost per lb
Soybean meal	44	$328	2.3	$0.37
Liquid protein (urea)	32	154	3.13	0.24
Range cubes	20	198	5.0	0.49½
Alfalfa hay	17	100	6.0	0.29

TABLE 25-2   Ethanol and By-Product Output of Livestock Feeds

Feedstock	Average ethanol output			Average by-product output (feed ingredient)
	gal/ton	gal/bu	gal/ac	
Corn (56 lb/bu)	93	2.6	260	16.8 lb/bu @ 28% protein
Wheat (60 lb/bu)	87	2.6	84	20.7 lb/bu @ 35% protein
Grain sorghum (56 lb/bu)	100	2.8	146	16.8 lb/bu @ 28% protein
Sugar beets (16.5% sugar)	26.8	. . .	354	264 lb/ton @ 8% protein
Potatoes	29	. . .	300	296 lb/ton @ 23% protein

how to handle the whole stillage from the still to the point of feeding this by-product to livestock."

The simplest, cheapest method is to move the whole stillage from the still to some feeding operation as directly, quickly, and efficiently as possible. This usually means daily feeding of all—or virtually all—the wet stillage produced. The stuff doesn't keep very long, especially in warm weather.

"This procedure would involve whole stillage storage tanks at the distillery or at the feeding site, with at least holding capacity for 1.5 times the expected tank truck capacity or half of the daily whole stillage output of the plant, whichever is greater," adds McEll-hiney. "Storage tanks used for this material should be hopper-bottomed, equipped with a recirculating and load-out pump, and should probably be insulated and possibly heated to prevent freezing in cold weather."

The cost of such tanks would range from 20 cents per gallon capacity for used tanks in good condition to more than 50 cents per gallon capacity for new tanks—plus freight, installation, and plumbing costs.

Kevin Eilks, who manages a crop-and-cattle farm in central Missouri, has adopted what appears to be a workable system for feeding whole stillage.

Eilks pumps the whole stillage out of his ethanol plant and runs it through a perforated auger, of the type often used to clean seed. Much of the solid material is strained out of the liquid and is then top-dressed on hay or silage in cattle-feeding bunks. The liquid, which contains up to half of the nutrients in the stillage, is pumped into a tank wagon and hauled to feed catfish that grow in large irrigation reservoirs.

"If it weren't for the cattle and catfish, I wouldn't even consider making grain alcohol," says Eilks. "As it is, I don't know how long we can use the stillage liquid in catfish ponds. The stuff is pretty rich, and may cause excessive growth of pond weeds and algae. But so far, we haven't run into problems."

The alternatives to using the super-wet whole stillage directly begin to eat into the energy efficiency of an on-farm alcohol plant, but may be necessary (see Fig. 25-1). The first step in removing moisture is to use some mechanical method of separating liquids and solids. This process is complicated by the fact that as much as 40 percent of the nutrients are in solution in the water—and not everyone raises catfish, as does Kevin Eilks. About 60 percent of the total solids can be mechanically separated.

Depending on the system used, mechanical separation leaves a product that still contains from 50 to 85 percent moisture. Stationary or vibrating screens are used by many commercial processors, but are too expensive for most farm-sized plants. The moisture content of solids from screens is about 75 to 85 percent. Centrifuge devices separate liquids and solids by centrifugal force, reducing the moisture content of the product to about 50 percent. Press wheels and other mechanical presses can reduce the moisture content to about 70 percent.

Many alcohol-making farmers will likely rig up some sort of flow system—such as Eilks's auger strainer—to reduce moisture content as much as possible, without big equipment

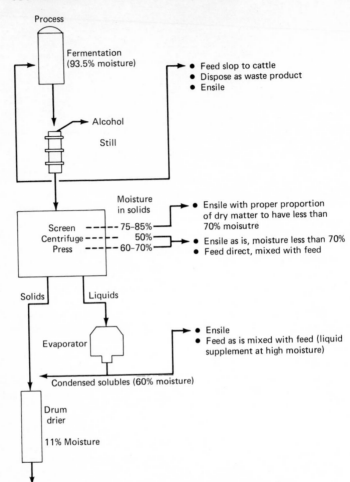

Process

Fermentation
(93.5% moisture)
→ • Feed slop to cattle
   • Dispose as waste product
   • Ensile

→ Alcohol

Still

Moisture
in solids
→ • Ensile with proper proportion
     of dry matter to have less than
     70% moisutre

Screen   ---- 75–85%
Centrifuge ----   50%
Press   ---- 60–70%
→ • Ensile as is, moisture less than 70%
   • Feed direct, mixed with feed

Solids   Liquids

Evaporator
→ • Ensile
   • Feed as is mixed with feed (liquid
     supplement at high moisture)

← Condensed solubles (60% moisture)

Drum
drier

11% Moisture

Distillers' dried grains ————→ • Use as feed ingredient

FIG. 25-1   Several options exist for treating, storing, and using the by-products of ethanol distillation. *(Courtesy: Robert M. George, Agricultural Engineer, University of Missouri.)*

costs and operating expenses. With a product that contains in the neighborhood of 65 percent moisture, there is more leeway in handling and feeding. Stillage mechanically separated to that moisture content can be ensiled, either by itself or along with forage materials, or fed directly with less water to handle and haul. However, in warm weather, the stuff will still spoil in short order.

Ensiling partially dried stillage is complicated by the fact that the distilling process is continuous, whereas most silos are designed for batch filling. Continuous ensiling would require planning and special management to accumulate the by-products and mix them with drier feeds for ensiling. A possibility is to use an air-tight structure where the material is filled at the top and unloaded at the bottom (see Fig. 25-2). The stillage would have to remain in the structure long enough for the ensiling process to be completed without oxygen being introduced.

A horizontal silo (pit, trench, or bunker) might be used if air is kept from the surface of the stillage as it is placed in the silo. This might be done with a plastic cover and a small fan attached to a perforated pipe to remove air from under the plastic.

Any distillery by-products with moisture content higher than 70 percent probably would need to be blended with dry materials to reduce the overall moisture. One possibility would be to mix 2000 lb of chopped dry cornstalks with about 4000 lb of whole stillage, to produce ensilage at 65 to 68 percent moisture.

A caution is in order at this point: These ideas on handling and ensiling whole stillage are largely untried and untested. It would be the better part of wisdom for readers to try different ensiling recipes on a small scale first.

After the mechanical screening more moisture can be removed by heat drying. Most commercial plants use drum-type dryers to bring the distillery by-product down to about 11 percent moisture, where it will keep very well in conventional storage. But drying is expensive—in both dollars and units of energy. To aid in drying, heat exchangers could be used to recover waste heat from the distilling process. But considerable "makeup" energy would be needed, with the hazard that the whole alcohol-making operation could be thrown into an energy-deficit condition.

One of the better uses of wet stillage is to feed it as a partial protein supplement in rations of young, growing cattle—either after separation, or with the raw, wet product right out of the still. There's a problem with consumption, since cattle would have to consume about 16 lb of the 92 to 94 percent moisture stillage to get 1 lb of dry matter—only 28 percent of which is protein.

Iowa State University scientists have calculated that the stillage from a plant operated for 100 days to produce 4000 gal of alcohol would amount to 438,710 lb of material that

FIG. 25-2 Whole stillage drier than about 70 percent moisture content can be ensiled most conveniently in a top-loading, bottom-unloading airtight structure.

is 93.8 percent water. If each calf consumed 1 lb of stillage dry matter daily as a source of protein supplement, 273 head of cattle would be needed. (Each calf consumes 16.1 lb of wet stillage each day for 100 days = 1610 lb of stillage per calf = 1610 lb × 273 calves = about 439,000 lb.)

It might be possible to feed higher levels of stillage, as a replacement for protein supplement and for part of the grain in cattle rations. However, there is little research to support this practice. Again, the high moisture content of whole stillage and the need to feed it quickly (before it spoils) are the big problems. In research at the University of Kentucky, yearling cattle gained from 1.5 to 1.9 lb daily when fed nothing but whole stillage and 5 lb of hay. These cattle consumed an average of 20.6 gal (about 175 lb) of stillage daily, according to report. Apparently, feeding the dry hay partially countered the considerable laxative effect of feeding that much wet stillage.

In most cases, whole stillage probably should be limited to supplying no more than half of the dry matter in beef rations, perhaps less than that in dairy rations. Cattle may have trouble consuming 2 percent of their body weight in dry matter daily, if they take in huge amounts of water. There's also some evidence that prolonged feeding of very high moisture feeds may do damage to the animals' urinary system.

It's understandable: A yearling steer weighing 900 lb, on a finishing ration, will consume about 20 lb of dry matter daily—the equivalent of about 2¼ percent of his body weight. If 15 percent of the total dry matter is made up of roughage, and another 5 percent comes from vitamins, minerals, and other additives, that leaves 16 lb of dry matter to be supplied by grain and stillage. If stillage at 10 percent dry matter is to make up half of that, the steer will need to consume 80 lb to get 8 lb of dry matter.

The advantage of draining, straining, or squeezing out some of the moisture in whole stillage is evident.

Homer Sewell, cattle-feeding specialist at the University of Missouri, has formulated rations for beef cattle, using dried distillers' grains with solubles (DDGS) to supply part of the nutrients. The value of whole stillage dry matter is computed on the basis that it substitutes for soybean meal and corn as a source of protein and energy. The growing ration for a 500-lb steer is shown in Table 25-3.

While whole stillage can be fed to cattle (if you have enough cattle to take the stillage from a day's alcohol production), there are some practical limitations to feeding the

TABLE 25-3 Growing Ration for a 500-lb Steer

	Dry matter, lb	Protein, lb	TDN*, lb	Fiber, lb	Ca, lb	P, lb
Daily requirement	12.7	1.34	8.3	0.040	0.040	0.035
Ingredients (as fed, lb):						
Grass hay 5.00	4.50	0.35	2.34	1.53	0.017	0.011
Corn 3.00	2.58	0.26	2.40	0.06	0.001	0.008
Stillage 61.40	4.30	1.25	3.60	0.42	0.004	0.017
Limestone 0.07	0.06	—	—	—	0.023	—
Total 69.47	11.44	1.86	8.34	2.01	0.045	0.036

Stillage replaces:

1.0 lb soy meal @ 10¢	= 10¢
3.63 lb corn @ 4.5 ¢	= 16
Value of stillage	= 26¢

This makes the 4.3 lb of dry matter in the stillage worth about 6.05¢ per lb.

*Total digestible nutrients.

by-product to swine and poultry. The high water content, low concentration of nutrients, and poor balance of amino acids—especially lysine—reduce the value in rations for single-stomached animals.

DDGS can be used in swine rations if the limiting amino acids are supplemented. Note the nutrient comparison of DDGS with corn and soybean oil meal in Table 25-4.

Keep in mind that this comparison is with *dried* distillers' grains. The fact that hogs are limited in their capacity to utilize liquids affects the performance of raw, whole stillage. If a 100-lb pig needs 5 lb of dry feed a day, he'd have to consume at least 50 lb of whole stillage to get it. Obviously, the pig would have some trouble consuming half of his body weight each day, and lesser amounts would not provide the nutrients needed.

Some whole stillage probably can be fed to larger finishing hogs, and particularly to gestating sows. How much has not been determined by research as yet, however. Because of the limited lysine in whole stillage, the product probably should be held to about 20 percent of the daily dry-matter intake.

Because of the spoilage hazard and the need to limit-feed whole stillage (whereas most hogs are fed by automated or self feeders), alcohol by-products may be more trouble than they are worth in swine rations—at least until more is learned about how to feed them.

There's another coproduct of alcohol manufacture—one you seldom hear much about, as far as its usefulness is concerned. In big commercial alcohol plants, carbon dioxide (CO_2) is salvaged and marketed. Right now, there is not much commercial demand for small amounts of CO_2 generated during the fermentation of alcohol in farm-scale plants. But nearly a third of each bushel of corn is converted to CO_2 by fermentation yeasts, and there is a potential use for this by-product right on the farm. Plants use CO_2 and give off oxygen, in a reverse of an animal's respiratory process. A greenhouse built adjacent to an alcohol plant could make the CO_2 a useful coproduct of farm ethanol production (see Fig. 25-3). It shouldn't be much trouble to capture the gas and move it into a vegetable-growing greenhouse. Along with the "used" heat from the distilling column, CO_2 might grow bumper crops of January tomatoes.

ANAEROBIC DIGESTION BY-PRODUCTS

The most thought-of use for effluent from anaerobic biogas digesters is as plant nutrients, to replace commercial fertilizer on crop- and pastureland. But there's growing interest in feeding the high-protein residue from digesters; it's being done in California, Colorado, Nebraska, and Oklahoma. And dairymen use dried residue for bedding.

We'll get to the feeding of sludge shortly. But first, let's look at the value of a digester's by-product as fertilizer. Most solids not converted to biogas settle out in the digester vessel as a high-moisture sludge. Depending on the raw materials used and the efficiency of digestion, this by-product contains nitrogen, phosphorus, potassium, and several trace elements needed for plant growth.

Nitrogen is especially important, because of its vital role in plant nutrition and also because of the high energy cost associated with manufactured nitrogen fertilizers. About

TABLE 25-4 Nutrient Comparison of DDGS with Corn and Soy Meal

Nutrient content	Corn	Soy meal	DDGS
Crude protein, %	9.00	45.0	27.0
Lysine, %	0.25	2.9	0.6
Fiber, %	2.20	7.3	9.1
Digestible energy, kcal/lb	1600	1520	1600

FIG. 25-3 Carbon dioxide, a major fermentation by-product, is not often utilized in small farm-scale alcohol operations. But it could be piped off as an aid to plant growth in a greenhouse. *(Photo: U.S. Department of Agriculture.)*

half of the nitrogen in digested sludge is in the ammonium form, readily available to plants. Digested manures contain this range of nitrogen percentages, on a dry-weight basis:

Hog manure	6.0–9.2
Cow manure	2.7–5.1
Poultry manure	5.2–9.1

Another happy feature of digested manure sludge: It doesn't have the nose-twitching odor of raw manure from a pit or holding tank. A farmer can spread the by-product from his methane digester and stay on friendly terms with his downwind neighbors.

There's more interest all the time in feeding the effluent from digesters, too. A. G. Hashimoto, at the U.S. Meat Animal Research Center, Clay Center, Nebraska, notes that a feedlot steer can produce about $20 worth of methane and about $40 or more worth of protein feed (at $60 per metric ton of dry solids) in one year. The higher potential credit for the protein than for the biogas itself is a big reason interest has perked up in feeding digested effluent.

"During anaerobic fermentation, between 40 and 50 percent of the influent dry matter is converted to methane and carbon dioxide, while little total ash or total nitrogen is lost," says Hashimoto.

However, the relative proportion of ammonia to total nitrogen increases from 27 percent in the influent to about 48 percent in the effluent. If all the nitrogen in the effluent could be recovered and used as a feed supplement, it would have a crude protein content of about 60 percent.

Perhaps the most pronounced impact fermentation has on the nitrogen fraction is the enhancement of total essential amino acid concentration. However, when the effluent is centrifuged or pressed (to eliminate excess moisture), about half of the amino acids are lost in solution as the water is drawn off.

Other nutrients show similar losses when digester effluent is centrifuged. Mixing the liquid effluent with other, drier ration elements can eliminate the need for expensive separation equipment and retain all the effluent nutrients in the bargain. But here again, as with wet distillers' grains, the water content of the material limits the amount of effluent that can be added to rations.

Although nearly half of the nitrogen in digester effluent is in the ammonium form (not readily available to animals), ruminants can utilize this product in much the same way they utilize urea and other nonprotein nitrogen forms.

In feeding trials at the Nebraska station, Hashimoto compared three groups of cattle on different rations, based on cracked corn and hay. One group received no supplemental protein, one group received soybean oil meal, and the third group received digester effluent. The first two groups of cattle had nearly the same daily gain, but for some reason, steers fed the digester effluent failed to gain quite as fast. There was no apparent difference in carcass quality, and taste panel members rated steaks from cattle fed the effluent as equal to those from steers fed the soybean meal ration.

"We aren't sure what the problem was in this first trial," says Hashimoto. "There doesn't seem to be any differences in taste and aroma of the rations with the effluent, so we don't think palatability is the trouble. We're repeating the feeding trials; we're convinced that we can replace virtually all of the plant protein in cattle rations with digester effluent."

When the bugs (no pun intended) are worked out of using effluent as a cattle feed supplement, it may well turn out that *biogas* is the by-product. The high-protein feed may be worth up to three times the value of the gas.

VEGETABLE OIL MEAL

North Dakota field trials indicate that sunflower seed, fed whole to cattle, has about the same energy value as barley, provided that not more than 3 lb per head is fed daily. In swine rations, the useful content of sunflower seed is similar to barley also, but only when used at less than 10 percent of the ration and mixed with grain before grinding.

The value of sunflower oil meal that remains from a farm extraction plant depends to a large extent on whether the seed is dehulled before extraction. Also, screw-press expellers, such as those used for most farm-scale oil extraction operations, remove only 75 to 80 percent of the oil. The balance of the oil, while not available for diesel fuel, is nevertheless a valuable feed ingredient.

About a 41 percent protein meal with low fiber can be expected when oil is extracted from sunflower seed that has been dehulled. If the seed is not dehulled before extracting the oil, the protein content is less—in the 28 to 29 percent range—and the fiber content is higher.

At present protein prices, 41 percent sunflower oil meal for swine is worth about $150 per ton, compared with soybean meal. The reason is that sunflower meal, like distillers' grain, is low in lysine—an essential amino acid that must be supplemented in hog rations. For cattle, the value of the meal is higher—about $200 per ton, compared with soybean meal. The value of 28 percent protein meal is about $100 per ton for swine and $135 to $140 in cattle rations.

Similar feeding values could be expected with the by-products from soybean and peanut oil extraction, although figures are not available on these farm energy products.

Vegetable oil meal also has some value as a soil additive, although less than its feeding value, in most cases. It's difficult to assign a value to sunflower meal as an aid to soil tilth, plowing ease, and erosion control, although these benefits come with any plant residue returned to the soil. However, with decomposition in the soil, the plant food content of the meal is released, and this by-product does have a measurable value, as a replacement for commercial fertilizer.

Current fertilizer prices for nitrogen, phosphorus, and potassium (in the author's area) are 18, 24, and 12 cents per pound, respectively. Table 25-5 shows nutrient values of sunflower oil meal when it replaces commercial fertilizer at these prices.

These values for meal as fertilizer fall short of the benefit of this by-product in livestock rations, and no charges for hauling, handling, or applying the material on the field have been included.

It's not difficult to see that the success of a farm fuel plant—whether alcohol, methane, or vegetable oil—depends heavily on planning the best use of coproducts, as well as of the fuel itself. With whole stillage worth $155 to $160 per ton (dry-matter basis), digester effluent worth $60 or more per ton, and vegetable oil meal worth up to $180 per ton, it's entirely possible that the by-product may be more valuable than the fuel produced.

TABLE 25-5   Nutrient Values of Sunflower Oil Meal

Element	Whole-seed meal		Dehulled-seed meal	
	lb/ton	$/ton	lb/ton	$/ton
Nitrogen (N)	86	15.45	138	24.84
Phosphorus (P_2O_5)	46	11.04	82	19.68
Potassium (K_2O)	24	2.88	60	7.20
Total		29.37		51.72

SOURCE: North Dakota State University.

Metric and Other SI Units for Agriculture

Back in 1960, the countries of the world formally adopted the International System of Units and identified six standard units from which all measurements can be derived. The system is referred to as the *SI system* or the *metric system,* and it hasn't affected your farm operation much . . . yet. But it may.

We all resist change, and the SI (or metric) system has not been warmly welcomed everywhere in the United States. But it's actually easier to use than our old awkward English system, once you understand the six basic standards of measurement:

Meter   is the standard unit of length, first defined by the French government nearly 200 years ago.

Kilogram   is the standard unit of mass (weight), originally defined as the weight of 1000 cubic centimeters of pure water.

Second   is the basic unit of time, and is already in use in the United States.

Degree Celsius   is the measure of temperature.

Ampere   is the standard measure of electrical current flow and is already in use in the United States.

Candela   is the measure of light intensity.

All other units of measurement can be derived from these six basic standards. For instance, speed is distance per unit of time and has as its basic unit meters per second. There's also a change in energy measurement with the SI system; energy is measured in calories instead of British thermal units (Btu). One calorie is the amount of energy needed to raise one gram of water by one degree Celsius, whereas one Btu is the energy required to raise one *pound* of water by one degree *Fahrenheit.* Rounded off, there are about 250 calories in 1 Btu.

Distance and mass or weight units in the SI system use the decimal, or base 10, system of accumulation. Prefixes designate multiples and divisions of the basic unit, as shown in Table A-1. For example, 1 kilometer is 1000 meters. The English system has no consistent base. There are 12 inches per foot, 3 feet per yard, 5½ yards per rod, 5280 feet per mile, etc. Once you have the standard measurements and the prefix terms down pat, the metric system is easier to use than the English system.

The three measurements of the SI system that will affect you most will be temperature

TABLE A-1   Prefixes Designating
Multiples and Divisions of Basic Units in
the SI System

Prefix	Multiply by:
Mega-	1,000,000
Myria-	10,000
Kilo-	1,000
Hecto-	100
Deca-	10
Deci-	$1/10$
Centi-	$1/100$
Milli-	$1/1000$
Micro-	$1/1,000,000$

(Celsius), length (meter) and mass or weight (kilogram). Table A-2 is a conversion table of English to SI measurements, in case you need to refer to it while you're still learning to "think metric."

The basic unit of length or distance is the meter, just longer than a conventional yardstick (about 39 in). A 40- by 200-ft building is a 12.2- by 61-m building.

The kilometer (1000 m) is replacing the mile as the measure of longer distances. The speedometer on your pickup probably has scales for both miles per hour and kilometers per hour on the dial. The distance from Des Moines to Boston is 1280 mi, or 2061 km—which may make the trip seem longer.

Most measurements now specified in inches will be changed to centimeters ($1/100$ m). A standard 2 by 4 will be 5.08 by 10.16 cm—or, what is more likely, the standard will be a 5- by 10-cm board, and lumber will be sold by cubic centimeters, rather than by board feet.

Smaller measurements of length will be specified in millimeters ($1/1000$ m). In fact, spark plug sizes are already stated in millimeters. If you own a vehicle or piece of farm equipment built overseas, you probably have seen metric measurements in specifications for parts.

Area, the product of length times width, is measured in square meters in the SI system.

TABLE A-2   Conversion of English Measurements to SI Measurements

Multiply:	By:	To get:
Inches (in)	2.54	Centimeters (cm)
Feet (ft)	0.3	Meters (m)
Miles (mi)	1.61	Kilometers (km)
Square inches (in²)	6.45	Square centimeters (cm²)
Square feet (ft²)	0.09	Square meters (m²)
Square yards (yd²)	0.81	Square meters (m²)
Acres (ac)	0.39	Hectares (ha)
Cubic inches (in³)	16.39	Cubic centimeters (cm³)
Cubic feet (ft³)	0.028	Cubic meters (m³)
Cubic yards (yd³)	0.76	Cubic meters (m³)
Pounds per square inch (psi)	704.0	Kilograms per square meter (kg/m²)
Pounds per square inch (psi)	70.4	Grams per square centimeter (g/cm²)
Ounces (oz)	28.4	Grams (g)
Pounds (lb)	454.0	Grams (g)
Tons*	0.91	Tonnes (t)
Btu (British thermal units)	0.252	Kilocalories (kcal)

*Reference here is to a short ton, which is equal to 2000 lb.

Roofing, carpeting, sheet plastic, and other materials now sold by the square foot or square yard will be sold by the square meter. For measurement of smaller areas, square centimeters ($\frac{1}{100}$ m²) or square millimeters ($\frac{1}{1000}$ m²) will be used.

Land area, now expressed in acres, will be measured in hectares. A hectare is the area contained in a square whose sides are one hectometer (100 m) long; it is equal to approximately 2.5 acres.

In the metric system, the liter is a standard measure of volume (length times width times height). One liter is equal to one cubic decimeter (1000 cm³). Milk, fuel oil, and gasoline —along with many other liquids—will be sold by the liter (booze already is sold that way). A liter is slightly larger than a quart. Smaller quantities are measured in cubic centimeters (milliliters); most vaccines, antibiotics, and other medicines have already been changed to the SI system.

Most bulky items, including farm crops, will be sold by weight, rather than by volume. But some materials, such as sand, gravel, or concrete will be measured in cubic meters. About 1.3 cubic yards are contained in a cubic meter.

Pressure, or force per unit of area, is measured in kilograms per square meter or grams per square centimeter, rather than in pounds per square inch.

Items measured in pounds in the English system are measured in kilograms. Livestock, grain, cotton, and most other farm commodities will be sold by the kilogram. Smaller items, such as trace minerals and some veterinary supplies, will be measured in grams and milligrams—some even in micrograms.

Larger masses will generally be measured in tonnes (1000 kg or 2200 lb). Fertilizer, limestone, and other bulk materials will be handled in tonnes, also called *metric tons.*

In the SI system, temperature is measured on the Celsius scale, with its 100° interval between the boiling point and freezing point of water. On this scale, 0°C (32°F) represents the point at which water freezes, and 100°C (212°F) represents the boiling point of water at atmospheric pressure at sea level. To convert Fahrenheit to Celsius, use the equation $C = (F - 32) \times 0.56$.

That's all there is to it. Your 30-in corn rows will be spaced 76.2 cm apart. A quarter-mile-long field will be 1609 m long. If your pickup is well-tuned, it should be getting about 8 km/l.

We'll all probably have to give up some favorite old sayings when SI metric sets in for good, though. "A miss is as good as 1.61 kilometers" just doesn't have the same ring as "A miss is as good as a mile."

Tables, Tips, and Rules of Thumb

TABLE B-1 Table of Equivalents

Liquid measure:

31½ gallons	=	1 barrel
2 barrels	=	1 hogshead
1 gallon	=	231 cubic inches
1 cubic foot (water)	=	62.5 pounds; 7.43 gallons
1 gallon (water)	=	8.34 pounds
240 gallons (water)	=	1 ton; 35.9 cubic feet

Cubic measure (volume):

1728 cubic inches	=	1 cubic foot
27 cubic feet	=	1 cubic yard
2150.4 cubic inches	=	1 bushel
128 cubic feet	=	1 cord (wood)

Square measure (area):

144 square inches	=	1 square foot
9 square feet	=	1 square yard
160 square rods	=	1 acre
43,560 square feet	=	1 acre
640 acres	=	1 square mile

Surface and volume measure:

Area of rectangle	=	length $\times$ width
Area of triangle	=	base $\times$ ½ perpendicular height
Area of circle	=	diameter $\times$ diameter $\times$ 0.7854
Diameter of circle	=	radius $\times$ 2
Circumference of circle	=	diameter $\times$ 3.1416
Volume of cube	=	width $\times$ length $\times$ height
Volume of cylinder	=	area of base $\times$ height
Volume of sphere	=	diameter $\times$ diameter $\times$ diameter $\times$ 0.5236
Volume of cone	=	area of base $\times$ ⅓ height

TABLE B-2   Common and Scientific Names for Chemicals and Formulas

Popular name	Chemical name	Chemical formula
Grain alcohol	Ethyl alcohol	C_2H_5OH
Wood alcohol	Methyl alcohol	CH_3OH
Baking soda	Sodium bicarbonate	$NaHCO_3$
Benzol	Benzene	C_6H_6
Ether	Ethyl ether	$(C_2H_5)_2O$
Glauber salt	Sodium sulfate	$Na_2SO_4 \cdot 10H_2O$
Marsh gas	Methane	CH_4
Potash	Potassium carbonate	K_2CO_3
Rochelle salt	Potassium sodium tartrate	$KNaC_4H_4O_6 \cdot 4H_2O$

Heating Degree Days

The heating-degree-day method of estimating heating loads has been in general use by the heating industry for 30 years, and is a fairly simple way to figure how much supplemental heat will be needed. Heating degree days are determined by subtracting the average outside temperature over a 24-hour period from a base of 65°F. Normally, heat is not required when the outside temperature averages 65°F or above.

For example, if the average temperature for October 30 is 45°F, the degree days for that particular date would be 20 (65 − 45 = 20). Heating degree days are cumulative. To find the degree days for a month or year, simply add together all the degree days for that period.

TABLE B-3   Degree Days per Month for Selected U.S. Locations

Location	July	Aug.	Sept.	Oct.	Nov.	Dec.	Jan.	Feb.	Mar.	Apr.	May	June
Birmingham, Ala.	0	0	6	93	363	555	592	462	363	108	9	0
Mobile, Ala.	0	0	0	22	213	357	415	300	211	42	0	0
Anchorage, Alaska	245	291	516	930	1284	1572	1613	1316	1293	879	592	315
Flagstaff, Ariz.	46	68	201	558	867	1073	1169	991	899	651	437	180
Phoenix, Ariz.	0	0	0	22	234	415	474	328	217	75	0	0
Little Rock, Ark.	0	0	9	127	465	716	756	577	434	126	8	0
Eureka, Calif.	270	257	258	328	414	499	546	470	505	438	372	285
Oakland, Calif.	53	50	45	127	309	481	527	400	353	255	180	90
San Diego, Calif.	6	0	15	37	123	251	313	249	202	123	84	36
Denver, Colo.	6	9	117	428	819	1035	1132	938	887	558	288	66
New Haven, Conn.	0	12	87	347	648	1011	1097	991	871	543	245	45
Wilmington, Del.	0	0	51	270	588	927	980	874	735	387	112	6
Miami Beach, Fla.	0	0	0	0	0	40	56	36	9	0	0	0
Tallahassee, Fla.	0	0	0	28	198	360	375	286	202	36	0	0
Atlanta, Ga.	0	0	18	127	414	626	639	529	437	168	25	0
Boise, Idaho	0	0	132	415	792	1017	1113	854	722	438	245	81
Chicago, Ill.	0	0	81	326	753	1113	1209	1044	890	480	211	48
Springfield, Ill.	0	0	72	291	696	1023	1135	935	769	354	136	18
Fort Wayne, Ind.	0	9	105	378	783	1135	1178	1028	890	471	189	39
Evansville, Ind.	0	0	66	220	606	896	955	767	620	237	68	0
Des Moines, Iowa	0	9	99	363	837	1231	1398	1163	967	489	211	39

Location	July	Aug.	Sept.	Oct.	Nov.	Dec.	Jan.	Feb.	Mar.	Apr.	May	June
Topeka, Kans.	0	0	57	270	672	980	1122	893	722	330	124	12
Louisville, Ky.	0	0	54	248	609	890	930	818	682	315	105	9
New Orleans, La.	0	0	0	19	192	322	363	258	192	29	0	0
Portland, Me.	12	53	195	508	807	1215	1339	1182	1042	675	372	111
Baltimore, Md.	0	0	48	264	585	905	936	820	676	327	90	0
Boston, Mass.	0	9	60	316	603	983	1088	972	846	513	208	36
Detroit, Mich.	0	0	87	360	738	1088	1181	1058	936	522	220	42
Minneapolis, Minn.	22	31	189	505	1014	1454	1631	1380	1166	621	288	81
Jackson, Miss.	0	0	0	65	315	502	546	414	310	87	0	0
St. Louis, Mo.	0	0	60	251	657	936	1026	848	704	312	121	15
Billings, Mont.	6	15	186	487	897	1135	1296	1100	970	570	285	102
Lincoln, Neb.	0	6	75	301	726	1066	1237	1016	834	402	171	30
Ely, Nev.	28	43	234	592	939	1184	1308	1075	977	672	456	225
Concord, N.H.	6	50	177	505	822	1240	1358	1184	1032	636	298	75
Atlantic City, N.J.	0	0	39	251	549	880	936	848	741	420	133	15
Albuquerque, N. Mex.	0	0	12	229	642	868	930	703	595	288	81	0
Buffalo, N.Y.	19	37	141	440	777	1156	1256	1145	1039	645	313	99
Raleigh, N.C.	0	0	21	164	450	716	725	616	487	180	34	0
Bismarck, N.D.	34	28	222	577	1083	1463	1708	1442	1208	645	329	117
Akron, Ohio	0	9	96	381	726	1070	1138	1016	871	489	202	39
Cincinnati, Ohio	0	0	54	248	612	921	970	837	701	336	118	9
Oklahoma City, Okla.	0	0	15	164	498	766	868	664	527	189	34	0
Portland, Oreg.	25	28	114	335	597	735	825	644	586	396	245	105
Philadelphia, Pa.	0	0	60	291	621	964	1014	890	744	390	115	12
Pittsburgh, Pa.	0	9	105	375	726	1063	1119	1002	874	480	195	39
Providence, R.I.	0	16	96	372	660	1023	1110	988	868	534	236	41
Columbia, S.C.	0	0	0	84	345	577	570	470	357	81	0	0
Sioux Falls, S.D.	19	25	168	462	972	1361	1544	1285	1082	573	270	78
Nashville, Tenn.	0	0	28	158	495	732	778	644	512	189	40	0
Amarillo, Tex.	0	0	18	205	570	797	877	664	546	252	56	0
Dallas, Tex.	0	0	0	62	321	524	601	440	319	90	6	0
Houston, Tex.	0	0	0	6	183	307	384	288	192	36	0	0
Salt Lake City, Utah	0	0	81	419	849	1082	1172	910	763	459	233	84
Burlington, Vt.	28	65	208	539	891	1349	1513	1333	1187	714	353	90
Richmond, Va.	0	0	36	214	495	784	815	703	546	219	53	0
Seattle, Wash.	50	47	129	329	543	659	738	599	577	396	242	99
Spokane, Wash.	9	25	168	493	879	1083	1231	980	834	531	288	135
Charleston, W.Va.	0	0	63	254	591	865	880	770	648	300	96	9
Milwaukee, Wis.	43	47	174	471	876	1252	1376	1193	1054	642	372	135
Cheyenne, Wyo.	19	31	210	543	924	1101	1228	1056	1011	672	381	102

Measuring Water Flow in Pipes

When you're pumping water for irrigation or piping water for a hydroelectric system, you need to know how much water you're using to make the most economical and intelligent use of it. Overirrigation adds unnecessary cost and creates soil drainage problems. Underirrigation wastes labor and cuts crop quality and yield.

Water is measured a number of ways and in several different units:

Acre-inch   is a measure of the volume of water. One acre-inch (ac-in) is the volume of water necessary to cover one acre with one inch of water, or the amount of water falling on one acre in a one-inch rain. One acre-inch equals 3630 ft³ or 27,154 gal.

Acre-foot   is also a measure of the volume of water. One acre-foot (ac-ft) is the volume of water needed to cover one acre with one foot of water. One acre-foot equals 43,560 ft³, 12 ac-in, or 325,851 gal.

Gallons per minute   is a measure of the flow rate of water. One gallon per minute (gpm) is the rate of a pump, stream, or pipeline delivering a volume of one gallon of water in one minute. One gallon is exactly 231 in³.

Cubic feet per second   is also a measure of the flow rate of water. One cubic foot per second (cfs) is a rate of flow equivalent to a stream one foot wide and one foot deep flowing at a velocity of one foot per second.

Here are the equivalent relationships of those measurements:

$$1 \text{ cfs} = 450 \text{ gpm} = 1 \text{ ac-in/hour} = 2 \text{ ac-ft/day}$$

A good water manager in an irrigation system needs to know the volume of water pumped, depth of water applied, number of acres covered, and how long it takes to apply the water. Any one of the four can be computed if the other three are known.

For example, if you operate a pump discharging 1350 gpm and spend 120 hours irrigating a 90-acre field, what average depth of water is applied?

$$450 \text{ gpm} = 1 \text{ ac-in/hour}$$
$$1350 \div 450 = 3 \text{ ac-in/hour}$$
$$3 \text{ ac-in for 120 hours} = 360 \text{ ac-in}$$
$$360 \text{ ac-in} \div 90 \text{ acres} = 4 \text{ in}$$

The average depth of the water is 4 in.

Pump manufacturers provide rating curves that can be used to estimate flow rates if the pump is new or in good condition. These curves are inaccurate for older pumps because wear cuts pump efficiency.

Install a pressure gauge on the discharge side of the pump and a vacuum gauge on the suction side. Use a tachometer to measure pump rpm. Pump discharge depends on the pump speed and the combined suction plus discharge head against which the pump is operating. Find the flow rate for that total head on the pump rating curves. For deep-well turbine pumps, estimate the lift and friction loss through the pump column. Add these two amounts to the discharge pressure gauge reading to obtain total head.

A plumb bob can be used to estimate flow rate by the trajectory method. Attach a plumb bob to the end of a yardstick or rod, as shown in Fig. B-1, so that the distance from the bottom edge of the stick to the point is 8 in. Mark the board or rod in half-inch increments, starting at the plumb bob end.

Place the board or rod on top of the pipe with the plumb bob extended past the flow of water. Pull back slowly until the end of the plumb bob touches the top of the main stream of water, not just the spray or foam. Read the flow rate corresponding to the measured B value from Table B-4. Use this procedure to measure discharge from pipes flowing full.

If a pipe flows only partially full, measure the empty section of the pipe as shown in Fig. B-2, and use Table B-5 to find the approximate discharge. This method works only with level pipes, of course.

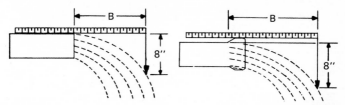

FIG. B-1 Plumb bob method of measuring water flow rate from pipes. *(University of Missouri.)*

TABLE B-4   Flow Rate from Pipes Flowing Full

Inside diameter, in	Area, in²	Flow rate (in gpm) when horizontal distance B is:								
		12 in	14 in	16 in	18 in	20 in	22 in	24 in	26 in	30 in
2	3.14	48	56	64	73	80	88	96	105	120
2½	4.91	75	88	101	112	125	138	150	163	188
3	7.07	108	126	144	162	180	199	216	234	270
4	12.57	193	224	256	288	320	353	385	417	481
5	19.64	301	350	400	451	501	551	601	651	750
6	28.27	432	505	576	659	720	793	864	936	1080
8	50.27	769	896	1025	1153	1280	1409	1538	1663	1922
10	78.54	1201	1400	1600	1803	2002	2200	2400	2604	3000
12	113.10	1730	2020	2302	2592	2884	3176	3460	3755	4320

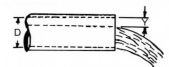

FIG. B-2 Measuring flow from partially full pipe. Take readings in tenths of pipe diameter.

TABLE B-5   Flow Rate from Pipes Flowing Partially Full

Y reading*	Flow rate (in gpm) when inside diameter is:				
	4 in	6 in	8 in	10 in	12 in
0.1	142	334	579	912	1310
0.2	128	302	524	824	1185
0.3	112	264	457	720	1034
0.4	94	222	384	605	868
0.5	75	176	305	480	689
0.6	55	130	226	355	510
0.7	37	88	152	240	345
1.0	0	0	0	0	0

*Y readings are in tenths of inside diameter of pipe.

TABLE B-6   Liquid Manure Production

Animal	Daily production per animal
Dairy cattle	11 gal or 1.5 ft³
Beef cattle	5½ gal or 0.75 ft³
Swine	2 gal or 0.25 ft³
Poultry	0.033 gal

TABLE B-7   Liquid Manure Storage Requirements

Numbers and kinds of animals	Storage period, days	Holding tank capacity, gal
60 dairy cattle	30	20,000
60 dairy cattle	90	60,000
100 beef cattle	60	33,000
100 beef cattle	90	50,000
250 hogs	60	30,000
250 hogs	90	45,000
1000 layers	365	12,000

TABLE B-8   Holding Pit Sizes and Capacities

Dimensions, ft	Capacity, ft³	Capacity, gal
15 × 100 × 6	9000	67,000
20 × 40 × 8	6400	48,000
20 × 40 × 10	8000	60,000
20 (diameter) × 10	3140	23,550
30 (diameter) × 10	7970	53,025

Solar Radiation

Insolation, or the amount of solar energy at any one place, is measured most often in *langleys.* One langley equals one gram-calorie per square centimeter per minute, and is measured with an instrument called a *pyranometer.*

The average amount of solar radiation reaching the earth's atmosphere per minute (called the *solar constant*) is just under 2 langleys. This is the equivalent of 442.4 Btu/(h)(ft²), or 1395 W/m².

The amount of solar radiation received on the earth's surface at any point is affected by the density of the atmosphere and the sun's angle of declination. The National Oceanic and Atmospheric Administration (NOAA) measures solar radiation daily at about 130 locations. The maps in Figs. B-3 to B-14 show mean daily solar radiation by month. By referring to these maps and doing some guesswork or interpolation for your area, you can come up with a rough estimate of the solar radiation. To put the information into usable form, you'll need to convert gram-calories per square centimeter to Btu's per square foot. To find Btu's per square foot for solar radiation falling on the earth's surface, multiply the figure given in langleys by 3.69.

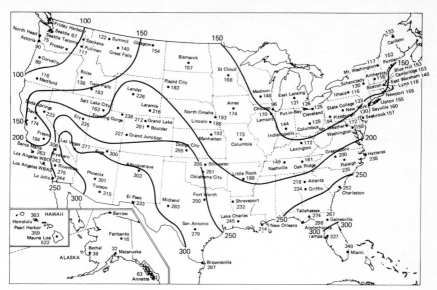

FIG. B-3 Mean daily solar radiation for January, langleys.

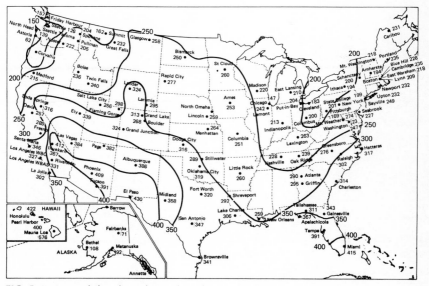

FIG. B-4 Mean daily solar radiation for February, langleys.

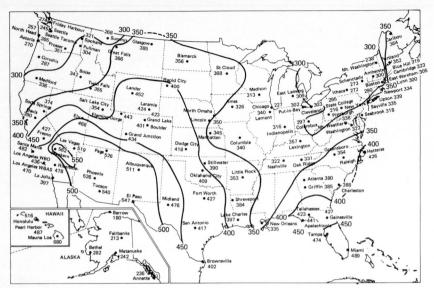

FIG. B-5 Mean daily solar radiation for March, langleys.

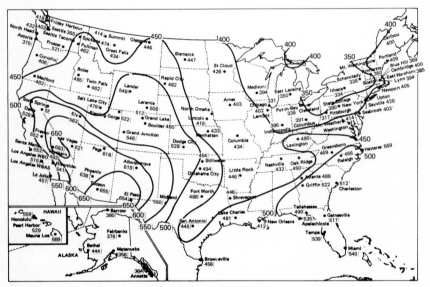

FIG. B-6 Mean daily solar radiation for April, langleys.

B–8

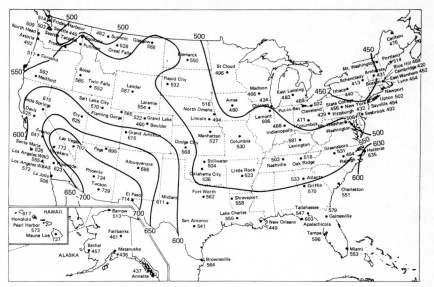

FIG. B-7 Mean daily solar radiation for May, langleys.

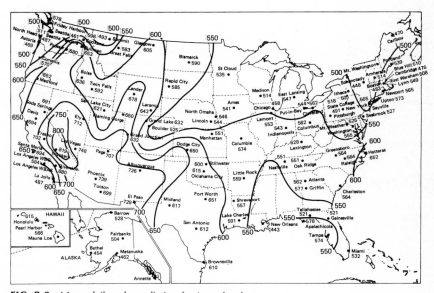

FIG. B-8 Mean daily solar radiation for June, langleys.

B–9

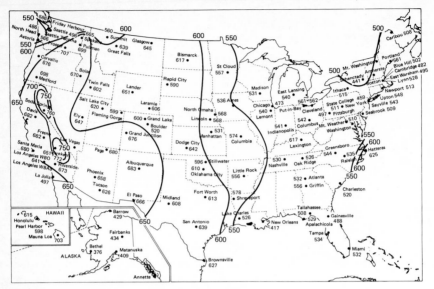

FIG. B-9 Mean daily solar radiation for July, langleys.

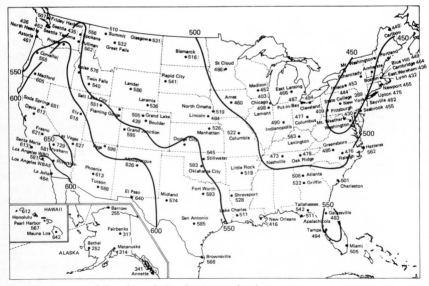

FIG. B-10 Mean daily solar radiation for August, langleys.

B–10

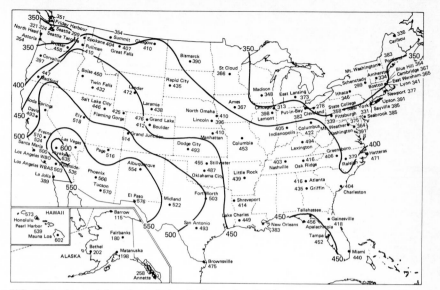

FIG. B-11 Mean daily solar radiation for September, langleys.

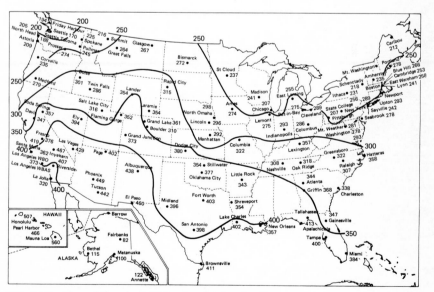

FIG. B-12 Mean daily solar radiation for October, langleys.

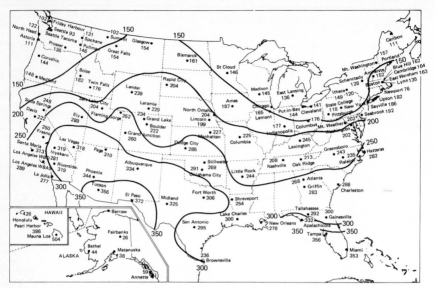

FIG. B-13 Mean daily solar radiation for November, langleys.

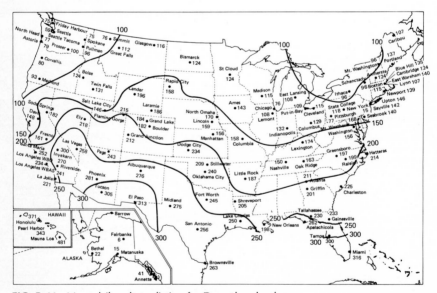

FIG. B-14 Mean daily solar radiation for December, langleys.

For More Energy Information

Alternative Energy—General

Alternate Sources of Energy
Don Marier (ed.)
Route 2
Milaca, MN 56353

A good general periodical, covering a wide variety of alternative energy sources.

Appropriate Technology Sourcebook
Ken Darrow and Rick Pam, 305 pp.
Volunteers in Asia, Inc.
VITA
3706 Rhode Island Avenue
Mount Rainier, MD 20822

Energy Directory and Bibliography
Richard Berman and Irmgard Hunt, 27 pp.
New York State Alliance to Save Energy
36 West 44th St.
New York, NY 10036

Energy for Rural Development
Advisory Committee on Technology Innovation, 306 pp.
National Academy of Sciences
Washington, DC 20523

Farm Energy Magazine
200 West Towers
1200 35th Street
West Des Moines, IA 50265

Monthly publication, produced by the Iowa Corn Promotion Board, focusing mainly on farm-produced fuels.

Handbook of Homemade Power
Staff of Mother Earth News (eds.), 374 pp.
P.O. Box 70
Hendersonville, NC 28739

Rather dated roundup of simple technology approaches to energy production and conservation.

Introduction to Appropriate Technology
R. J. Congdon (ed.), 205 pp.
Rodale Press, Inc.
Emmaus, PA 18049

Producing Your Own Power
Staff of Organic Gardening, 322 pp.
Rodale Press, Inc.
Emmaus, PA 18049

Village Technology Handbook
Volunteers in Technical Assistance, 350 pp.
VITA
3706 Rhode Island Ave.
Mount Rainier, MD 20822

Your Energy-Efficient House
Anthony Adams, AIA, 120 pp.
Garden Way Publishing Company
Charlotte, VT 05445

Alcohol Fuels

Auto Fuels of the 1980s
Jack Frazier
Solar Age Press
Indian Hills, WV

Biohazard of Methanol in Proposed New Uses
Herbert S. Posner
National Institute of Environmental Health Science
P.O. Box 12233
Park, NC 27709

Brown's Alcohol Motor Fuel Cookbook
Michael H. Brown
Desert Publications
Cornville, AZ

Fermentation and Enzyme Technology
D. I. C. Wang, et al, 135 pp.
John Wiley & Sons, Inc.
New York, NY

Forget the Gas Pump—Make Your Own Fuel
Jim Wortham and Barbara Whitener, 84 pp.
Love Street Books
P.O. Box 58163
Louisville, KY 40258

Fuel from Farms—A Guide to Small-Scale Ethanol Production
SERI Consultants, 150 pp.
U.S. Department of Energy
P.O. Box 62
Oak Ridge, TN 37830

Future Automotive Fuels: Prospects, Performances, Perspective
Joseph Colucci and Nicholas Gallopoulas (eds.)
Plenum Books
227 West 17th Street
New York, NY 10011
Report of symposium held at General Motors Research Lab.

International Symposium on Alcohol Fuel Technology
U.S. Department of Energy
P.O. Box 62
Oak Ridge, TN 37830
Report of seminar held at Wolsburg, Federal Republic of Germany.

The Lure of Still Building
Gibat and Gibat
Rutan Publishing Co.
P.O. Box 3585
Minneapolis, MN 55403

Makin' It on the Farm
Micki Nellis, 88 pp.
American Agriculture Movement
P.O. Box 100
Iredell, TX 76649

Making Alcohol Fuel—Recipes and Procedure
Lance Crombie, 40 pp.
Rutan Publishing
P.O. Box 3585
Minneapolis, MN 55403

Making Fuel in Your Backyard
Jack Bradley, 63 pp.
Biomass Resources
Wenatchee, WA 98801

Manual for the Home and Farm Production of Alcohol Fuel
Stephen Mathewson, 208 pp.
J. A. Diaz Publications
Los Banos, CA 93635

Methanol and Other Ways around the Gas Pump
John Ware Lincoln, 134 pp.
Garden Way Publishing Company
Charlotte, VT 05445

Methanol as a Fuel
Seminar report; vol. 1, 36 pp.; vol. 2, 118 pp.
Swedish Methanol Development Co.
Stockholm, Sweden

Methanol Technology and Application in Motor Fuels
J. K. Paul, 85 pp.
Noyes Data Corp.
Mill Road and Grand
Park Ridge, NJ 07656

Biogas—Methane, Producer Gas

Bioconversion of Agricultural Wastes
W. J. Jewell, et al., 321 pp.
National Technical Information Center
U.S. Department of Commerce
Springfield, VA 22161

Bio-gas Plant: Generating Methane from Organic Wastes
Ram Bux Singh, 93 pp.

Gas Research Station
Ajitmal, Etawah, UP, India

Biological Reclamation of Solid Wastes
Clarence G. Golueke, 249 pp.
Rodale Press, Inc.
Emmaus, PA 18049

Cellulose as a Chemical and Energy Resource
M. Mandels, 82 pp.
John Wiley & Sons, Inc.
New York, NY

The Complete Biogas Handbook
D. House (ed.)
Route 2, Box 259
Aurora, OR 97002

Compost, Fertilizer and Biogas Production from Human and Farm Waste in the People's Republic of China
Michael G. McGarry (ed.), 94 pp.
Internation Development Research Centre
P.O. Box 433
Murray Hill Station
New York, NY 10016

Energy, Agriculture and Waste Management
William J. Jewell (ed.), 540 pp.
Ann Arbor Science
P.O. Box 1425
Ann Arbor, MI 48106

Energy and Agriculture in the Third World
Arjun Makhijani, 168 pp.
Ballinger Publishing Company
17 Dunster St.
Cambridge, MA 02138

Energy and Economic Assessment of Anaerobic Digesters and Biofuels for Rural Waste
T. A. Abeles, D. F. Freedman, L. A. DeBaere, and D. A. Ellsworth, 175 pp.
Oasis 2000
University of Wisconsin Center
Rice Lake, WI 54868

Energy Potential through Bio-Conversion of Agricultural Wastes
John L. Burford and Fred Varani, 196 pp.
Bio-Gas of Colorado
342 E. Third St.
Loveland, CO 80537

Energy Primer: Solar, Water, Wind and Biofuels
Richard Merrill and Thomas Gage, 256 pp.
Dell Publishing
1 Dag Hammerskjold Plaza
New York, NY 10017

Methane Digesters for Fuel Gas and Fertilizer
John Fry and R. Merrill, 46 pp.
New Alchemy Institute—West
Box 2206
Santa Cruz, CA 95063

Methane Generation from Human, Animal and Agricultural Wastes
National Research Council, 131 pp.
National Academy of Science
2101 Constitution Ave.
Washington, DC 20418

Methane on the Move: A Discussion of Small Anaerobic Digesters
95 pp.
Bio-Gas of Colorado
342 E. Third St.
Loveland, CO 80537

Methane: Planning a Digester
Peter-John Meynell, 180 pp.
Schocken Books
200 Madison Ave.
New York, NY 10016

Practical Building of Methane Power Plants
L. John Fry, 96 pp.
1223 North Nopal St.
Santa Barbara, CA 93103

Three Cubic Meter Bio-Gas Plant: Construction Manual
Volunteers in Technical Assistance, 25 pp.
VITA
3706 Rhode Island Ave.
Mount Rainier, MD 20822

Uses of Agricultural Wastes: Food, Fuel, Fertilizer
Peter J. Catania (ed.), 371 pp.
Canadian Plains Research Center
University of Regina
Regina, Saskatchewan S4S OA2

Wastewater Engineering: Collection, Treatment, Disposal
Metcalf and Eddy, Inc., 600 pp.

McGraw-Hill Book Company
New York, NY 10020

Building Design

*Alternate Natural Energy Sources in Build-
ing Design*
Albert J. Davis and Robert P. Schubert, 252 pp.
Van Nostrand Reinhold Company
450 West 33rd St.
New York, NY 10001

*American Building: The Environmental
Forces That Shape it*
James Marston Fitch, 349 pp.
Schocken Books
200 Madison Ave.
New York, NY 10016

*The Architecture of the Well-Tempered En-
vironment*
Reyner Banhom, 295 pp.
University of Chicago Press
5801 Ellis Avenue
Chicago, IL 60637

Buildings and the Environment
Robert Goodland (ed.), 215 pp.
The Cary Arboretum & New York Botanical
Garden
Millbrook, NY 12545

Design for a Limited Planet
Norma Skurka, 215 pp.
Ballantine Books
201 East 50th St.
New York, NY 10022

*Design with Climate: Bioclimatic Approach
to Architecture*
Victor Algyay, 236 pp.
Princeton University Press
41 Williams St.
Princeton, NJ 08540

Energy, Environment and Building
Philip Steadman, 287 pp.
Cambridge University Press
32 East 57th St.
New York, NY 10022

Energy Use for Building Construction
B. M. Hannon, et al., 185 pp.

U.S. Department of Commerce
Springfield, VA 22161

Environmental Design Primer
Tom Bender, 207 pp.
Schocken Books
200 Madison Ave.
New York, NY 10016

Farm Builders Handbook
Robert J. Lytle, 288 pp.
McGraw-Hill Book Company
New York, NY 10020

The Home for Man
Barbara Ward, 297 pp.
W. W. Norton & Company
500 Fifth Avenue
New York, NY 10036

Low-Cost Energy-Efficient Shelter
Eugene Eccli, 408 pp.
Rodale Press, Inc.
Emmaus, PA 18049

Midwest Plan Service
Department of Agricultural Engineering
Iowa State University
Ames, IA 50010

Excellent source of plans and technical guides
for all kinds of farm buildings and facilities.

*Solar Energy: Fundamentals in Building De-
sign*
Bruce Anderson, 374 pp.
McGraw-Hill Book Company
New York, NY 10020

Electrical Power—General

National Electrical Code
Published by the National Board of Fire Underwriters;
see local electrical suppliers for copy.

*Power Generation: Resources, Hazards,
Technology and Costs*
Philip Hill, 402 pp.
MIT Press
Cambridge, MA

Engineering Principles

American Society of Agricultural Engineers
St. Joseph, MI 49085

Technical papers and specifications on a variety of farm structures and equipment.

Architecture and Energy
Richard G. Stein, 322 pp.
Anchor Books/Doubleday
Garden City, NY

Biotecture II, Solar Grown Homes
Rudolf Doernach, 92 pp.
D-7277
Wildberg 4, Germany

Environmental Technologies in Architecture
Bertram Y. Kinzey, Jr., and Howard M. Sharp, 788 pp.
Prentice-Hall, Inc.
Englewood Cliffs, NJ 07632

Handbook of Fundamentals
ASHRAE
American Society of Heating, Refrigerating and Air-Conditioning Engineers, Inc.
New York, NY 10017

An Introduction to Statistical Methods and Data Analysis
L. Ott
Duxbury Press
North Scituate, MA

Measurement Systems: Application and Design
E. O. Doebelin
McGraw-Hill Book Company
New York, NY 10020

Standard Handbook for Mechanical Engineers
Theodore Baumeister, et al.
McGraw-Hill Book Company
New York, NY 10020

Heat Exchangers

Heat Pipes
D. D. Dunn and D. A. Reay

Pergamon Press
New York, NY

Heat Transfer
A. J. Champman
Macmillan Company
New York, NY

Hydroelectric Power

Hydro Electric Handbook
W. P. Creager and J. D. Justin
John Wiley & Sons, Inc.
New York, NY

Hydroelectric Engineering Practices
J. Guthrie Brown
Gordon and Breach
New York, NY

Design and Construction of Overshot Water Wheel
50 pp.

Low Cost Development of Small Water Power Sites
Hans Hamm, 50 pp.

Small Michelli (Banki) Turbine: A Construction Manual
55 pp.
All from VITA
3706 Rhode Island Ave.
Mount Rainier, MD 20822

Solar

The National Solar Heating and Cooling Information Center
P.O. Box 1607
Rockville, MD 20850

Serves as a clearing house for information on solar heating, cooling, and power generation. Various publications available.

Application of Solar Technology to Today's Energy Needs
vol. 1, 525 pp.; vol. 2, 756 pp.
Superintendent of Documents
U.S. Government Printing Office
Washington, DC 20402

Applications of Solar Energy for Heating and Cooling
R. C. Jordan and B. Y. H. Liu
ASHRAE, Inc.
345 East 47th St.
New York, NY 10017

Barriers and Incentives to Solar Energy Developments
Arnold R. Wallenstein, 103 pp.
Northeast Solar Energy Center
Cambridge, MA

The Buy-Wise Guide to Solar Heat
Floyd Hickok, 121 pp.
Hour House
St. Petersburg, FL

Building a Durable Low-Cost Solar Collector
Paul F. Larsen, 16 pp.
Loveland, CO

Building and Using Our Sun-Heated Greenhouse
Helen Nearing, 148 pp.
Garden Way Publishing
Charlotte, VT 05445

Building for Self-Sufficiency
R. Clarke, 296 pp.
Universe Books
New York, NY 10016

Build-It Book of Solar Heating Projects
William F. Foster, 195 pp.
Tab Books
Blue Ridge Summit, PA 17214

Cloudburst: A Handbook of Rural Skills and Technology
V. Marks (ed.), 126 pp.
Cloudburst Press
Brackendale, British Columbia

Designing and Building a Solar House
D. Watson, 240 pp.
Garden Way Publishing
Charlotte, VT 05445

Do-It-Yourselfers Guide to Modern Energy-Efficient Heating and Cooling Systems
J. Traister, 280 pp.
Tab Books
Blue Ridge Summit, PA 17214

Energybook No. 1: Natural Sources and Backyard Applications
J. Prenis (ed.), 112 pp.
Running Press
Philadelphia, PA 19103

The Food and Heat Producing Solar Greenhouse
Rich Fisher and Bill Yanda, 160 pp.
John Muir Publications
P.O. Box 613
Santa Fe, NM 87501

Homegrown Sundwellings
Peter Van Dresser, 132 pp.
The Lightning Tree
P.O. Box 1873
Santa Fe, NM 87501

How To Build a Solar Heater
Ted Lucas, 236 pp.
Ward Ritchie Press
Pasadena, CA 91103

How To Use Solar Energy in Your Home and Business
Ted Lucas
Ward Ritchie Press
Pasadena, CA 91103

Land-Use Barriers and Incentives to the Use of Solar Energy
Paul Spivak, 35 pp.
Solar Energy Research Institute
Golden, CO

Passive Design Ideas for the Energy Conscious Architect
90 pp.
National Solar Heating and Cooling Information Center
P.O. Box 1607
Rockville, MD 20850

Passive Solar Buildings: A Compilation of Data and Results
R. P. Stromberg, et al., 71 pp.
National Technical Information Service
U.S. Department of Commerce
Springfield, VA 22161

Solar Access Law
Gail Boyer Hayes, 303 pp.
Ballinger Publishing Company
17 Dunster St.
Cambridge, MA 02138

Solar Age Catalogue
Editors of *Solar Age* magazine, 232 pp.
Brick House Publishing Co.
Harrisville, NH 03450

*The Solar Decision Book: Guide to Making
  a Sound Investment*
Richard H. Montgomery, et al.; 312 pp.
Dow Corning Corp.
Midland, MI

Solar Dwelling Design Concepts
AIA Research Corp., 146 pp.
U.S. Department of Housing and Urban Development
Washington, DC

Solar Energy: Fundamentals in Building Design
Bruce Anderson, 374 pp.
McGraw-Hill Book Company
New York, NY 10020

*Solar Energy Systems: Survey of Materials
  Performance*
L. F. Skoda and L. W. Masters, 110 pp.
National Bureau of Standards
U.S. Department of Commerce
Springfield, VA 22161

Solar Grain Dryer
20 pp.
VITA
3706 Rhode Island Ave.
Mount Rainier, MD 20822

The Solar Greenhouse Book
James C. McCullagh (ed.), 328 pp.
Rodale Press, Inc.
Emmaus, PA 18049

Solar Stills
20 pp.
VITA
3706 Rhode Island Ave.
Mount Rainier, MD 20822

Solar Water Heater
15 pp.
VITA
3706 Rhode Island Ave.
Mount Rainier, MD 20822

Solar Heated Buildings of North America
William Shurcliff, 295 pp.

Brick House Publishing Co.
Harrisville, NH 03450

*Solar Heating Design by the F-Chart
  Method*
William A. Beckman, et al., 200 pp.
John Wiley & Sons, Inc.
One Wiley Drive
Somerset, NJ 08873

The Solar Home Book
Bruce Anderson, 298 pp.
Brick House Publishing Co.
Harrisville, NH 03450

Solar Homes and Sun Heating
G. Daniels, 178 pp.
Harper and Row, Inc.
New York, NY 10022

Solar Law: Present and Future, with Proposed Forms
Sandy F. Kraemer, 364 pp.
P. O. Box 1235
Colorado Springs, CO 80901

SUN: A Handbook for the Solar Decade
Stephan Lyons (ed.), 364 pp.
Friends of the Earth
124 Spear St.
San Francisco, CA 94105

Underground Structures

Earth Sheltered Housing Design
The Underground Space Center, 310 pp.
Department of Civil and Mechanical Engineering
University of Minnesota
Minneapolis, MN 55455

Underground Designs
Malcolm Wells, 87 pp.
P.O. Box 183
Cherry Hill, NJ 08002

Wind Energy Conversion Systems

A Buyer's Guide to Wind Power
Real Gas & Electric Company, Inc.
P.O. Box F
Santa Rosa, CA 95402

Catch the Wind: A Book of Windmills and Windpower
Dennis Landt, 114 pp.
Four Winds Press
New York, NY

Energy from the Wind: Annotated Bibliography
Barbara Burke and Robert N. Merony
Engineering Research Center
Colorado State University
Fort Collins, CO 80523

Energy Primer: Solar, Water, Wind and Biofuels
Richard Merrill and Thomas Gage, 256 pp.
Dell Publishing Co.
1 Dag Hammerskjold Plaza
New York, NY 10017

The Feasibility of Using Wind Power to Pump Irrigation Water
Vaughn Nelson, et al., 58 pp.
Department of Physics
West Texas State University
Canyon, TX 79016

Fundamentals of Wind Energy
Nicholas P. Cheremisinoff, 170 pp.
Ann Arbor Science
P.O. Box 1425
Ann Arbor, MI 48106

The Generation of Electricity by Wind Power
E. W. Golding, 332 pp.
Halsted Press
605 Third Ave.
New York, NY 10016

Handbook of Solar and Wind Energy
Floyd Hickok, 125 pp.
Cahners Books
221 Columbus Ave.
Boston, MA 02116

Harnessing the Wind for Home Energy
Dermot McGuigan, 134 pp.
Garden Way Publishing
Charlotte, VT 05445

The Homebuilt Wind Generated Electricity Handbook
Michael A. Hackleman, 194 pp.
Peace Press, Inc.

3828 Willat Ave.
Culver City, CA 90230

Power from the Wind
Palmer C. Putnam, 244 pp.
Van Nostrand Reinhold Co.
450 West 33rd St.
New York, NY 10001

Simplified Wind Power Systems for Experimenters
Jack Park, 80 pp.
Helion
Box 4301
Sylmar, CA 91342

Technician's and Experimenter's Guide to Using Wind Power
Richard E. Pierson
Parker Books
West Nyack, NY

Wind and Windspinners
Michael A. Hackleman and David House, 140 pp.
Peace Press, Inc.
3828 Willat Ave.
Culver City, CA 90230

Wind Energy
Ben Wolff and Hans Neyer, 82 pp.
Franklin Institute Press
P.O. Box 2266
Philadelphia, PA 19103

Wind Energy Report
Wind Publishing Company
189 Sunrise Highway
Rockville Centre, NY 11570

Monthly publication.

Wind Engineering
Multi-Science Publishing Company, Ltd.
The Old Mill
Dorset Place
London, E1S 1DJ, England

Quarterly publication.

Wind Power
Daniel M. Simmons, 300 pp.
Noyes Data Corp.
Mill Road at Grand Avenue
Park Ridge, NJ 07656

Wind Power Access Catalog
Editors of *Wind Power Digest*
Bristol, IN 46507

Wind Power and Other Energy Options
David Rittenhouse Inglis, 312 pp.
The University of Michigan Press
P.O. Box 1104
Ann Arbor, MI 48106

Wind Power Book
Jack Park,
Cheshire Books
514 Bryant St.
Palo Alto, CA 94301

Wind Power Digest
American Wind Energy Assn.
Bristol, IN 46507

Quarterly publication.

*Wind Power for Farms, Homes and Small
Industry*
Jack Park and Dick Schwind
National Technical Information Service
U.S. Department of Commerce
Springfield, VA 22161

Wind Power for Your Home
George Sullivan, 127 pp.
Cornerstone Library
New York, NY

Wind Technology Journal
American Wind Energy Assn.
P.O. Box 7
Marston Mills, MA 02648

Quarterly publication.

Windmills and Watermills
John Reynolds, 196 pp.
Praeger Publications
200 Park Avenue
New York, NY 10017

*Windmills for Water Lifting and Generating
Electricity on Farms*
E. W. Golding
Food and Agriculture Organization
Via delle Terme di Caracalla
Rome, Italy

Winds and Wind Systems Performance
Carl G. Justus, 120 pp.

Franklin Institute Press
P.O. Box 2266
Philadelphia, PA 19103

Wood Heat

Axes and Chainsaws
Rockwell Stephens, 30 pp.
Garden Way Publishing
Charlotte, VT 05445

Buying and Installing a Wood Stove
Charles Self, 31 pp.
Garden Way Publishers
Charlotte, VT 05445

Chimney and Stove Cleaning
Christopher Curtis, et al., 27 pp.
Garden Way Publishers
Charlotte, VT 05445

The Chimney Brush
Eva Horton, 41 pp.
Kristia Associates
Box 1118
Portland, ME 04104

The Complete Book of Heating with Wood
Larry Gay, 128 pp.
Garden Way Publishers
Charlotte, VT 05445

Firewood
M. Michaelson, 157 pp.
Gabriel Books
P.O. Box 224
Mankato, MN 56001

How To Build an Oil Barrel Stove
Ole Wik, 24 pp.
Alaska Northwest Publishing Co.
130 Second Avenue, South
Edmonds, WA 98020

*How To Select, Cut and Season Good Fire-
wood*
John Vivian, 28 pp.
Customer Service Dept.
Stihl, Inc.
P.O. Box 5514
Virginia Beach, VA 23455

The New, Improved Wood Heat
John Vivian, 428 pp.

Rodale Press, Inc.
Emmaus, PA 18049

A Resource Book on the Art of Heating with Wood
Kristia Associates, 64 pp.
Box 1118
Portland, ME 04104

Save $$ on Firewood
Charles Self, 30 pp.
Garden Way Publishers
Charlotte, VT 05445

The Wood Burner's Encyclopedia
Jay Shelton and Andrew B. Shapiro, 155 pp.
Vermont Crossroads Press
Waitfield, VT 05673

Wood Energy
Mary Twitchell, 144 pp.
Garden Way Publishers
Charlotte, VT 05445

Wood Furnaces and Boilers
Larry Gay, 31 pp.
Garden Way Publishers
Charlotte, VT 05445

Miscellaneous

Energy Savers Catalog
Editors of *Consumers Guide,* 160 pp.
G. P. Putnam's Sons
200 Madison Avenue
New York, NY 10016

Fuels and Their Combustion
F. H. Haslam and R. P. Russell
McGraw-Hill Book Company
New York, NY 10020

Saving Home Energy
Rachel Snyder
Colorado Solar Energy Assn.
P.O. Box 3608
Estes Park, CO 80517

Starting Your Own Energy Business
Avrom Ben David, et al.
Institute for Local Self Reliance
1717 18th St., N.W.
Washington, DC 20009

Using Land to Save Energy
Corbin Crews Harwood, 336 pp.
Ballinger Publishing Company
17 Dunster
Cambridge, MA 02138

Directory of Manufacturers And Technical Service Firms (U.S. and Canadian)

Alcohol Fuels

ACR Process Corp.
808 South Lincoln
Urbana, IL 61801

Distillation plants, 20 gph and larger

Agri-Stills of America
3550 Great Northern Ave.
Springfield, IL 62707

Alcohol plants and equipment

Alcohol Plant Supply Co.
P.O. Box 248
Sherwood, OR 97140

Alcohol fuel and methane plants in Pacific Northwest

Alternative Energy Limited
Route 1
Colby, KS 67701

Design and construction of farm-size alcohol plants

Anheuser-Busch, Inc.
721 Pestalozzi St.
St. Louis, MO 63118

Fermentation yeasts

Aquaterra Biochemicals Corp.
P.O. Box 496

Lancaster, TX 75146

Glucoamylase enzymes

Arbor Sales Company
P.O. Box 6
Des Moines, IA 50301

Custom-made column plates

Bechtel Corporation
50 Beale St.
San Francisco, CA 94119

Large alcohol plants

W. A. Bell
P.O. Box 105
Florence, SC 29503

Components for alcohol plants

Biocon, Inc.
261 Midland Ave.
Lexington, KY 40507

Fermentation yeasts, enzymes

Cole-Parmer Instruments Co.
7425 N. Oak Park Ave.
Niles, IL 60648

Gauges, components for alcohol distillation plants

Conrad Industries
Box 130

Bonaparte, IA 52620
Farm-sized stills

Dorr-Oliver
77 Havemeyer Lane
Stamford, CT 06904

Separators, milling equipment

Double-A Products
P.O. Box 1107
Albert Lea, MN 56007

Farm-sized distillation plants

Easy Engineering
3353 Larimer St.
Denver, CO 80205

20-gph alcohol plants; tours of plant, workshops

Energy Independence Corp.
Montrose, MN 54363

20-gph turnkey alcohol plants, operating instruction

Enzyme Development Corp.
210 Plaza
New York, NY 10001

Starch conversion enzymes

Ethanol International, Inc.
1704 Third St.
Brookings, SD 57007

Alcohol plants, farm-scale to large commercial size

Farm Energy Systems
P.O. Box 2573
Pocatello, ID 83201

80-gal batch alcohol plants

Fermco Biochemicals, Inc.
2638 Delton Lane
Elk Grove Village, IL 60007

Enzymes

Flexi-Liners
5940 Reeds Road
Mission, KS 66202

Fermentation tank liners

Fort Wayne Dairy Equipment Co.
Box 269
Fort Wayne, IN 46801

Tanks, components for alcohol cooking and fermentation

Glitsch, Inc.
P.O. Box 226227
Dallas, TX 75266

Skid-mounted distillation units

Great Northern Equipment Co.
3550 Great Northern Ave.
Springfield, IL 62707

Small-scale alcohol plants

Grove Engineering
1714 Gervais Ave.
North St. Paul, MN 55109

Starch conversion enzymes

International Fuel Systems
1820 West 91st Place
Kansas City, MO 64114

Farm-scale alcohol plants

Jacobson Machine Works
2445 Nevada Ave., N.
Minneapolis, MN 55427

Hammer mills, grinders, processing equipment

Kargard Industries
Marinette, WI 54143

Stainless steel tanks

KBK Industries
East Highway 96
Rush Center, KS 67575

Fiberglass fermentation tanks

Louisville Drying Machinery
Div. of Custom Dryers
232 E. Main
Louisville, KY

Distillers' grain dryers

MRC Energy Systems
Route 2, Box 399
Plymouth, IN 46563

Alcohol plants

Marlin Car Care, Inc.
P.O. Box 1009
Marlin, TX 76661

Alcohol distillation plants

Miles Laboratories
P.O. Box 932
Elkhart, IN 46515

Starch conversion enzymes

Novo Laboratories, Inc.
59 Danbury Road
Wilton, CT 06897

Enzymes

Protectoseal Co.
225 West Foster Ave.
Bensonville, IL 60106

Alcohol production components

Jim Pufahl
Route 2, Box 99
Milbank, SD 57252

Mild and stainless steel distillation columns

R. B. Industries
Riverdale, MI 48877

Small-scale (20-gal) stills

Renewable Energy Systems
P.O. Box 1134
Lebanon, PA 17042

Farm still hardware and kits

Rohm and Haas Company
Independence Mall, West
Philadelphia, PA 19105

Enzymes

Schmitt Energy Systems
Route 2
Hawkeye, IA 52147

Small-scale alcohol plants

Schwartz Services International
230 Washington St.
Mt. Vernon, NY 10551

Enzymes

Semplex
P.O. Box 12276
Minneapolis, MN 55412

Yeasts and enzymes

Seven Energy Corp.
3760 Vance

Wheat Ridge, CO 80033

Farm alcohol plants and accessories

David Vendergriend
Sheldon, IA

Distilling columns

Wenger Manufacturing
Sabetha, KS 66534

Cooking and fermentation equipment

Duane Wessels, Inc.
305 East Main St.
Cedar Falls, IA

Semiautomated alcohol plants

Archie Zeithamer
Route 2, Box 63
Alexandria, MN 56308

Farm-scale alcohol fuel plants and equipment

Alternative Energy—General

Advanced Energy Services
800 Business 70E
Columbia, MO 65201

Ceramic insulation

Design Insulation Systems, Inc.
77 13th Avenue, N.E.
Minneapolis, MN 55413

Insulation materials

GLECON Insulation Products
2909 East 79th St.
Cleveland, OH 44104

Insulation products and accessories

Insultek Corporation
82 Crestwood Road
Rockaway, NJ 07866

Insulated copper pipe

Loren Cook Company
2015 E. Dale St.
Springfield, MO 65803

Ventilating fans

Paul Mueller Company
P.O. Box 828

Springfield, MO 65804

Heat exchangers and heat transfer materials

Sto-Cote Products, Inc.
Drawer 310
Richmond, IL 60071

Polyethylene duct tubing for heating, cooling

Biogas Materials and Equipment

AgFerm Corporation
P.O. Box 55604
Indianapolis, IN 46205

Digester design and construction

Agri-Lines Corporation
P.O. Box 248
Parma, ID 83660

Electric, PTO-driven manure pumps

Alternative Energy Systems
305 Basseron
Vincennes, IN 47524

Digesters and components

Anaerobic Energy Systems, Inc.
P.O. Box 1477
Bartow, FL 33830

Digester design and construction

Bienergy Organizers, Inc.
P.O. Box 5715
Baltimore, MD 21218

Methane digesters and supplies

Biogas of Colorado
5611 Kendall Court
Arvada, CO 80002

Skid-mounted methane digesters

The DeLaval Separator Company
Poughkeepsie, NY 12602

Manure-handling equipment

Perennial Energy, Inc.
P.O. Box 15
Dora, MO 65637

Design and technical services; kits; components; biogas-fired engine-generator systems

Schaffer & Roland, Inc.
130 N. Franklin St.
Chicago, IL 60606

Digester design and construction

Vaughan Company, Inc.
364 Monte-Elam Road
Montesano, WA 98563

Manure-handling pumps

Building Materials and Services

ADA Enterprises, Inc.
Box 151
Freeborn, MN 56032

Tenderfoot livestock floors

Bonanza Buildings, Inc.
Box 9
Charleston, IL 61920

Post-frame metal buildings

Capital Building Systems, Inc.
Box 830
Huron, SD 57350

Metal buildings and components

Central Steel Tube Co.
Clinton, IA 52732

Welded and drawn steel tubing

Circle Steel Corp.
Circle Park
Taylorville, IL 62568

Metal buildings and bins

Confinement Livestock Systems, Inc.
P.O. Box 497
Eldora, IA 50627

Livestock housing and facilities

Dess Company
P.O. Box 342
Morton, IL 61550

Heating and ventilating equipment

Empire-Detroit Steel Div.
Dover, OH 44622

Steel and fiberglass panels

Gifford-Hill and Co., Inc.
West 601 Main Avenue
Spokane, WA 99201

Metal roofing and siding

Mid-West Company
221 N. LaSalle St.
Chicago, IL 60601

Welded steel tubing

The Panel Clip Company
P.O. Box 423
Farmington, MI 48024

Fasteners, truss building jigs, and hardware

W. H. Porter, Inc.
P.O. Box 1112
Holland, MI 49423

Insulated livestock buildings

Solarcrete Corp.
7505 Sussex Drive
Florence, KY 41042

Structural insulated concrete

Wood Connector Products, Inc.
Box 175
Gilbert, IL 60136

Wood fasteners

Control Equipment

Ammark Corp.
12-22 River Road
Fairlawn, NJ 07410

Differential and set-back thermostats

Anabil Enterprizes, Inc.
525 S. Aqua Clear Dr.
Mustang, OK 73064

Differential thermostats and sequence controllers

B.E.S.T., Inc.
Route 1, Box 106
Necedah, WI 54646

Solid-state DC-to-AC inverters

Dalen Products, Inc.
201 Sherlake Dr.

Knoxville, TN 37922

Thermal motors

Heat Motors, Inc.
635 W. Grandview Ave.
Sierra Madre, CA 91024

Thermal motors

Heliotrope General
3733 Kenora Dr.
Spring Valley, CA 92077

Differential thermostats and heating system controllers

Natural Power, Inc.
New Boston, NH 03070

Differential thermostats

Real Gas and Electric Co.
278 Borham Ave.
Santa Rosa, CA 95402

Synchronous inverters

Solar Control Corp.
5721 Arapahoe Road
Boulder, CO 80303

Solar controllers

Spectrex Co.
Bragg Hill Road
Waitsfield, VT 05673

Electrical meters

Westberg Manufacturing, Inc.
3400 Westach Way
Sonoma, CA 95476

Temperature sensors and alarms

Crop-Drying Equipment

Aerovent Fan and Equipment, Inc.
929 Terminal Road
Lansing, MI 48906

Grain dryers and aeration fans

Agri-Products Div.
Butler Mfg. Co.
7400 East 13th St.
Kansas City, MO 64126

Crop-drying controls

Beard Industries
Highway 28 West
Frankfort, IN 46041

Grain dryers

Behlen Manufacturing Co.
P.O. Box 569
Columbus, NE 68601

Grain dryers, fans, and bins

Chief Industries, Inc.
W. Highway 30
Grand Island, NE 68801

Grain dryers

Middle States Mfg., Inc.
P.O. Box 788
Columbus, NE 68601

Crop-drying controls

Richland Building Systems, Inc.
Highway 14 S.E.
Richland Center, WI 53581

Solar grain drying and storage

Stormor, Inc.
P.O. Box 198
Fremont, NE 68025

Residue-fired drying equipment

Electrical Equipment—General

Arco Electric
Box 278
Shelbyville, IN 46176

Phase converters

Dayton Electric Mfg. Co.
5959 Howard St.
Chicago, IL 60648

Motors and controllers

Gould, Inc.
10 Gould Center
Rolling Meadows, IL 60083

Storage batteries

W. W. Grainger, Inc.
5959 West Howard St.
Chicago, IL 60648

Motors, switching relays, controllers, and instruments

Prestolite Battery
Div. of Eltra Corp.
P.O. Box 649
Port Huron, MI 48060

Storage batteries

Ronk Electrical Industries, Inc.
1145 East State St.
Nokomis, IL 62075

Phase converters and transfer switches

Steelman Electric Mfg., Inc.
P.O. Box 1461
Kilgore, TX 75662

Phase converters and pump controls

Union Carbide Corp.
Battery Products Div.
270 Park Avenue
New York, NY 10017

Storage batteries

Engineering Firms

ABP Consultants, Ltd.
P.O. Box 321
Fredericton, New Brunswick E3B 4YP

Planning and design of farm structures

ACR Process Corp.
808 Lincoln
Urbana, IL 61801

Alcohol manufacturing plant design

Addink Engineering
7415 South Hampton Rd.
Lincoln, NE 68506

Irrigation design and feasibility

Bechtel Corporation
50 Beale St.
San Francisco, CA 94119

Alcohol plant design

Bohler Brothers of America, Inc.
1625 W. Belt North
Houston, TX 77043

Alcohol fermentation plants

Canadian Bio Resources Engineering
Suite 205, 17619 96th Ave.

Surrey, British Columbia, V37 4W2

Confined livestock housing and irrigation

Thomas G. Carpenter
6065 Sandgate Road
Columbus, OH 43229

Power and energy applications

Day and Zimmerman, Inc.
1818 Market St.
Philadelphia, PA 19103

Alcohol plant design and construction

Raphael Katzen Associates
1050 Delta Avenue
Cincinnati, OH 45208

Alcohol plant design and engineering

A. G. McKee Associates
10 South Riverside Plaza
Chicago, IL 60606

Alcohol plant engineering

Thomas R. Miles
P.O. Box 216
Beaverton, OR 97005

Energy and biomass planning and design

Perennial Energy, Inc.
P.O. Box 15
Dora, MO 65637

Biogas design and construction

Stone and Webster Engineering Corp.
One Penn Plaza
New York, NY 10001

Alcohol plant and manufacture

Tudor Engineering Company
149 N. Montgomery St.
San Francisco, CA 94105

Hydroelectric site evaluation, planning services, and feasibility studies

World Energy Company
Route 5, Box 251
Carthage, MO 64836

Energy development and engineering

Engine Modification

Arrakis Propane Conversions
Route 2, Box 96C
Leslie, AR 72645

Conversion from gasoline to propane

Consumers Solar Electric Power Corp.
5811 Uplander Way
Culver City, CA 90230

Conversion from gasoline to hydrogen-based fuel

Dual Fuel Systems, Inc.
720 W. Eighth St.
Los Angeles, CA 90017

Conversion from gasoline to methane and other biomass fuels

IMPCO
16916 Gridley Place
Cerritos, CA 90701

Conversion from gasoline to methane and other biomass fuels

M & W Gear
Route 47 South
Gibson City, IL 60936

Kits to burn alcohol-water mix in diesel engines

Moss Fuel Master, Inc.
Route 1, Box 150
Sioux Center, Iowa 51250

Kits to convert gasoline engines to hydrogen-based fuels

Hydroelectric Equipment

Gilbert Gilkes and Gordon, Ltd.
Kendal, England

Impulse turbine-generators

Independent Power Developers
Box 1497
Noxon, MT 59853

High- and low-head hydroturbines

Short Stoppers Electric
Route 4, Box 471 B

Coos Bay, OR 97420

Impulse and reaction turbine-generators

Small Hydroelectric of Canada, Ltd.
Box 54
Silverton, British Columbia VOG 2BO

Kits and components for small-scale private hydro systems

Small Hydroelectric Systems and Equipment
15220 S.R. 530
Arlington, WA 98223

Peltech impulse turbines, for high-head installations; bronze, steel, and aluminum castings; site engineering

The James Leffel and Company
Springfield, OH 45501

Hydroturbines of all sizes and custom castings; Hoppes self-contained turbine and generator units, from 1 to 10 kW

Irrigation and Water Pumping

Berkeley Pump Co.
P.O. Box 2007
Berkeley, CA 94702

Irrigation pumps

D & P Manufacturing
Box 97A
Forest City, NC 28043

Hydraulic rams

Defco, Inc.
325 North Dawson Drive
Camarillo, CA 93010

Drip irrigation systems

Delmhorst Instrument Co.
Boonton, NJ 07005

Soil-moisture testing instruments

Multi-Fittings, USA
P.O. Box 7216
Waco, TX 76710

PVC pipe and fittings

Rife Ram and Pump Works
Box 367

Millburn, NJ 07041

Hydraulic rams

The Toro Company
5825 Jasmine St.
Riverside, CA 92504

Irrigation sprinkler heads

Solar Equipment

American Acrylic Corp.
173 Marine St.
Farmingdale, NY 11735

Light-transmitting panels

Applied Energy Systems
3007 E. Jackson St.
Broken Arrow, OK 74012

Concentrating collectors

Babson Brothers Co.
2100 S. York Rd.
Oak Brook, IL 60521

Solar Surge milking parlor and solar water heaters

Bensun Solar Corp.
P. O. Box 612
Cape Canaveral, FL 32920

Solar collectors

Berdon, Inc.
711 Olympic Blvd.
Santa Monica, CA 90401

Light-transmitting panels

Colorado Sunworks
P. O. Box 455
Boulder, CO 80306

Collectors

Consumers Solar Elec. Power Corp.
5811 Uplander Way
Culver City, Calif. 90230

Gallium aluminum arsenide solar cells; concentrating collectors

Contemporary Systems, Inc.
68 Charlonne St.
Jaffrey, NH 03452

Solar collector fans

Energy Materials, Inc.
3300 S. Tamarac, Suite E105
Denver, CO 80231

Phase-change storage

Fal Bel Energy Systems
Box 6
Greenwich, CT 06830

Flat-plate collectors

Fiberglass Plastics, Inc.
7395 N. W. 34th Ct.
Miami, FL 33147

Light-transmitting panels

Glasteel, Inc.
1727 Buena Vista St.
Duarte, CA 91010

Light-transmitting panels

Grundfos Pumps Corp.
2555 Clovis Ave.
Clovis, CA 93612

Hot-water pumps

Idaho Chemicals Industries
P.O. Box 7866
Boise, ID 83707

Light-transmitting panels

International Solarthermics Corp.
Box 397
Nederland, CO 80466

Self-contained detached systems

Kalwall Corp.
P.O. Box 237
Manchester, NH 03105

Flat-plate collectors and fiberglass-reinforced plastic

Li-Cor, Inc.
Box 4425
Lincoln, NE 68504

Solar data recording equipment

Megatherm
803 Taunton Ave.
East Providence, RI 02914

Phase-change storage

MEMglass, Inc.
6501 Redgrack Lane

Knoxville, TN 37919

Tempered glass

PPG Industries, Inc.
One Gateway Center
Pittsburgh, PA 15222

Flat-plate collectors and solar panels

PSI Energy Storage Div.
1533 Fen Park Drive
Fenton, MO 63026

Phase-change storage

Ramada Energy Systems, Inc.
1421 S. McClintock
Tempe, AZ 85281

Solar collectors

Reichold
Reinforced Plastics Div.
P.O. Box 81110
Cleveland, OH 44181

Plastic and fiberglass panels

Resolite Division
Route 19
Zelienople, PA 16063

Solar panels

Rheem Water Heater Div.
City Investing Company
7600 S. Kedzie Ave.
Chicago, IL 60652

Solar water heaters

A. O. Smith Corp.
P. O. Box 28
Kankakee, IL 60901

Solar water heaters

Solar Energy Systems
5825 Green Ridge Dr., N.E.
Atlanta, GA 30328

Pumps and controllers

Solar Pathways, Inc.
3710 Highway 82
Glenwood Springs, CO 81601

Solar measuring equipment

Solarmatic
Div. of OEM Products, Inc.
Route 3, Box 295

Dover, FL 33527

Phase-change storage

Solaron Corp.
4850 Olive St.
Commerce City, CO 80022

Solar heating and cooling systems

Sun Stone Solar Energy Equipment
Box 941
Sheboygan, WI 53081

Blowers and pumps

Sun-Wall, Inc.
Box 9723
Pittsburgh, PA 15229

Flat-plate collectors

Sunwater Company
1112 Pioneer Way
El Cajon, CA 92020

Solar water heaters and space heaters

Sunworks, Inc.
669 Boston Post Road
Guilford, CT 06437

Flat-plate collectors

Surplus Center
1000 West "O" St.
Lincoln, NE 68501

Parts for fans; water and wind
equipment

Texxor Corp.
9910 North 48th St.
Omaha, NE 68152

Phase-change storage

Thomason Solar Homes, Inc.
6802 Walker Mill Rd., S.E.
Washington, DC 20027

Solaris fluid systems

Thorolyte Fiberglass, Inc.
8969 S.E. 58th St.
Portland, OR 97206

Solar panels

Tranter Manufacturing, Inc.
735 East Hazel St.
Lansing, MI 48909

Heat exchangers and solar absorbers

U.S. Solar Corp.
6407 Ager Road
West Hyattsville, MD 20782

Copper flat-plate collectors

Valmont Industries
Valley, NE 68064

Phase-change storage materials

Vistron Corp., Filon Div.
12333 S. Van Ness St.
Hawthorne, CA 90250

Light-transmitting panels

Standby Generators

BPS
P.O. Box 171
Balfour, NC 28706

Engine-driven generators

Katolight
P.O. Box 3229
Mankato, MN 56001

PTO-driven alternators and engine-driven generators

Onan
1400 73rd Avenue, N.E.
Minneapolis, MN 55432

Electric plants, generators, and engines (gasoline and diesel)

Winco
7850 Metro Parkway
Minneapolis, MN 55420

PTO-driven alternators and diesel and gasoline engine-driven generators

Wind Energy Equipment

Aeolian Energy
Route 4
Ligonier, PA 15658

Four-blade turbines, 8 kW

Aeolian Kinetics
Box 100
Providence, RI 02901

Wind data recording instruments

Aerolectric
13517 Winter Lane
Cresaptown, MD 21502

Wind turbines

Aircraft Components
North Shore Drive
Benton Harbor, MI 49022

Anemometers

American Wind Turbine, Inc.
1016 Airport Road
Stillwater, OK 74074

High-speed turbines, 2 kW

Automatic Power, Inc.
Pennwalt Corp.
205 Hutcheson St.
Houston, TX 77003

French-built Aerowatt wind plants,
24 to 4100 W

Bendix Environmental Science Div.
1400 Taylor Ave.
Baltimore, MD 21204

Anemometers

Budgen and Associates
72 Broadview Avenue
Pointe Claire 710, Quebec

Wind energy components

Climet Instruments Co.
P.O. Box 1165
Redlands, CA 92373

Anemometers

Dakota Wind and Sun, Ltd.
Box 1781
Aberdeen, SD 57401

Wind turbines

Danforth
Div. of Eastern Co.
Portland, ME 04103

Anemometers

Davis Instruments Co., Inc.
513 East 36th St.
Baltimore, MD 21218

Anemometers and recording devices

Dominion Aluminum
3570 Hawkestone Rd.

Mississauga, Ontario L5C 2V8

Wind turbines

Dwyer Instruments, Inc.
P.O. Box 373
Michigan City, IN 46360

Anemometers

Dynergy Corp.
Box 428
Laconia, NH 03426

15-ft Darrieus turbines, 3.3 kW

Energy Alternatives
Box 223
Leverett, MA 01054

Wind plants

Entertech Corp.
P.O. Box 420
Norwich, VT 05055

Wind turbines, 1.5 kW

Grumman Energy Systems
4175 Veterans Memorial Highway.
Ronkonkoma, NY 11779

25-ft turbines, 20 kW

Gurnard Mfg. Co.
100 Airport Road
Beverly, MA 01915

Wind turbine blades

Independent Power Developers
Box 618
Noxon, MT 59853

Wind plants

Kahl Scientific Instrument Corp.
Box 1166
El Cajon, CA 92022

Anemometers and recording devices

Kedco, Inc.
9016 Aviation Blvd.
Inglewood, CA 90301

Wind plants

Mehrkam Energy Development Co.
Route 2, Box 179
Hamburg, PA 19526

Wind turbines and components

Meteorology Research, Inc.
Box 637
Altadena, CA 91001

Anemometers

Millville Windmills
P.O. Box 32
Millville, CA 96062

25-ft turbines, 10 kW

North Wind Power Co.
P.O. Box 315
Warren, VT 05764

Wind turbines, 2 kW

Pinson Energy Corp.
P.O. Box 7
Marstons Mills, MA 02648

Vertical-axis turbines, 2 kW

Product Development Institute
508 S. Byrne Road
Toledo, OH 43609

Wind turbines

Real Gas and Electric Company, Inc.
P.O. Box F
Santa Rosa, CA 95402

Wind turbines and controllers

Sencenbaugh Wind Electric
P.O. Box 11174
Palo Alto, CA 94306

12-ft turbines, 1000 W

Senich Corp.
Box 1168
Lancaster, PA 17609

Turbine blades

R. A. Simerl Instrument Div.
238 West St.
Annapolis, MD 21401

Wind-measuring equipment

Taylor Instruments
Arden, NC 28704

Anemometers

TRW Enterprises
72 W. Meadow Lane
Sandy, VT 04070

Vertical-axis turbines

Texas Electronics, Inc.
5529 Redfield St.

Dallas, TX 75209

Wind-measuring equipment

Unarco-Rohn
6718 W. Plank Rd.
Peoria, IL 61656

Towers for wind machines

WTG Energy Systems, Inc.
Box 87
Angola, NY 14006

Wind turbines

Westberg MFG, Inc.
3400 Westach Way
Sonoma, CA 95476

Wind-measuring equipment

Robert E. White Instruments, Inc.
33 Commercial Wharf
Boston, MA 02110

Anemometers

Winco
7850 Metro Parkway
Minneapolis, MN 55420

Battery-charger plants

Wind Engineering Corp.
Box 5936
Lubbock, TX 79417

Wind turbines, 25 kW

Wind Power Systems, Inc.
P.O. Box 17323
San Diego, CA 92117

Wind turbines, 6 kW

Windmill Water Pumpers

Aeromotor Water Systems
Div. of Braden Industries
P.O. Box 1364
Conway, AR 72032

Bowjon Co.
2829 Burton Ave.
Burbank, CA 91504

Dempster Industries, Inc.
P.O. Box 848
Beatrice, NE 68310

The Heller-Aller Company
Perry and Oakwood
Napoleon, OH 43545

KMP
P.O. Box 441
Earth, TX 79031

O'Brock Windmill Sales
Route 1
North Benton, OH 44449

Wadler Mfg. Co.
Galena, KS 66739

Wood-Burning Equipment

All-Nighter Stove Works
80 Commerce St.
Glastonbury, CT 06033

Wood stoves

Aquaheater Corp.
Box 815S
Clark, CO 80428

Wood-fired water heaters

Ashley Wood Heaters
1604 17th St., S.W.
Sheffield, AL 35660

Circulating heaters and furnaces

Cawley Stove Co.
27 N. Washington St.
Boyertown, PA 19512

Air-tight cast-iron stoves

Hampton Technologies Corp., Ltd.
Box 2277, 126 Richmond St.
Charlottetown, Prince Edward Island, Can.

Wood-burning furnaces

King Products Div.
P. O. Box 128
Florence, AL 35630

Wood and coal circulators

Longwood Furnace Corp.
Gallatin, MO 64640

Dual-fuel furnaces

G & S Mill Co.
Otis St.
Northborough, MA 01532

Wood-burning furnaces

Mohawk Industries, Inc.
P. O. Box 71
Adams, MA 01220

Wood and coal stoves

Multi-Fuel Energy Systems
2185 N. Sherman Dr.
Indianapolis, IN 46218

Dual-fuel furnaces

National Stove Works
Box 640
Cobleskill, NY 12043

Stoves and furnaces

New Hampshire Wood Stoves, Inc.
38 Haywood St.
Greenfield, MA 01301

Air-tight wood stoves

Preston Distributors
Whidden St.
Lowell, MA 01852

Wood and coal stoves

Riteway Manufacturing
Box 6
Harrisonburg, VA 22801

Dual-fuel furnaces

Therm-Kon Products
207 E. Mill Road
Galesville, WI 54630

Wood and coal stoves and furnaces

U.S. Stove Co.
Box 151
South Pittsburg, TN 37380

Wood stoves

Miscellaneous Energy Equipment

Anachron
Box 8860
Portland, OR 97208

Chemical chimney cleaners

Anchor Tools and Wood Stoves
618 N.W. Davis
Portland, OR 97209

Chimney-cleaning brushes

Billings Energy Corp.
2000 E. Billings Ave.
Provo, UT 84601

Hydrogen generators and storage tanks

Moss Fuel Master, Inc.
Route 1, Box 150
Sioux Center, IA 51250

Hydrogen generators

Preston Distributors
14 Whidden St.
Lowell, MA 01852

Chemical flue cleaners

Edmund Scientific
1875 Edscorp Bldg.
Barrington, NJ 08007

Photovoltaic cells

APPENDIX E

Abbreviations Used in This Book

A, amp	amperes	Hz	Hertz
AC	alternating current	in	inch
ac-ft	acre-feet	kcal	kilocalorie
ac-in	acre-inch	kg	kilogram
Ah	ampere-hour	km	kilometer
Btu	British thermal unit	kW	kilowatt
bu	bushel	kWh	kilowatt hour
°C	degrees Celsius	L	liter
cfm	cubic feet per minute	lb	pound
cfs	cubic feet per second	LP gas	"liquid petroleum" (propane)
cm	centimeter	m	meter
cwt	hundredweight	mA	milliampere
d	day	mi	mile
DC	direct current	MHz	megahertz
DDGS	dried distillers' grains with solubles	min	minute
		mph	miles per hour
°F	degrees Fahrenheit	m/s	meters per second
ft	foot	MW	megawatt
g	gram	mV	millivolt
gal	gallon (U.S.)	psi	pounds per square inch
g-cal	gram-calories	psig	pounds per square inch, gauge
gph	gallons per hour	rpm	revolutions per minute
gpm	gallons per minute	s	second
h	hour	TDN	total digestible nutrients
ha	hectare	V	volt
hL	hectoliter	W	watt
hp	horsepower	Wh	watthour
hp-h	horsepower hours		

APPENDIX F

Glossary

Absorber Plate   Part of a solar collector that absorbs solar radiation, converts it to heat energy, and releases it to the working fluid.

Absorptance   The ratio between the radiation absorbed by a surface and the total energy falling on that surface, usually expressed as a percentage.

Active System   A solar system requiring fans or pumps to move the working fluid.

Aerobic Bacteria   Microorganisms that require oxygen.

Alcohol   A class of chemical compounds, composed of carbon, hydrogen, and oxygen, that can be burned as fuel. Ethanol and methanol are two main alcohols being considered for fuel use.

Alternating Current (AC)   Electric current which changes direction of flow at regular intervals, normally making 60 cycles per second.

Ambient Temperature   Outdoor or surrounding temperature.

Ampere   The unit of rate of flow in an electric circuit.

Ampere-Hour   Unit of electrical charge, equal to the quantity of electricity flowing in one hour past any point in a circuit.

Anaerobic Bacteria   Microbes that can live without oxygen.

Angle of Incidence   The angle between the sun's rays and a line perpendicular to the surface on which the sunlight falls.

Anhydrous   Containing virtually no water.

Average Wind Speed   The mean wind speed over a specified period of time.

Backplate   Back of a solar collector; the part farthest from the sun.

Bedplate   A baseplate for supporting a wind-energy system or component.

Biogas   The gaseous product obtained by anaerobic fermentation or pyrolysis of organic materials.

Biomass   Any organic material that can be used to produce food, feed, fiber, or energy.

Breakeven Costs   The cost at which the price of a system's product equals the price of an equivalent product from another source; also used to describe the point at which fixed and variable costs are equaled by production.

British Thermal Unit (Btu)   A unit of measurement that provides a way to compare energy available. One Btu is the amount of heat energy required to raise the temperature of one pound of water one degree Fahrenheit.

Busbar Price   The price of electricity at a generating plant.

Calorie (gram-calorie)   Unit of measurement used to measure heat energy in the SI system. One calorie is the amount of heat energy required to raise the temperature of one gram of water one degree Celsius.

Capital Costs   Investment or ownership costs.

Cellulosic Material   Fibrous plant material.

Centrifugal Pump   A high-speed pump that pushes water with a rotating impeller.

Chord   The distance from the leading to trailing edge of an airfoil.

Collector   Device to receive and absorb solar energy and convert it to heat.

Condensate   Liquid formed when a vapor condenses.

Conduction   Heat transfer through or between bodies in physical contact.

Conifers   Order of trees that usually keep their leaves through winter and reproduce by means of cones.

Convection   Heat transfer by fluid motion.

Cord   A unit of volume measurement, commonly used to describe the amount of wood that will occupy 128 cubic feet of space.

Cover   Collector part that admits solar radiation to the absorber, shields it from heat loss, and reduces long-wave radiation.

Creosote   A sticky, odorous distillate of wood tar that forms on chimneys as a result of incomplete combustion.

Cut-in Speed   Speed at which a wind machine is activated, as the wind speed increases.

Deciduous   Order of trees that shed leaves in autumn.

Desiccant   A drying agent that can remove moisture from other substances.

Destructive Distillation   Process to produce fuel from cellulosic materials by heating in the absence of oxygen, decomposing the material, then distilling the resulting vapors.

Detention Time   The average time that a material remains in a system.

Differential Thermostat   An automatic switch that makes or breaks contact when a temperature difference between two points exceeds or falls below the setpoint.

Diffuse Radiation   Sunlight scattered by particles in the atmosphere; solar energy available on a cloudy day.

Digestion   The process by which complex organic molecules are broken down into simpler compounds.

Direct Current (DC)   Electric current that flows in one direction.

Direct Gain   Solar heating by direct exposure to sunlight.

Downwind   Opposite from the direction from which the wind is blowing.

Efficiency   A measurement of how much of the energy applied to a device is utilized in useful work, expressed as a percentage.

Electrolysis   The process of decomposing a compound by passing an electric current through it.

Emittance   The fraction, between 0 and 1, that indicates the tendency of a material to radiate or emit energy of a specified wavelength.

Energy   The ability to do work.

Energy Conversion   The change of one form of energy into another form of energy; for example, the conversion of the wind's kinetic (mechanical) energy into electricity.

Ethanol   Grain (or ethyl) alcohol made by fermenting simple sugars with microorganisms and enriching the alcohol fraction by distillation.

Eutectic Salt   Materials with a relatively low melting point and large heat of fusion.

Feedstock   The raw material from which alcohol can be made.

Fermentation   Chemical changes caused in materials containing sugars or starches by the action of enzymes produced by living organisms.

Fixed Collector   Collector that does not track the sun.

Fluid   Any liquid or gas.

Fossil Fuels   Natural fuels formed from prehistoric plant and animal remains.

Freestanding   Self-supporting; not mounted on or part of another structure.

Fiberglass-Reinforced Plastic   Polyester resin imbedded with glass fibers.

Gasohol   A blend of 10 percent anhydrous alcohol and 90 percent unleaded gasoline.

Gigawatt   A measure of electrical power, equal to 10^9 watts.

Glazing   The transparent cover on a solar collector or greenhouse.

Head   Vertical distance of a water source above the point of use; pressure exerted by water.

Heat Exchanger   A device to transfer heat from one fluid to another without direct contact between the two fluids.

Horsepower   A measure of the rate of doing mechanical work, equal to 33,000 foot-pounds or 754.2 watts.

Hydrolysis   Enzyme or acid activity to break down the fibrous elements of cellulose.

Infrared Radiation   Radiation with wavelengths longer than 0.7 micrometers; also called long-wave radiation.

Insolation   Sunlight or solar energy.

Inverter   A device for converting direct current into alternating current.

Kilowatt   A unit of electrical power equal to 1000 watts.

Kilowatt-Hour   A measure of energy, equal to the consumption of 1000 watts for one hour.

Latent Heat   The energy absorbed or released by a material when it changes phase.

Latitude   Distance in degrees of arc north or south of the equator on the earth's surface.

Methane   The simplest hydrocarbon, consisting of one carbon atom and four hydrogen atoms; its common name is *marsh gas.*

Methanol   Wood (or methyl) alcohol, made from wood or carbonaceous materials by destructive distillation.

Natural Convection   Natural heat transfer caused by the density difference between hot and cold fluids.

Ohm   The unit of resistance in electrical circuits.

Opaque   Not transparent; admitting no light.

Passive Solar System   A solar system that relies on natural movement of the working fluid or on direct exposure to sunlight.

Penstock   Conduit that connects a water source to a turbine or waterwheel.

pH   The expression of acidity or alkalinity of a substance, measured on a scale from 0 to 14.

Phase-change Material   A material that stores energy as latent heat.

Reflectance   The fraction, between 0 and 1, of incident radiation that reflects off a surface.

Rectifier   A device that converts alternating current into direct current.

Retrofit   Adaptation of a technological innovation to an existing structure.

R value   The resistance of a material to heat flow.

Sensible Heat   The energy applied to raise the temperature of a material or the energy removed to cool it.

Shortwave Radiation   High-energy radiation with wavelengths shorter than three micrometers.

Solar Altitude   Angle between the sun's rays and a horizontal surface.

Solar Constant   The average amount of solar energy available on a sun-following surface outside the earth's atmosphere.

Solar Time   Time based on the position of the sun.

Specific Heat   The amount of heat required to raise the temperature of a unit mass of material by one degree.

Still   Apparatus used to distill or remove water from alcohol.

Summer Solstice   Longest day of the year in the northern hemisphere; first day of summer.

Sun-Tracking Device   A device that maintains a position in relation to the sun.

Tilt Angle   The angle between a collector surface and a horizontal surface.

Torque   The movement of a force around an axis.

Total Solids   The weight of the solid matter remaining after a sample is dried to constant weight at 103°C.

Ultraviolet Light   Radiation with wavelengths just shorter than those of visible light.

Venturi Effect   The increase in velocity of a flow of fluid that is created by a constriction, as in a flume or tube.

Volatile Acids   Volatile portion of solids heated at 550°C; difference between total solids content and the ash remaining after ignition at 550°C.

Volt   The unit of pressure in an electrical circuit.

Voltage Regulator   A device used to maintain constant generated voltage.

Watt   The unit of measurement for work done in an electrical device (watts = volts × amperes).

Winter Solstice   Shortest day of the year in the northern hemisphere; first day of winter.

Working Fluid   Air or liquid that removes heat from the solar collector absorber.

Index

Absorber plate, **5**-4
Adsorption, **14**-23
Ag-Rain, Inc., **12**-5
Alcohol (*see* Ethanol)
Anaerobic bacteria, **16**-1
Anemometer, **19**-4
Azeotrope, **14**-10

Battelle Memorial Institute, **12**-7
Biogas, **16**-1 to **16**-14
   to fuel engine-generator, **24**-2
   production of, **16**-4, **16**-8
   properties of, **16**-2
   from pyrolysis, **16**-11 to **16**-14
   uses of, **16**-10 to **16**-11
Biogas digesters, design of, **16**-5 to **16**-8
Biomass:
   by-products of, **25**-2
   direct combustion of, **11**-2
   national potential of, **13**-5 to **13**-7
Breimyer, Harold F., xi
Bruwer, J. J., **15**-2
Btu (British thermal unit), xvii, **3**-6
Budgets (costs and returns):
   for electricity, **18**-1 to **18**-2
   for farm-grown fuels, **13**-8 to **13**-9
   grain-drying, **10**-5 to **10**-6
   solar-heating, **5**-17 to **5**-19
   for solar hot water, **6**-4
   for vegetable oils, **15**-6 to **15**-7
Buildings:
   livestock, **2**-3, **3**-1 to **3**-8
      energy for heating, **3**-3 to **3**-4
         for "warm" confinement, **3**-3
      heat loss in, estimating, **3**-4 to **3**-6
   shading effect of, **5**-2
Bundy, Dwaine, **4**-1
Bureau of Alcohol, Tobacco, and Firearms, **14**-13, **22**-2

By-products, **24**-1, **25**-1 to **25**-9

Calcium chloride hexahydrate, **5**-16
Circulating pumps, **6**-5 to **6**-8
Clark, R. N., **12**-10
Co-generation of energy, **18**-3, **23**-2
Commodity Credit Corporation, **23**-3
Community Services Administration, **23**-4
Conservation of energy, **1**-1 to **1**-7, **2**-1 to **2**-9
Contractors:
   general, **22**-4
   solar energy, **5**-19 to **5**-20, **6**-9
Cord (firewood measurement), **7**-5
Corn cobs as fuel, **11**-3 to **11**-4
Crop drying, **8**-3 to **8**-4
   (*See also* Grain drying)
Crop varieties, **1**-5
Cropping, energy-efficient, **8**-1 to **8**-6

Dairy operations:
   energy conservation in, **1**-7, **2**-3
   to recover heat from milk, **2**-5
Dessicant grain drying, **10**-9
Diesel fuel, **15**-1 to **15**-9
Diesel fuel substitutes (*see* Vegetable oils)
Differential thermostat, **6**-2
Digesters, biogas, design of, **16**-5 to **16**-8
Disaccharides, **14**-9
Drain-back water-heating system, **6**-1 to **6**-3
Dried distillers' grains, **25**-6 to **25**-7

Economics (*see* Budgets)
Eilks, Kevin, **25**-3
Electricity:
   conservation of, **1**-6

Electricity (*Cont.*):
  costs of, **18**-1 to **18**-2
  emergency generators, **21**-4 to **21**-6
  heating, **19**-9
  from hydroelectric power (*see* Hydro-
    electric power)
  induction generator, **19**-9
  inverters, **19**-8
  rectifiers, **19**-8
  requirements of, **18**-4
  from solar power, **21**-1 to **21**-4
    (*See also* Solar energy)
  from wind power (*see* Wind power)
Electrolysis, **17**-2
Energy:
  co-generation of, **18**-3, **23**-2
  conservation of, **1**-1 to **1**-7, **2**-1 to **2**-9
  costs of, **2**-1, **3**-7, **18**-2
  requirements for building heat, **3**-3 to
    3-8
  thermal, **24**-2, **24**-3
  (*See also* Hydroelectric power; Solar en-
    ergy; Wind power)
Energy Security Act, **23**-3
Energy tax credit, **25**-1 to **25**-3
Entrainment distillation, **14**-10
Enzymes, **14**-12
Ethanol, **14**-1 to **14**-30
  comparison with gasoline, **14**-5
  denaturing, **14**-13
  distillation of, **14**-13
  feedstocks, **14**-8 to **14**-10
  fermentation of, **14**-2 to **14**-12
  history of, **14**-1 to **14**-2
  properties of, **14**-2 to **14**-4
  still design, **14**-14 to **14**-19
  storage of, **14**-26
Ethanol by-products, **25**-3
Eutectic salts, **5**-16
Expeller, screw-press, **15**-5

Farmers Home Administra-
    tion (FmHA), **23**-3
Feedstocks, **14**-8 to **14**-10
Fertilizer, **8**-2 to **8**-3
  green manure, **8**-3
  nitrogen, **8**-2 to **8**-3, **25**-7
  solar-manufactured, **8**-2
Field operations to save fuel, **1**-3 to **1**-4,
    8-4 to **8**-5
Firewood, **7**-2 to **7**-9

Fischer, James, **24**-2
Francis turbine, **20**-9
Frisby, James C., **8**-4
Fuel:
  biogas (*see* Biogas)
  from corn cobs, **11**-3 to **11**-4
  costs of, **3**-6 to **3**-8
  diesel, **15**-1 to **15**-9
  energy content of, **13**-2
  estimating needs for, **3**-6 to **3**-8, **13**-3 to
    13-4
  ethanol (*see* Ethanol)
  firewood, **7**-2 to **7**-9
  hydrogen (*see* Hydrogen)
  solar equivalents, **5**-19 to **5**-21
    (*See also* Solar energy)
  storage of, **1**-5, **13**-2, **14**-26 to **14**-28
  vegetable oils (*see* Vegetable oils)
Fulhage, Charles, **24**-2

Gasohol, **14**-28 to **14**-29
Gebhardt, Maurice, **8**-4
Generators, standby electric, **21**-4 to **21**-6
George, Robert, **9**-1
  illustration, **25**-4
Geothermal heating and cooling, **7**-9
Glauber's salt, **5**-16
Grain alcohol (*see* Ethanol)
Grain drying:
  bins for, **9**-5 to **9**-8
  budget for, **10**-5 to **10**-6
  combination, **9**-4 to **9**-5, **10**-3
  with crop wastes, **11**-1 to **11**-4
  dessicant, **10**-9
  fuel-saving, **1**-4, **9**-2 to **9**-5
  low-temperature, **9**-1 to **9**-8
  natural-air, **9**-2
  solar, **10**-1 to **10**-10
  (*See also* Crop drying)
Grain(s):
  dried distillers', **25**-6 to **25**-7
  safe storage moistures for, table, **9**-3

Hall, Marvin, **5**-7
Harley, Chuck, **10**-6
Hashimoto, A. G., **25**-8
Heat exchangers:
  air-to-water, **6**-3 to **6**-4
  in dairy buildings, **2**-4, **6**-5 to **6**-6
  in ethanol stills, **14**-14

Heat exchangers (*Cont.*):
  in swine buildings, **2**-3 to **2**-4
  in wood stoves, **7**-8
Heat loss in buildings, estimating, **3**-4 to **3**-6
Heat pumps, solar-assisted, **5**-21
Heating buildings, energy for, **3**-3 to **3**-8
Heating degree days, **3**-4, **B**-2
Heid, Walter G., **10**-3
Hofman, Vern, **15**-5
Hydroelectric power, **20**-1 to **20**-10
  head, measuring, **20**-7 to **20**-8
  pump-back storage, **20**-8
  site evaluation, **20**-2 to **20**-3
Hydroelectric Systems and Equipment, Inc., **20**-7
Hydrogen, **17**-1 to **17**-6
  from electrolysis, **17**-2
  energy potential of, **17**-2 to **17**-5
  storage of, **17**-6

Injectors (diesel engine), **15**-3
Insulation, estimating needs of, **4**-2 to **4**-7
  K value, **4**-6
  R value, **4**-3 to **4**-5
  types of, **4**-3 to **4**-5
  values for various materials, table, **4**-4
Investment tax credit, **2**-2, **23**-1, **23**-2
Iowa State University, **14**-22, **25**-3
Irrigation cost cutting, **1**-5, **12**-1 to **12**-12
  methods, **12**-2 to **12**-3
    center-pivot sprinklers, **12**-3
    drip-trickle system, **12**-3
    surface, **12**-3
  pumps, **12**-4
  solar-powered, **12**-6 to **12**-10
  timing, **12**-4 to **12**-5
  wind-powered, **12**-10 to **12**-12

James Leffel and Company, **20**-9

K value, **4**-6
Kansas State University, **5**-18, **16**-11, **25**-2
Kaplan propeller, **20**-9

Ladisch, Michael, **14**-24
Landers, Ted, **16**-7
Langleys (solar measurement), **5**-2, **B**-6

Livestock:
  confinement housing for (*see* Buildings, livestock)
  cooling, **7**-9
  feeding, **25**-2 to **25**-3
  water supply for, **2**-6 to **2**-9
Lorenzen, John, **17**-2
Low-temperature grain drying, **9**-1 to **9**-8

M&W Gear Company, **15**-8
McEllhiney, Robert, **25**-2
McKenzie, Bruce, **8**-3
Mash (ethanol), **14**-12
Meador, Neil F., **22**-4, **24**-2
Metal hydrides, **17**-6
Methane (*see* Biogas)
Middaugh, Paul, **14**-16
Midwest Plans Service, illustrations, **4**-7, **9**-6, **10**-8
Milk, energy to cool, **2**-5
Monosaccharides, **14**-8

National Alcohol Fuel Still Certification Board, **14**-14
National Bureau of Standards, **23**-4
National Public Utilities Regulatory Act, **18**-2
Natural-air grain drying, **9**-2
Nebraska-type swine building, **2**-2
New Mexico State University, **12**-7
Nilsson, Christer, **4**-3
Nimmermark, Sven, **4**-3
North Dakota State University, **15**-5

Odometer, **19**-4
Ohio State University, **6**-11
OPEC (Organization of Petroleum Exporting Countries), xii, **25**-1

Paul Mueller Company, graph, **2**-5
Peltech turbine, **20**-7
Pelton wheel, **20**-9
Pesticides, **8**-5 to **8**-6
Phillips, Richard E., **3**-3, **4**-14
Photovoltaic cells, **21**-1 to **21**-4
Pogue, Gene, **5**-18
Polysaccharides, **14**-9
Prier, Leslie, **5**-14

Purdue University, **14**-24
Pyrolysis, producing gas by, **16**-11 to **16**-14

Quad (quadrillion Btu), xvii

R value, **4**-3 to **4**-5
Radio Corporation of America, **21**-3
Reduced tillage, **8**-4 to **8**-5

Sallvik, Krister, **4**-3
Schneeberger, Kenneth, xiii
Screw-press expeller, **15**-5
Sewell, Homer, **25**-6
Sisson, Donald, **12**-5
Small Business Administration (SBA), **23**-1
Snow, Gale, **6**-5
Soderholm, L. H., **19**-10
Solar energy:
  collectors, **5**-1 to **5**-21, **6**-1 to **6**-5, **10**-1 to **10**-6
    active, **5**-4
    bare-plate, **5**-13
    cover glazing, **5**-10 to **5**-12
    efficiency of, **5**-5 to **5**-8, **5**-18
    flat plate, **5**-4
    with heat pumps, **5**-21
    passive, **5**-5
    portable, illustrated, **10**-4
  contractors, choosing, **5**-19 to **5**-20, **6**-9
  declination, **5**-2
  heat storage, **5**-13, **6**-5
  pond, **6**-11 to **6**-12
  water heaters, types of, **6**-1 to **6**-5
Solar-generated electricity, **21**-1 to **21**-4
Solar grain drying, **10**-1 to **10**-10
Solar-powered irrigation, **12**-6 to **12**-10
Solar-Surge system, illustrated, **6**-10
South Dakota State University, **14**-16
Spray cooling, **7**-10 to **7**-12
Standby electric generators, **21**-4 to **21**-6
Starches, **14**-8
Stillage as livestock feed, **25**-2 to **25**-5
Stormor, Inc., **11**-1
Sugars, **14**-9
Sunflower oil, **15**-3
Swine confinement buildings, **2**-3, **24**-2
  (*See also* Buildings, livestock)
Synchronous inverter, **19**-8 to **19**-9
Synfuels, xiv

Tank, freeze-proof, for stock watering, **2**-6 to **2**-9
Taxes:
  energy credit, **23**-1 to **23**-3
  Gasohol exemption, **23**-3
  investment credit, **2**-2, **23**-1, **23**-2
  special fuel, **23**-2
Thermal energy, **24**-2, **24**-3
Tractor horsepower, xvii
Tractors, fuel-saving, **1**-4, **8**-5

U.S. Army Corps of Engineers, **20**-1
U.S. Department of Agriculture, xviii, **23**-1
U.S. Department of Commerce, **19**-4, **23**-1
U.S. Department of Energy, xviii, **23**-1, **23**-4
U.S. Meat Animal Research Center, **25**-8
University of Idaho, **12**-4
University of Illinois, **15**-7,
  illustration, **10**-4
University of Missouri, **2**-6, **9**-1, **24**-2
University of Nebraska, **12**-7, **21**-3
Uranium-235, xii

Vegetable oil meal (as livestock feed), **25**-9 to **25**-10
Vegetable oils (as diesel fuel substitutes), **13**-7, **15**-1 to **15**-9
  costs of, **15**-6 to **15**-7
  production of, **15**-5 to **15**-6
Velocity head rod, **20**-4 to **20**-5
Ventilation, **4**-1
  air inlet sizing, **4**-10 to **4**-12
  controllers for, **4**-10, **4**-13 to **4**-14
  energy-conserving, **1**-7
  fans for, **4**-9 to **4**-10
  purposes of, **3**-3
  requirements for, **4**-7 to **4**-12
  troubleshooting systems in, **4**-14 to **4**-16
  winter minimum, tables, **3**-4, **4**-9

Water heaters, solar, **6**-1 to **6**-5
Water power (*see* Hydroelectric power)
Weirs (for measuring stream flow), **20**-5
Wessels, Duane, **14**-21, **14**-29
Wick, Emil, **14**-22
Wicklow, Donald T., **14**-25
Wind power, **19**-1 to **19**-12
  generators, **19**-6 to **19**-9
    battery storage system, **19**-7 to **19**-8

Wind power (*Cont.*):
  heating system, **19**-9 to **19**-11
  plant design, **19**-5 to **19**-6
  site evaluation, **19**-3 to **19**-5
  turbines (rotors), **19**-5 to **19**-6
    Darrieus rotor, **12**-10 to **12**-11
    horizontal-axis, **19**-6
    vertical-axis, **12**-10, **19**-6
  water pumpers, **19**-11 to **19**-12
Wind-powered irrigation, **12**-10 to **12**-12

Windfall Profits Tax Act, **23**-1
Wood heat, **7**-1 to **7**-9
Woodburning furnaces, **7**-3 to **7**-8
Woodburning safety, **7**-9

Yeasts in fermentation, **14**-2, **14**-12 to **14**-13

Zone heating, **6**-6 to **6**-7

About the Author

James D. Ritchie, a leading farm authority, has had more than twenty-five years of experience in the agriculture field, both as an editor and a journalist.

One of James Ritchie's most recent books is *Successful Alternative Energy Methods,* which was a featured selection of the Playboy Book Club. Mr. Ritchie has served as an editor for *Today's Farmer* and *Farm Journal.* He is also well-respected for his numerous articles in such publications as *Progressive Farmer, Successful Farming, BEEF, Farm Building News, National Hog Farmer,* and *PIG International.*

A graduate of the University of Missouri's School of Journalism, Mr. Ritchie is a resident of Versailles (pronounced "ver-sales"), Missouri.

RITCHIE, JAMES D.
    SOURCEBOOK FOR FARM ENERGY
ALTERNATIVES.